Abiotic Stress Tolerance in Crop Plants

Edited by :

Dr. M. Prakash
Professor
Faculty of Agriculture, Annamalai University
Annamalai nagar - 608 002

&

Dr. K. Balakrishnan
Professor of Crop Physiology
Agricultural College & Research Institute
Madurai – 625 104

SSPH

SATISH SERIAL PUBLISHING HOUSE
403, Express Tower, Commercial Complex, Azadpur, Delhi-110033 (India)
Phone : 011-27672852, Fax : 91-11-27672046
E-mail : info@satishserial.com, hkjain1975@yahoo.com
Website : www.satishserial.com

Published by :

SATISH SERIAL PUBLISHING HOUSE

403, Express Tower, Commercial Complex, Azadpur, Delhi-110033 (INDIA)
Phone : 011-27672852 Fax : 91-11-27672046
E-mail : info@satishserial.com, hkjain1975@yahoo.com

ISBN 978-93-81226-92-6

Typeset at: Harminder Kharb *for* Laxmi Art Creation # 9971212252
Printed at: Salasar Imaging Systems, Delhi-110035

PREFACE

Stress environment is one of the major limiting factors affecting the crop productivity. Crop stress may be defined as any factor that affects growth and/or yield below the limit imposed by plant genotype and available solar radiation. Crop stress can be classified into two categories namely, biotic and abiotic stress affecting the crop in various aspects. Biotic stress is the stress caused by biotic factors like pest and disease attack on crop plants. Globally, over 70 per cent of natural disasters are related to weather and climate. Variability in weather and climatic elements, especially rainfall is the principal source of fluctuation in food production. The major abiotic stresses encountered by agriculture sector are drought, floods/water logging, soil and water salinity, soil acidity and nutrient deficiency which are known to negatively impact agricultural production. Almost 80 million hectares of our land is rainfed. About 25 m ha of soils with pH less than 5.4 are critically degraded and have very less productivity. These stresses can reduce productivity of crops by as much as 65% in India depending on the crop, region and severity of the stresses. Effect of these and other environmental stresses on crops could intensify further as the environment continue to change and as the global population rises, forcing agriculture to expand into areas currently considered unsuitable for crop production.

The focus of this book, therefore, has been to elucidate mechanism of stress tolerance, management strategies to overcome their adverse effect and to sustain crop productivity under different stress environments. This book has the latest information on different aspects of stress contributed by broad range of researchers, from Crop physiologists to Agronomist and Plant breeders. We hope that this book will be highly valuable to researchers, students of Post Graduate Studies in appropriate disciplines in Agricultural sciences.

In conclusion, we would like to say sincere thanks to all the contributing authors for their great work.

M. Prakash
K. Balakrishnan

CONTENTS

Chapter 1

Molecular Genetic Approaches for Improvement of Drought Tolerance in Rice

MONIKA DALAL[1] AND VISWANATHAN CHINNUSAMY[2]

[1] National Research Centre on Plant Biotechnology, Pusa Campus, New Delhi-110012

[2] Division of Plant Physiology, Indian Agricultural Research Institute, New Delhi-110012

Summary

Sustainability of rice production is essential for food security of Asia, which is the largest producer and consumer of rice in the world. Drought is the major factor limiting grain yield especially under rainfed areas. To maximize production under these conditions, traits for drought tolerance need to be combined with traits that contribute to yield in drought stress environments. These traits include maximization of water uptake through deep root system, minimization of water loss through stomata, leaf rolling, osmotic adjustment, epidermal conductance, and epicuticular wax, etc. Depending on the nature of drought stress, plants may employ some or all of these mechanisms. Significant progress has been made in physiological dissection of complex drought tolerance mechanisms into component traits, and mapping of the quantitative trait loci (QTLs) for these traits. The QTLs have been shown to contribute from 5% to 66% of the phenotypic variation in component traits. QTLs with large effect (51%) have also been identified for yield under reproductive drought stress. Fine mapping and identification of gene(s) underlying the QTLs will help in pyramiding of QTLs by Marker Assisted Selection (MAS) and Marker Assisted Recurrent Selection (MARS) breeding. Transcriptome engineering is emerging as an important tool to combat abiotic stresses.

Some of the effector genes or transcriptional regulators when overexpressed in rice conferred reproductive stage drought tolerance under field conditions. Advancement in omics coupled with high through put precision phenomics and efficient use of diverse genetic resources will increase the pace of our understanding of the molecular basis of drought tolerance and development of crop varieties and hybrids with enhanced water use efficiency and drought tolerance.

Introduction

Rice (*Oryza sativa*) is the most important food crop of the world. It is grown in more than hundred countries and is staple food for more than half of the world's population. In 2011, from the total harvested area of 164 million hectares, about 722 million tons of rice was produced worldwide (FAO Stats). Of the total rice production, 90% of rice is produced in Asia. Rice is grown in three major environments; irrigated, rainfed lowland (periodically flooded by rain), and rainfed upland (dryland, unirrigated) system. The irrigated lowland rice provides 75% of the world's rice production and plays an important role in food security of Asian continent. The upland rice and rainfed lowland rice, which together account for about 48% of the rice cropping area, contribute only 20% and 4% respectively, to the world's total rice production. The unpredictable rainfall is the major reason for such low productivity in rainfed lowland and upland ecosystems. On the contrary, the irrigated rice consumes about 40% of the world's irrigation water and 30% of the world's developed freshwater resources (www.irri.org). Fresh water availability is decreasing and by 2050, many of the river basins will reach the water scarcity line. Due to shortage of water, it will be difficult to sustain production even from irrigated rice cultivation. To meet the increasing demand for food in the time of climate change, it has become imperative to devise technologies for growing rice with less water, without any compromise on production. For combining grain yield under drought stress conditions Salekdeh *et al.* (2009) proposed a conceptual framework for drought phenotyping based on the Passioura equation. According to this equation, yield is expressed as the product of WU (amount of water used), WUE (the conversion of WU to dry mass gain) and HI (the fraction of dry mass gain converted to grain). Based on this equation the phenotypic traits associated with different factors of yield under drought include deep roots, seedling vigour, osmotic adjustment, transpiration, leaf wax, stay-green, panicle extrusion, stem carbohydrates, grain size , grain number, and grain weight etc.

In the past two decades, efforts have been made to map QTLs for these physiological traits contributing to drought stress tolerance and direct selection for yield *per se* and several genes have been characterized in rice for imparting drought tolerance in rice. The chapter presents an overview of studies carried out towards development of drought tolerant rice through molecular breeding and transgenic approach.

Mapping for Physiological Traits

Root traits

The roots play pivotal role in water and nutrient uptake, and act as sensors of water and nutrient status of the soil. Under rainfed conditions, possession of a deep and thick root system and other root morphological traits that allow access to water and nutrients from deeper soil layers is one of the most promising traits for improving drought tolerance. Moreover, the ability of plants to cool their canopy temperature through transpiration needs efficient functioning of the root water uptake system. The actual measurement of root morphological and other traits that contribute to enhanced water uptake under field drought conditions require a whole crop season in addition to other resources. Moreover, the plasticity and difficulty in phenotyping this underground trait becomes a limiting factor to be included in the breeding program. Hence, efforts were made to identify the quantitative trait loci (QTL) for various root parameters such as root length, thickness, penetration etc. Once QTLs are tagged with molecular markers, these markers can be used to screen and select the desired plant type (with desired QTLs) from the segregating population even in the seedling stage in the lab conditions.

For identifying QTLs related to root traits in rice, mapping populations derived primarily from cross between lowland *indica* rice and upland *japonica* rice varieties have been exploited (Table 1). The lowland rice, which is grown in flooded conditions has shallow root system and more susceptible to drought while upland-adapted cultivars have deeper root systems and has been used as source of root traits for lowland rice. In rice alone, more than 675 QTLs controlling 29 parameters for root traits have been mapped on 12 different chromosomes (Courtois *et al.* 2009). In most of the earlier studies, either same populations were analysed by different groups or atleast one of the parent was common across some populations (Table 1). The number of QTLs identified varied from 3-56 which could explained 1.1 % to 38 % of phenotypic variation in different root traits. The meta analysis of the root QTL data revealed six major QTL concentration areas on chromosomes 1, 2, 3, 7, 9 and 11. Of this, about 30 QTLs for 12 root traits are located in the interval between 15 Mb and 20 Mb on chromosome 9 (Courtois et *al* 2009).

Table 1 : Studies on root QTLs for drought tolerance in rice

S. No	Genetic Background	Characters	Population	Environment	Reference
1	Co39xMorobere kan	Root thick ness, root: Shoot ratio, root dry weight / tiller, deep root dry weight/tiller, maxi. root length, root Penetration	203 RIL, F7 And 52 RILs F7	Pots in Glass house, Wax layers	Champoux *et al.* (1995) Ray *et al.* (1996) Lilley *et al.* (1996)
2	BalaxAzucena	Root penetration root Length, root thick ness, root ; shoot ratio, deep Root weight	205 RILs F6	soil chambers and Wax layers	Price *et al.* (2000, 2002) MacMillan *et al.* (2006)
3	IR64xAzucena	Root thickness, maxi. root length, total root weight, deep root weight, deep root weight per tiller, and root to shoot ratio	105 DH/and 90 DH	Wax layers, PVC Cylinders	Zheng *et al.* (2000) Yadav *et al.* (1997) Venuprasad *et al.* (2002)

4	IR1552xAzucena	Seminal root length, adventitious root number, Lateral root length, and number	150 RILs F10	solution culture, paper culture, Sandy soil culture in cylindrical pots	Zhang *et al.* (2001b) Zheng *et al.* (2003)
5	CT9993xIR62266	Root Penetration, root pulling force	154 DH	Wax layers	Zhang *et al.* (2001a)
6	Zhenshan97xIR AT109	Root depth, root volume, drought induced root growth	180 RILs at F9/F10	PVC pipes, Rain out shelter	Yue *et al.* (2006)
7	IAC 165xCo39	Root length, thickness and root dry weight	125 RILs F7	PVC tubes green house	Courtois *et al.* (2003)
8	IR64xKinandang Patong	Deeper rooting	117 RILs	Basket method in green house and upland Field	Uga *et al.* (2011)

Recently a set of 117 recombinant inbred lines (RILs) derived from a cross between the lowland cultivar IR64 (shallow rooting), and the upland cultivar Kinandang Patong (deep rooting) were evaluated by basket method and a major QTL *Dro 1* (Deeper rooting 1) was identified in rice. Located on chromosome 9, this QTL explained 66.6% of the total phenotypic variance in RDR (the proportion of total roots that elongated through the basket bottom) in the RILs and appeared to play an important role in increased deep rooting under upland field conditions (Uga *et al* 2011). Efforts have also been made to introgress the rice QTLs controlling root traits from Azucena, an upland *japonica* into an Indian upland *indica* rice variety, Kalinga III (Steele *et al* 2006). Five segments from Azucena on different chromosomes were targeted for introgression out of which four segments carried QTLs for improved root morphological traits (root length and thickness). The target segment on chromosome 9, significantly increased root length under both irrigated and drought stress treatments. The study was further extended and introgressed lines and the recurrent parent were analysed for six years in upland fields in Eastern India and yields were recorded. The combination of the four QTLs (2, 7, 9, and 11) significantly increased yield by 1 t ha^{-1} under favourable field trials. However, there were no significant differences between any of the lines for yield or maturity in unfavourable conditions. The most successful line from this breeding programme PY 84 has been released in India as Birsa Vikas Dhan 111 (Steele *et al.*, 2013).

Water relations

Water relations of a plant are indicators of a plant's capacity to tolerate drought. The component traits for water relations include relative water content, leaf rolling, osmotic adjustment, stomatal conductance, canopy temperature etc. Osmotic adjustment (OA) is an inheritable trait and it refers to active accumulation of solutes during water stress in plants, which allows maintenance of higher turgor potential. OA has been associated with delayed leaf rolling, and leaf senescence and yield under water stress in rice (Hsiao et al 1984). Plants loose water to the atmosphere through stomata during transpiration and by direct evaporation of water from epidermal cell surface. Transpiration through stomata accounts for majority of the water loss. Stomatal pores allow CO_2 influx for photosynthetic carbon fixation and water loss via transpiration to the atmosphere. Thus, the rate of transpiration and photosynthesis depends upon the plants ability to regulate its stomatal pores. Transpiration also helps in reducing the canopy temperature of the plant. Under water stress condition when transpiration rate exceeds the water uptake by plant, the relative water content of leaves decreases, resulting in loss of turgor. Leaf rolling is an adaptation to conserve internal water by reducing the transpirational losses under drought stress conditions. In spite of being important physiological markers for defining plants water status, most of these traits except canopy temperature, could not be efficiently utilized for screening and selection in breeding for drought tolerance. The reason is that most of these traits vary with the physiology of the plant throughout the day. Hence, estimation of these component traits under field conditions and across a population in a time frame is very difficult and laborious task. The identification of QTLs and genes responsible for these traits would simplify the screening and exploitation of these traits for increasing drought tolerance of rice. Lilley and Ludlow (1996) screened 61 diverse rice lines and 8 *Oryza* spp accessions and found that rice genotypes exhibit a greater degree of variation for osmotic adjustment (0.4 to 1.5 MPa) and lethal osmotic potential (-4 to -10 MPa) (dehydration tolerance). The genotypes with *Japonica* background showed poor dehydration tolerance and low osmotic adjustment, while *Indica* lines had greater dehydration tolerance and greater osmotic adjustment (Lilley and Ludlow, 1996). Of the 61 genotypes, Co39, a lowland *Indica* cultivar and Moroberekan, an upland *Japonica* cultivar were selected, and RILs of 52 were evaluated for osmotic adjustment (Lilley *et al.* 1996). A single QTL on chromosome 8, which explained 32-34% of osmotic adjustment was identified. Later Zhang *et al.* (2001a) used 154 doubled haploid lines derived from CT9993 X IR 62266 and identified total 5 QTLs for osmotic adjustment which together could account 31% of variance in OA. The individual contribution of these varied from 8-13%. Interestingly, one of the QTL *oa 8.1* identified in this study was located in the same genomic region as the OA QTL detected by Lilley et al (1996). For introgression of OA, an advanced back-cross population of IR62266-42-6-2X IR60080-46A was developed (Robin *et al.* 2003). One hundred and fifty BC3F3 families were evaluated for osmotic adjustment and 14 QTL for OA were identified which together explained 58% of phenotypic variability. The QTL on chromosome 8 (*oa8.1*) was found consistent across different populations in the three studies on OA. Moreover, this region is syntenic to a QTL for OA in both wheat (Lilley *et al.* 1996) and barley (Teulat *et al.* 1998). Thus, this region appears to be important for OA and needs to be analysed in detail.

Price *et al.* (1997) screened a population of 178 F2 plants from Cross between Azucena and Bala and identified a QTL for slow leaf rolling on chromosome 1, two QTLs associated with resting stomatal conductance on chromosomes 3 and 12 and QTLs for the rate of stomatal closure were found on chromosomes 3 and 7. The analysis of these traits in a doubled-haploid population of 105 lines derived from the cross IR64/Azucena of rice revealed 11 QTLs for leaf rolling, 10 for leaf drying, 11 QTLs for relative water content and 10 for relative growth rate under stress (Courtois *et al.* 2000). Though number of QTLs varied across populations, yet some were relatively, consistent across genetic backgrounds. A large effect QTL, *rwc 9.1* was identified in one of the three trials conducted on a doubled-haploid (DH) population of 154 rice lines from the cross CT9993-5-10-1-M/IR62266-

42-6-2 (Babu *et al.* 2003). This single QTL explained up to 59% of phenotypic variation for leaf RWC under stress. In the same trial, a QTL ct2.1, for canopy temperature was also detected which contributed 47% to the phenotypic variation. The QTLs for leaf rolling and leaf drying were of minor effects in the range of 8.5 to 21%.

An important factor in water relations is stomatal frequency that determines the stomatal conductance, water loss and carbon dioxide gain. The analysis of relationship between the stomatal frequency and the photosynthetic ability in a wide range of rice genotypes has revealed that higher stomatal frequencies of *indica* rices are related to high photosynthetic abilities as compared to *japonica* rice (Weng and Chen 1988). The study of the stomatal frequency in a cross between an *indica* rice (high stomatal frequency) and a *japonica* rice (lower stomatal frequency) showed genetic linkage with an isozyme of esterase gene (Weng and Chen, 1988). The genetic relations between adaxial and abaxial stomatal frequencies was studied in a population of backcross inbred lines of Nipponbare (*japonica*) X Kasalath (*indica*) rice genotypes. Of the four QTLs controlling adaxial and abaxial stomatal frequencies, QTL for adaxial stomatal frequency overlapped that for abaxial stomatal frequency on chromosome 3 of rice. The high correlation ($r = 0.660$, $P < 0.01$) between these alleles contributed by Kasalath suggested that the same gene may pleiotropically control the stomatal frequency on both surfaces of the leaf (Ishimaru *et al.* 2001).

The stress hormone, Abscisic acid (ABA) is one major factor implicated for leaf drying (senescence) under severe drought. ABA synthesized by root under receding soil water conditions or by leaf when the transpiration exceeds water uptake act as a signal to control stomatal responses (Christmann *et al.* 2007). The first study for identification of QTL for ABA accumulation was carried out in detached and partially dehydrated rice leaves, using a population of F_2 plants generated from the lowland × upland cross IR20 (high-ABA) × 63–83 (low-ABA) by Quarrie *et al.* (1997). Total ten QTL for ABA accumulation were identified on chromosome 2, 3, 4 ,6, 7, and 9 in rice under drought stress conditions which together accounted for 55% of the phenotypic variation.

Yield Under drought stress

In Asian countries, rainfed rice is generally planted in the monsoon season. Due to bimodal pattern of rainfall, the rice crop faces three types of stresses viz. early stress, mild- intermittent stress and late stress (Fukai and Cooper 1995). Late stress that is stress at reproductive stage of rice causes the greatest reduction in grain yield. The correlation of component traits such as root traits, leaf rolling, OA with yield and yield components under stress condition was analysed by Babu *et al* (2003). Total 47 QTLs were identified for various water stress indicators, phenology, and production traits under control and water stress conditions in the field, which individually explained 5 to 59% of the phenotype variation. Moreover, root traits also showed positive correlations with yield and yield components under drought stress. For grain yield, four QTLs were identified. Out of these four QTLs, only one QTL, *gys 1.1* was common in two of the trials. Thus, in different studies multiple QTLs for yield and yield components were identified. Most of these QTLs identified were minor effect QTLs, showed inconsistency across environment, and has limited genetic background. Later in past few years, three major effect QTLs for grain yield were reported in four different studies. Bernier *et al.* (2007) analysed a population of 436 random F3-derived lines from a cross between upland rice cultivars Vandana and Way Rarem for yield under reproductive stage drought. The two year screening of the population in upland drought stress and non-stress conditions led to identification of a major QTL *qtl 12.1* on Chromosome 12 which explaining 51% of the genetic variance. The QTL was narrowed down to 10.2 cM region between SSR markers RM28048 (45.2 cM) and RM511 (55.5 cM). The QTL also explained a large proportion of the genetic variance for harvest index (35%), biomass yield (40%), and plant height (36%) and reduced the number of days to flowering under stress conditions while it did not affect any trait under non stress conditions. The authors suggested that effect on grain yield under stress appeared to be due to effects on panicle exsertion and spikelet fertility under stress and also its effect on plant water status. Therefore, the physiological effects of the locus were evaluated for water status in a subset of lines having similar crop growth duration but contrasting genotypes at *qtl12.1* under field and greenhouse conditions (Bernier *et al* 2009b). The *qtl12.1* was found to improve grain yield under drought mainly through a slight improvement (7%) in plant water uptake under water-limited conditions. The authors also carried out 21 field trials in a span of four years (2004-2007) at IRRI in the Philippines and at target environment of Eastern India (Bernier et al 2009a). The relative effect of the QTL on grain yield increased with increasing intensity of drought stress. It had no effect under well-watered conditions but showed an additive effect of more than 40% of the trial mean in the most severe stress treatments. The improvement in grain yield was consistent in nine out of ten direct-seeded upland trials with severe or moderate drought stress. However, there was no effect under well-watered aerobic conditions or under transplanted lowland conditions. These trials confirmed that *qtl 12.1* has a large and consistent effect on grain yield under upland drought stress conditions, in a wide range of environments.

A major QTL was reported on chromosome 1 (Kumar et al 2007), for grain yield under lowland reproductive stage drought stress in doubled haploid lines derived from the population CT9993/IR62266. The stress yield QTL found in the interval EM11_11–RG109 explained 32% of the genetic variation for yield under stress, but only 4% under non-stress conditions thus suggesting that it is related to stress tolerance and not to yield potential. Another QTL, DTY3.1 with large effect on yield was identified in an F4:5 population of 490 recombinant inbred lines (RILs) from the cross Apo/Swarna (Venuprasad et al 2009). This QTL had the largest and most consistent effects on yield under both lowland stress and aerobic conditions. It explained 31% and 15–28% of genetic variation in grain yield under severe stress in low land and in aerobic trials respectively.

To identify large and consistent QTLs for grain yield under reproductive drought stress in elite genetic background, Vikram et al (2011) studied three F3:4 mapping populations derived from crosses of N22 with Swarna, IR64, and MTU1010. Among these, N22 is a drought-tolerant cultivar and Swarna, IR64, and MTU1010 are high-yielding popular mega-varieties grown on millions of hectares in irrigated and rainfed ecosystems in India. These three populations were phenotyped under reproductive stage drought stress (RS) in the dry seasons of 2009 and 2010 at IRRI. A major QTL qDTY1.1, for grain yield under reproductive stage drought was identified on rice chromosome 1. Under RS, this QTL showed an additive effect of 29.3%, 24.3%, and 16.1% of mean yield in N22/Swarna, N22/IR64, and N22/MTU1010, respectively. Recently, QTL $qDTY_{12.1}$ was validated in backcross inbred line population derived from IR74371-46-1-1 × Sabitri where it explained 23.8% of the phenotypic variance under severe lowland drought (Mishra *et al.* 2013). Thus, qDTY12.1 showed a large effect against multiple recipient genetic backgrounds as well as under highly diverse upland and lowland rice ecosystems. A meta analysis of grain yield QTL revealed presence of qDTY12.1 allele in 85% of the lines from a panel of random drought-tolerant lines (Swamy et al 2011). Total 14 meta-QTL were obtained on seven chromosomes. The meta QTLs namely MQTL1.2, MQTL1.3, MQTL1.4, and MQTL12.1 were around 700 kb and corresponded to a reasonably small genetic distance of 1.8 to 5 cM and hence are suitable for use in marker-assisted selection (MAS). The genes underlying all the meta-QTL regions were also analysed by a comparative genomics approach. The identified candidate genes belong to broadly three groups involving stress-inducible genes, growth and development-related genes, and sugar transport-related genes. These genes underlying QTL can be characterized and exploited for improving drought stress tolerance of both lowland and upland rice varieties of rice through molecular breeding and transgenic approach.

Transgenic Approach for Enhancing Drought Tolerance

The transgenic approach for enhancing drought tolerance includes over/under expression of gene expression, which primarily confers water deficit stress tolerance at cellular levels. Initially plants were engineered for effector genes such as LEA proteins (Rohila *et al* 2002), Heat shock proteins (Sato and Yokoya 2007) and organic compatible solutes such as glycine betaine (Sakamoto et al 1998), proline (Su and Wu, 2004) and trehalose (Garg *et al* 2002) which conferred enhanced abiotic stress tolerance including drought. Since drought tolerance is a multigenic trait, need for gene pyramiding or multigene transgenic was realized. This led to the concept of transcriptome engineering or engineering of master regulators. These master regulators are the genes that code for stress sensors, signalling components or transcription factors that regulate several target genes and pathways involved in abiotic stress tolerance. Several genes belonging to family of protein kinases, transcription factors, such as APETALA2/Ethylene Responsive Element Binding Protein family, C-repeat Binding Factor/ Dehydration Responsive Element Binding factor family, MYB/MYC gene family, NAC family have been engineered in to rice for enhanced drought tolerance (Table 2). Despite of having so many genes validated for stress tolerance in rice, the progress in exploiting these genes for crop improvement has not been so encouraging. The reason being that most of these have been validated in early growth stages or under controlled conditions of green / glass house, and have not been tested for drought tolerance and yield in field conditions. However, there are few studies that have been successfully demonstrated the performance of transgenic plants in field conditions. Rice transgenics overexpressing stress responsive gene *SNAC1* (*STRESS-RESPONSIVE NAC 1*) were evaluated at reproductive stage under three drought-stressed conditions viz. refined paddy field under rain off shelter, natural drought-prone field and in PVC tubes. All of the *SNAC1*-overexpressing plants produced significantly higher (22.0 to 34.6%) spikelet fertility than the negative control under all three treatments. there was no yield penalty in the transgenic plant.

Table 2 : List of genes validated in rice for drought tolerance

Gene	Gene Function	Trait	Reference
NAC10 (NAM, ATAF1-2, and CUC2)	Transcription factor	Root-specific overexpression of *OsNAC10* increased the grain yield by 25% to 42% under stress in field	Jeong et al. 2010
DSM1 (drought-hypersensiti ve mutant1)	Raf-like MAPKKK, protein kinase	Tolerance to dehydration stress at the seedling stage	Ning *et al.* 2010
DSM2	-Carotene hydroxylase	Higher spikelet fertility, increased resistance to drought and oxidative stresses at seedling and reproductive stage	Du *et al.* 2010
AP37 (*Apetela 37*)	Transcription factor	*OsCc1:AP37* plants showed increased grain yield by 16% to 57% under severe drought conditions in field	Oh *et al.* 2009
SKIPa (Ski-interacting protein)	Transcriptional factor	Higher relative spikelet fertility and grain yield	Hou *et al.* 2009
bZIP23 (*Basic leucine zipper* 23)	Transcription factor	Tolerance to drought and high-salinity stresses and sensitivity to ABA	Xiang *et al.* 2008

CBF3 (C-repeat binding factor /DREB1A)	Transcription factor	High relative yield and relative spikelet fertility under drought stress in field	Xiao *et al.* 2008
SOS2 (Salt Overly sensitive 2)	Protein Kinase	High relative yield and relative spikelet fertility under drought stress in field	Xiao *et al.* 2008
NPK1 (Nicotiana Protein Kinase 1)	Protein Kinase	High relative yield and relative spikelet fertility under drought stress in field	Xiao *et al.* 2008
LOS5 (Low Osmotic stress 5)	Molybdenum cofactor (MoCo) sulfurase	High relative yield and relative spikelet fertility under drought stress in field	Xiao *et al.* 2008
ZAT10 (C2H2 type zinc finger)	Transcriptional repressor	High relative yield and relative spikelet fertility under drought stress in field	Xiao *et al.* 2008
NHX1 (Na^+/H^+ antiporter)	Vacuolar Transporter	High relative yield and relative spikelet fertility under drought stress in field	Xiao *et al.* 2008
OCPI1 (Oryza sativa chymotrypsin inhibitor-like 1)	Proteinase Inhibitor	Higher relative spikelet fertility and grain yield	Huang *et al.* 2007
NAC6 (NAM, ATAF1-2, and CUC2)	Transcription factor	Improved tolerance to dehydration, salt and blast disease but coupled with growth retardation and low productivity	Nakashima *et al.* 2007
HRDY (hard-D)	AP2/ERF-like transcription factor	Improved water use efficiency, adaptive increase in root biomass under drought stress.	Karaba *et al*. 2007

HVA-1	Group 3 LEA protein	Higher grain yield under stress	Xiao *et al.* 2007
OsCIPK12 (calcineurin B-like protein-interacting) protein kinase 1	Protein kinase	Drought tolerance at the vegetative stage	Xiang *et al.* 2007
SNAC1 (NAM, ATAF1-2, and CUC2)	NAC, transcription factor	Higher spikelet fertility, tolerance at reproductive and vegetative stage	Hu *et al.* 2006
P5CS	- 1-pyrroline-5-carboxylate synthase	Faster growth of shoots and roots of seedlings in comparison with non-transformed plants under drought and salt stress	Su and Wu 2004
MAPK5 (Mitogen-activated protein kinase 5)	Protein kinase	Overexpression positively regulate drought, salt, and cold tolerance and negatively modulate pathogenesis-related gene expression and broad-spectrum disease resistance	Xiong *et al.* 2003
otsA and *otsB*	Trehalose-6-phosphate synthase/phosphatase	Increased trehalose accumulation correlated with higher soluble carbohydrate levels and an elevated capacity for photosynthesis under both stress and nonstress conditions,	Garg *et al.* 2002
CDPK7 (Calcium-dependent Protein Kinase 7)	Protein kinase	Drought, salt and cold tolerance at seedling stage	Saijo *et al.* 2000
HSP17.1	Heat shock protein	Improved recovery after drought stress on rewatering in seedling stage	Sato and Yokoya 2007

The transgenic and control plants had similar performance for all the agromonic traits under well irrigated conditions (Hu *et al.* 2006). Subsequently there were four consecutive studies reported (Xiao *et al* 2007, Huang *et al.* 2007, Xiao *et al.* 2008, Hou *et al.* 2009) in which total nine genes were evaluated for drought tolerance in field and field like conditions. In one study rice transgenic plants carrying *OsLEA3-1* gene driven by three different promoters (*CaMV* 35S, *Actin1,HVA1-like*) were developed and observation on grain yield and spikelet fertility were used to evaluate drought resistance under field conditions. The homozygous T2 and T3 lines over-expressing *OsLEA3-1* under the control of the *CaMV* 35S or the drought-inducible *HVA1-like* promoter were shown to produce significantly higher yield than wild type (WT) under drought stress conditions over two year field trial (Xiao *et al.* 2007). In second study, transgenic rice plants harboring seven different genes (*CBF3, SOS2, NCED2, NPK1, LOS5, ZAT10,* and *NHX*1) driven by constitutive promoter *Actin1* and stress-inducible promoter of rice *HVA22* homolog were developed. These transgenic plants were evaluated based on relative yield and relative spikelet fertility as criteria for drought resistance under field conditions. The results revealed that constitutive and/or inducible overexpression of all the candidate genes except *NCED2* could contribute to the improvement of drought resistance at the level of yield per plant and/or spikelet fertility (Xiao *et al.* 2008). Similarly rice transgenics expressing chymotrypsin inhibitor, namely *Oryza sativa chymotrypsin inhibitor-like 1* (*OCPI1*) and a homolog of Ski-interacting protein (*SKIP*) in rice, *OsSKIPa,* were analysed for reproductive drought stress in paddy field (Huang *et al.* 2007, Hou *et al.* 2009). The transgenic lines had significantly higher relative spikelet fertility and grain yield than the WT under drought stress.

Jeong *et al.* (2010) demonstrated that overexpression of *OsNAC10* in rice driven by constitutive promoter *GOS2* and the root-specific promoter *RCc3* increased the plant tolerance to abiotic stresses such as drought, high salinity, and low temperature at the vegetative stage. The homozygous (T4 and T5) transgenic plants were evaluated for yield components under normal and field drought conditions for two cultivating seasons. The results revealed that root-specific overexpression of *OsNAC10* (*RCc3:OsNAC10*) increased the grain yield by 25% to 42% and by 5% to 14% over controls in the field under drought and normal conditions, respectively. Another remarkable increase in grain yield was reported when *AP37* gene under the control of constitutive promoter *OsCc1* was overexpressed in rice (Oh *et al.* 2009). The *OsCc1:AP37* plants showed increased grain yield by 16% to 57% over controls under severe drought conditions in field. Similar to that of RCc3:OsNAC10, transgenic plants with OsCc1:AP37 also showed tolerance to abiotic stresses in vegetative stage, thus taking care of two important stages of stress in rice crop.

Conclusion and Prospective

In agriculture, the relevance of drought tolerance is estimated by increased or sustained yield under drought stress condition. However, yield in itself is a complex trait, which makes development of drought tolerant plants even more challenging. Success in developing rice varieties with high grain yield under stress condition requires 1) diverse sources of drought tolerance, 2) high precision phenotyping; 3) QTLs with large and consistent effect on grain yield under stress and 4) introgression of these QTLs in to popular / local high yielding varieties. The available germplasm need to be phenotyped for different component traits and genetic base needs to be diversified. The exploitation of germplasm and genomic resources for crop improvement would benefit from emerging field of plant phenomics where functional characterization of genes and component traits such as water relations, root architecture, biomass etc. can be carried out in reliable, efficient and high through put manner. The development of an accurate and uniform phenotyping protocol will further enhance our understating of stress-responsive mechanisms. Till date many QTLs have been mapped for component traits of drought stress tolerance and yield. These QTLs have not been used in marker-assisted selection (MAS) breeding because these genetic studies were largely conducted in selected set of populations with few parents independent of breeding and have not been validated in different genetic background and target environments. The identification of large effect QTLs such as $qDTY_{12.1}$, which has been found to be consistent across stress environment, and genetic background, appears to be promising. Such large effect QTLs can then be dissected by fine mapping or by genomic tools and candidate gene can be identified and effectively used by molecular breeding or developing transgenics. The results from rice transgenics expressing transcription factors and signalling proteins, which have been successfully tested for grain yield at field, demonstrate that transcriptome engineering can be a promising method to engineer drought tolerance in rice. The enhanced drought tolerance of transgenic plants over expression genes under specific promoters such as *OsNAC10* with root specific promoter warrants use of drought-responsive, tissue-and organ-specific promoters for transgene expression. This will help in avoiding problems such as silencing and pleiotropic effects on plant.

The aerobic rice production system appears to be a promising method to increase water productivity of rice, understanding of the physiological and molecular basis of tolerance of rice of aerobic conditions is pivotal for sustaining rice production. The development of appropriate genotypes with enhanced nutrient uptake and use, nematode resistance, and weed competitiveness under water-deficit aerobic and upland conditions is the key to increasing grain yield under these environments. With the available rice genome sequence and high through put genomics and proteomics tools, functions of several rice genes hitherto known needs to be unravelled. Crop management practices such as tillage and nutrient and water management have a significant influence on drought resistance of crops. Hence, studies on the genetic basis of the interaction between crop management practices and drought resistance will be necessary.

References

Babu RC, Nguyen BD, Chamarerk V, Shanmugasundaram P, Chezhian P, Jeyaprakash P, Ganesh SK, Palchamy A, Sadasivam S, Sarkarung S, Wade LJ and Nguyen HT. 2003. Genetic analysis of drought resistance in rice by molecular markers association between secondary traits and field performance. *Crop Sci.* 43: 1457- 1469.

Bernier J, Kumar A, Ramaiah V, Spaner D and Atlin G. 2007. A Large-effect QTL for grain yield under reproductive-stage drought stress in upland rice. *Crop Sci.* 47: 507–518.

Bernier J, Kumar A, Venuprasad R, Spaner D, Verulkar S, Mandal NP, Sinha PK, Peeraju P, Dongre PR, Mahto RN and Atlin G. 2009a. Characterization of the effect of a QTL for drought resistance in rice, qtl12.1, over a range of environments in the Philippines and eastern India. *Euphytica.* 166: 207-217.

Bernier J, Serraj R, Kumar A, Venuprasad R, Impa S, Gowda RPV, Oane R, Spaner D and Atlin G. 2009b. The large-effect drought-resistance QTL qtl12.1 increases water uptake in pland rice. *Field Crops Res.* 110: 139-146.

Christmann A, Weiler EW, Steudle E and Grill E. 2007. A hydraulic signal in root-to-shoot signalling of water shortage. *Plant J.* 52: 167–174.

Courtois B, Ahmadi N, Khowaja F, Price AH, Rami JF, Frouin J, Hamelin C and Ruiz M. 2009. Rice root genetic architecture: meta-analysis from a drought QTL database. *Rice* 2:115-128.

Courtois B, McLaren G, Sinha PK, Prasad K, Yadav R and Shen L. 2000. Mapping QTL associated with drought avoidance in upland rice. *Mol Breed.* 6: 55–66.

Courtois B, Shen L, Petalcorin W, Carandang S, Mauleon R and Li Z. 2003. Locating QTLs controlling constitutive root traits in the rice population IAC 165 × Co39. *Euphytica* 134: 335-345.

Du H, Wang N, Cui F, Li X, Xiao J and Xiong L. 2010. Characterization of the beta-carotene hydroxylase gene DSM2 conferring drought and oxidative stress resistance by increasing xanthophylls and abscisic acid synthesis in rice. *Plant Physiol.* 154, 1304–1318.

Fukai S and Cooper M. 1995. Development of drought resistant cultivars using physio-morphological traits in rice. *Field Crops Res.* 40:67–86.

Garg AK, Kim JK, Owens TG, Ranwala AP, Choi YD, Kochian LV and Wu RJ. 2002. Trehalose accumulation in rice plants confers high tolerance levels to different abiotic stresses. *Proc Natl Acad Sci USA.* 99: 15898-15903.

Hou X, Xie K, Yao J, Qi Z and Xiong L 2009. A homolog of human skiinteracting protein in rice positively regulates cell viability and stress tolerance. *Proc Natl Acad Sci* USA. 106:6410-6415.

Hsiao TC, O'Toole JC, Yambao EB and Turner NC. 1984. Influence of osmotic adjustment on leaf rolling and tissue death in rice (*Oryza sativa* L.). *Plant Physiol.* 75:338–341.

Hu H, Dai M, Yao J, Xiao B, Li X, Zhang Q and Xiong L. 2006. Overexpressing a NAM, ATAF, and CUC (NAC) transcription factor enhances drought resistance and salt tolerance in rice. *Proc Natl Acad Sci. USA* 103: 12987–12992.

Huang Y, Xiao B and Xiong L. 2007. Characterization of a stress responsive proteinase inhibitor gene with positive effect in improving drought resistance in rice. *Planta* 226: 73–85.

Ishimaru K, Shirota K, Higa M and Kawamitsu Y. 2001. Identification of quantitative trait loci for adaxial and abaxial stomatal frequencies in *Oryza sativa. Plant Physiol. Biochem.* 39: 173-177

Jeong JS, Kim YS, Baek KH, Jung H, Ha SH, Choi YD, Kim M, Reuzeau C and Kim JK. 2010. Root-specific expression of OsNAC10 improves drought tolerance and grain yield in rice under field drought conditions. *Plant Physiol.* 153: 185–97.

Karaba A, Dixit S, Greco R, Aharoni A, Trijatmiko K R, Marsch-Martinez N, Krishnan A, Nataraja KN, Udayakumar M and Pereira A. 2007. Improvement of water use efficiency in rice by expression of *HARDY*, an Arabidopsis drought and salt tolerance gene. *Proc. Natl. Acad. Sci. USA,* 104: 15270–15275.

Kumar R, Venuprasad R, Atlin GN. 2007. Genetic analysis of rainfed lowland rice drought tolerance under naturally-occurring stress in eastern India: heritability and QTL effects. *Field Crops Res.* 103:42-52.

Lilley JM and Ludlow MM. 1996. Expression of osmotic adjustment and dehydration tolerance in diverse rice lines. *Field Crops Res.* 48: 185-197.

Lilley JM, Ludlow MM, McCouch SR and O'Toole JC. 1996. Locating QTL for osmotic adjustment and dehydration tolerance in rice. *J. Exp. Bot.* 47: 1427-1436.

MacMillan K, Emrich K, Piepho H-P, Mullins CE and Price AH. 2006. Assessing the importance of genotype x environmental interaction for root traits in rice using a mapping population II: conventional QTL analysis. *Theor Appl Genet.*113:953–64.

Mishra KK, Vikram P, Yadaw R B, Mallikarjuna Swamy BP, Dixit S, Sta Cruz MT, Maturan P, Marker S and Kumar. 2013. A qDTY12.1: a locus with a consistent effect on grain yield under drought in rice. *BMC Genet.* 14: 12.

Nakashima K, Tran LS, Van Nguyen D, Fujita M, Maruyama K, Todaka D, Ito Y, Hayashi N, Shinozaki K and Yamaguchi-Shinozaki K. 2007. Functional analysis of a NAC-type transcription factor OsNAC6 involved in abiotic and biotic stress-responsive gene expression in rice. *Plant J.* 51:617-630.

Ning J, Li X, Hicks LM and Xiong L. 2010. A Raf-like MAPKKK gene DSM1 mediates drought resistance through reactive oxygen species scavenging in rice. *Plant Physiol.* 152:876-890.

Oh SJ, Kim YS, Kwon CW, Park HK, Jeong JS and Kim JK. 2009. Overexpression of the transcription factor AP37 in rice improves grain yield under drought conditions. *Plant Physiol,* 150:1368-1379.

Price AH, Steele KA, Moore BJ and Jones RGW. 2002. Upland rice grown in soil filled chambers and exposed to contrasting water-deficit regimes. II. Mapping QTLs for root morphology and distribution. *Field Crop Res.* 76: 25–43.

Price AH, Steele KA, Moore BJ, Barraclough PB and Clark LJ. 2000. A combined RFLP and AFLP linkage map of upland rice used to identify QTLs for root-penetration ability. *Theor Appl Genet.*100:49–56.

Price AH, Young EM and Tomos AD. 1997. Quantitative trait loci associated with stomatal conductance, leaf rolling and heading date mapped in upland rice (*Oryza sativa*). *New Phytol.* 137: 83-91.

Quarrie SA, Laurie DA, Zhu J, Lebreton C, Semikhodskii A, Steed A, Witsenboer A and Calestani C. 1997. QTL analysis to study the association between leaf size and abscisic acid accumulation in droughted rice leaves and comparisons across cereals. *Plant Mol. Biol.* 35: 155-165.

Ray JD, Yu L, McCouch SR, Champoux MC, Wang G and Nguyen HT. 1996. Mapping quantitative trait loci associated with root penetration ability in rice. *Theor Appl Genet.* 92:627–36.

Robin S, Pathan MS, Courtois B, Lafitte R, Carandang S, Lanceras S, Amante M, HT Nguyen and Z Li. 2003. Mapping osmotic adjustment in an advanced backcross inbred population of rice. *Theor Appl Genet.* 107: 1288- 1296.

Rohila JS, Jain RK and Wu R. 2002. Genetic improvement of Basmati rice for salt and drought tolerance by regulated expression of a barley *Hva1*cDNA. *Plant Sci.* 163: 525–532.

Saijo Y, Hata S, Kyozuka J, Shimamoto K and Izui K. 2000. Over-expression of a single $Ca2^+$ dependent protein kinase confers both cold and salt/drought tolerance on rice plants. *Plant J.* 23: 319-327.

Salekdeh GH, Reynolds M, Bennett J and Boyer J. 2009. Conceptual framework for drought phenotyping during molecular breeding. *Trends Plant Sci.* 14:488-496.

Sakamoto A, Alia and Murata N. 1998. Metabolic engineering of rice leading to biosynthesis of glycinebetaine and tolerance to salt and cold. *Plant Mol Biol* 38: 1011-1019.

SatoY and Yokoya S. 2007. Enhanced tolerance to drought stress in transgenic rice plants overexpressing a small heat-shock protein, sHSP17.7. *Plant Cell Rep.* 27: 329–334.

Steele K A, Price A H, Shashidhar HE and Witcombe JR. 2006. Marker-assisted selection to introgress rice QTLs controlling root traits into an Indian upland rice variety. *Theor Appl Genet.* 112:08–221

Steele KA, Price AH,. Witcombe JR, Shrestha R, Singh BN, Gibbons JM and Virk DS. 2013. QTLs associated with root traits increase yield in upland rice when transferred through marker-assisted selection. *Theor Appl Genet.* (2013) 126:101–108.

Su J and Wu R. 2004. Stress-inducible synthesis of proline in transgenic rice confers faster growth under stress conditions than that with constitutive synthesis. *Plant Sci.*166: 941-948.

Swamy BPM, Vikram P, Dixit S, Ahmed HU and Kumar A. 2011. Meta-analysis of GY QTL identified during agricultural drought in grasses showed consensus. *BMC Genom.* 12:319.

Teulat B, This D, Khairallah M, Borries C, Ragot C, Sourdille P, Leroy P, Monneveux P and Charrier A. 1998. Several QTLs involved in osmotic adjustment trait variation in barley. *Theor Appl Genet*, 96:688–698.

Uga Y, Okuno K and Yano M. 2011. Dro1, a major QTL involved in deep rooting of rice under upland field conditions. *J Exp Bot.* 62: 2485–2494.

Venuprasad R, Dalid CO, Del Valle M, Zhao D, Espiritu M, Sta Cruz MT, Amante M, Kumar A and Atlin GN. 2009. Identification and characterization of large effect quantitative trait loci for grain yield under lowland drought stress in rice using bulk-segregant analysis. *Theor Appl Genet.* 120:177-190.

Venuprasad R, Shashidhar HE, Hittalmani S and Hemamalini GS. 2002. Tagging quantitative trait loci associated with grain yield and root morphological traits in rice (*Oryza sativa* L.) under contrasting moisture regimes. *Euphytica* 128: 293–300.

Vikram P, Swamy BPM, Dixit S, Ahmed HU, Sta Cruz MT, Singh AK and Kumar A. 2011. qDTY1.1, major QTL for rice GY under reproductive-stage drought stress with a consistent effect in multiple elite genetic backgrounds. *BMC Genet.* 12:89.

Weng JH and Chen CY. 1988. Stomatal frequency associated with an esterase band in rice genotypes, *Rice Genet News.* 5: 93–95.

Xiang Y, Huang Y and Xiong L. 2007. Characterization of stressresponsive CIPK genes in rice for stress tolerance improvement. *Plant Physiol.* 144:1416-1428.

Xiang Y, Tang N, Du H, Ye H and Xiong L. 2008. Characterization of OsbZIP23 as a key player of the basic leucine zipper transcription factor family for conferring abscisic acid sensitivity and salinity and drought tolerance in rice. *Plant Physiol.* 148:1938-1952.

Xiao B, Huang Y, Tang N and Xiong L. 2007. Over-expression of a LEA gene in rice improves drought resistance under the field conditions. *Theor Appl Genet.* 115:35-46.

Xiao BZ, Chenb X, Xiang CB, Tanga N, Zhanga QF and Xiong LZ. 2008. Evaluation of seven function-known candidate genes for their effects on improving drought resistance of transgenic rice under field conditions. *Mol Plant* 2: 73–83.

Yadav R, Courtois B, Huang N and McLaren G. 1997. Mapping genes controlling root morphology and root distribution in a doubled haploid population of rice. *Theor Appl Genet.* 94: 619–32.

Yue B, Xue W, Xiong L, Yu X, Luo L, Cui K, Jin D, Xing Y, and Zhang Q. 2006. Genetic basis of drought resistance at reproductive stage in rice: separation of drought tolerance from drought avoidance. *Genetics*.172:1213–28.

Zhang J, Zheng HG, Aarti A, Pantuwan G, Nguyen TT, Tripathy JN, Sarial AK, Robin S, Babu RC, Nguyen BD, Sarkarung S, Blum A and Nguyen HT. 2001a. Locating genomic regions associated with components of drought resistance in rice: comparative mapping within and across species. *Theor Appl Genet.* 103: 19-29.

Zhang WP, Shen XY, Wu P, Hu B and Liao CY. 2001b. QTLs and epistasis for seminal root length under a different water supply in rice (*Oryza sativa* L.) *Theor Appl Genet.* 103:118–123.

Zheng BS, Yang L, Zhang WP, Mao CZ, Wu YR, Yi KK, Liu FY and Wu P. 2003. Mapping QTLs and candidate genes for rice root traits under different water supply conditions and comparative analysis across three populations. *Theor Appl Genet.* 107: 1505-1515.

Zheng HG, Babu RC, Pathan MS, Ali L, Huang N, Courtois B and Nguyen HT. 2000. Quantitative trait loci for root-penetration ability and root thickness in rice: comparison of genetic backgrounds. *Genome* 3:53–61.

❏❏❏

Chapter 2

Global Warming on Photosynthesis and Crop Productivity

K. BALAKRISHNAN[1] AND R. BALASUBRAMANIAM[2]

[1] Professor & Head (Crop Physiology)

[2] Professor and Head (Agronomy)

Agricultural College and Research Institute, Madurai-625 104.

One of the major driving forces for global climate change is the rapid increase in greenhouse gas concentration in the atmosphere. Human activities have lead to a build up of greenhouse gases including water vapour, carbon dioxide, methane, nitrous oxide and chlorofluorocarbon (CFC) resulting in global climate changes. All these gases except water vapour are produced directly or indirectly by human activities such as the burning of fossil fuels, cutting of forests and grasslands for land clearing and use of CFC's in refrigerants, industrial processes and aerosol sprays.

When plants made the transition to land, some 420 million years ago (MYA), the atmosphere may have contained as much as 4,000 to 6,000 µmol CO_2 (Boger, 1980). The rise of vascular plants, and their resultant photosynthesis, was a major factor the decline of the atmospheric CO_2 concentration, which by 300 MYA may have dropped to present values. Thus, the atmospheric CO_2 concentration has fluctuated perhaps by as much as 20-fold over geological time, thus higher values are not a new phenomenon for plants (Bowes, 1996).

Anthropogenic activities such as fossil fuel burning and deforestation lead to the exponential rise in the concentration of atmospheric CO_2 at the rate of 1.8 µmol mol^{-1} per year and are expected to reach 550 ìmol mol^{-1} by 2050. According to Long *et al.* (2004), at present the CO_2 is higher than it was at any time in the past 36 million years. The fossil fuel puts 5.7 Gt, whereas deforestation adds 2.3 Gt of CO_2 in the atmosphere per year, accumulating about 8.0 Gt carbon per year with atmospheric level increasing by 0.5% per year. Of this about 4.7 Gt is being absorbed by oceanic and terrestrial biota per year and 3.3 Gt CO_2 remains in the atmosphere accounting to the rise in its concentration (Trans and Bakwin, 1995). The land biosphere appears to have gained carbon immediately after de glaciation and then leaked 10% of its carbon content ie., about 200 Gt C. Ocean is the largest recipient of C, it gained 150 Gt whereas, atmosphere retained 50 Gt of C (Ciais, 1995).

Agriculture provides both sources and sinks of green house gases (GHG). It has been estimated that 20% of the green house effect is atmospheric concentration of these green house gases which has increased since the 19th century by 30% for CO_2, 145% for methane and 15% for nitrous oxide (Houghton *et al.*, 1996). Per capita CO_2 emission from South Asia was 0.47 Tg in1995, whereas, its projected value is 0.90 Tg for the year 2050. The total regional emission of GHG (in CO_2, equivalent) is 3% of the global total. Its annual growth rate is as high as 6% for India, 8% for Bangladesh and 10% for Pakistan. Agricultural emission is 0.2 tC (tons of carbon) per person from India, 1.2 tC per person for global average and 5.0 tC per person from USA (IPCC, 2001) (Table 1). Lal *et al.* (1998) showed that US agriculture has contributed 116 of the total 1,596 million metric tonnes of carbon equivalents (MMTCE) (ie. 7.3% of US emissions). The global carbon cycle is made up of large carbon reservoirs and its flow is important to agriculture. Estimates of carbon pools indicate the total carbon content in the atmosphere as 750,000 (MMTCE); biomass 550,000 (MMTCE); soil organic matter (1,500,000 MMTCE) and oceans 38,000,000 (MMTCE) (Ocean being the largest pool for carbon (Eswaran *et al.*, 1993) (Table 2). Mineral soil has low organic carbon content compared to organic soil (Lal, 1997). Different cropping systems yielded different amounts of soil carbon, which shows that management options exist for controlling soil organic matter and improving soil carbon levels. Tillage results in a large decline in soil carbon due to significant flush of CO_2. Tillage perturbs the soil system and causes a shift in the gaseous equilibrium by releasing CO_2 due to enhanced oxidation of soil C and organic matter loss (Paustian *et al.*, 1997). Sustainable agriculture requires new technologies for efficient biomass C utilization. While a large portion of crop residue C is recycled to CO_2 through microbial decomposition due to tillage, a small portion remains as human that contributes to long-term sequestration in soil. Islam *et al.* (1999) suggested that troposphere CO_2 enrichment and well-watered conditions might favour an accumulation of low molecular weight and more aliphatic carbon, and ozone stress may favour high molecular weight and more aromatic carbon. The carbon from agricultural crop residues is a small fraction (1%) of the estimated total global C fixed in photosynthesis, however, this one is amenable to management. Torbert *et al.* (2000) examined the impact of elevated CO_2 on crop residue decomposition within agro ecosystems using cotton, wheat, sorghum and soybean with cotton. N mineralization reduced with elevated CO_2, however, C mineralization was found stimulated. Little difference was observed in CO_2 enrichment in soybean but C mineralization reduced in grain sorghum.

Agriculture plays an important role in mitigating the greenhouse gas emissions. It brings down 32% CO_2 emissions, 423% carbon offsets from bio-fuel production, 16% from reduced CH_4 emissions and 10% from reduced emissions of N_2O. Cole *et al.* (1996) indicated that agriculture could contribute to mitigation of climate change by combining appropriate land with best management practices to increase crop production. Lal (1997) suggested that a significant soil carbon sequestration comes from conservation tillage and crop residue management. An understanding of the critical management practices will bring enhanced soil and plant management and the development of new technologies of increased crop production efficiency with minimum impact of environmental quality.

Table 1: Carbon dioxide emissions in the atmosphere (tons of carbon/person).

Agriculture emission	0.2 TC
Global average of carbon emission	2.1 TC
Emission of carbon (USA)	5.0 TC

(IPCC, 2001).

Measurement of CO_2

The measurements of atmospheric CO_2 concentration began at South Pole Antarctica early in1957 and at Mauna Loa (Hawaii) in March 1958 by analyzing the air trapped in bubbles in polar rice cores. The ice core studies revealed at 20,000 years ago, CO_2 concentration was about 205 ìmol mol^{-1}. Its pre-industrial value was 280 μmol mol^{-1}.

Mauna Loa studies also observed an increase of 1.0 ìmol mol^{-1} CO_2 per year from 1959-1982. The 1958 value of atmospheric CO_2 was 316 ìmol mol^{-1} (Keeling *et al.*,1995) and in 1995-2000, it was recorded as high as 360 ìmol mol^{-1} (Collete, 1995) and 378-380 ìmol mol^{-1} in 2006-08 (Uprety *et al.*, 2007) (Table 2). It is currently increasing @ 1.5 to 1.8 ìmol mol^{-1} per annum (Keeling *et al.*, 1995) and likely to be doubled by the end of 21st century.

Table 2: Atmospheric concentration of CO_2.

Time	Concentration (ppm)
Some 20,000 years ago	205
Prior to 1860 (Last 10,000 years)	274-280
Prior to 1900	290
1958	316
1982	340
1990	350
1999	365
2003	371
2007-08	378-380

(Upretty and Reddy, 2008).

Plants are directly affected by rising atmospheric CO_2 concentration because it is the first molecular link from atmosphere to biosphere. It serves as substrate to photosynthetic carbon assimilation. There is concomitant decline in photorespiration process and alteration in stomatal activity (Idso, 1995), C_3 plants (wheat, rice, oilseeds, pulses etc.). respond to elevated CO_2 since they reduce the oxygenase activity of RuBP carboxylase oxygenase enzyme in plants. Lang *et al.* (2004) have clearly attributed that the elevated CO_2 induced photosynthesis to (i) competitive inhibition of the oxygenase activity of Rubisco, and (ii) acceleration of carboxylation because the CO_2 binding site is not saturated at the current CO_2. C_4 plants (sorghum, maize, sugarcane) show little or no photosynthetic response to elevated CO_2 because C_4 pathway is not competitively inhibited by O_2 and is completely CO_2 saturated. The differing responses of C_3 and C_4 species to CO_2 is likely to change their competitiveness and C_3 weeds are likely to become a problem in C_4 crops while C_3 crops may get advantage over C_4 weeds. This could affect the complexion of agro ecosystems. Natural vegetation is also expected to show significant response to rising CO_2, since 95% of the species of higher plants and 100% of the trees are C_3 plants to low CO_2 environment (Hall and Allen, 1993) and has lower CO_2 compensation point. Plants with C_4 photosynthetic pathway show negligible photosynthetic response to elevated CO_2 because the C_4 cycle increases the CO_2 concentration in bundle sheath cells to the point where very little photorespiration occurs and the Calvin cycle is nearly saturated with CO_2 (Ziska *et al.*, 1999). However, there is no consensus on the quantitative effects of increased CO_2 on plant processes and growth because of differences in response at different stages of growth, species of crops and growth limiting environmental factors.

Global Warming and CO_2

The main CO_2 contributors have been the developed countries like US and EU but now other developing countries like China are slowly replacing as the main polluters. However, the per capita emission reveals that the main emitters are the developed countries. If the pace of this emission moves, then it is expected the CO_2 concentration which is currently less than 400 parts per million might shoot above 800 if high emission continues by the end of this century (Table 3).

Table 3: Global CO_2 emission and the country share

Country	CO_2 Emission (tonnes)	% Share of world	Per capita emission (tonnes)
Australia	377	1.42	18.5
Brazil	334	1.26	1.8
China	5060	19.06	3.9
India	1149	4.33	1.1
Japan	1214	4.57	9.5
Russia	1544	5.81	10.8
USA	5841	22.00	19.7
EU	3271	12.32	8.5

(Chaudhury, 2010).

The main cause of the human-induced component of warming is the increased atmospheric concentration of greenhouse gases (GHGs) such as carbon dioxide (CO_2), which leads to warming of the surface and lower atmosphere by increasing the greenhouse effect. Greenhouse gases are released by activities such as burning of fossil fuels, land clearing and agriculture. The contribution of different sectors to the global warming or CO_2 emission are illustrated in Fig.1.

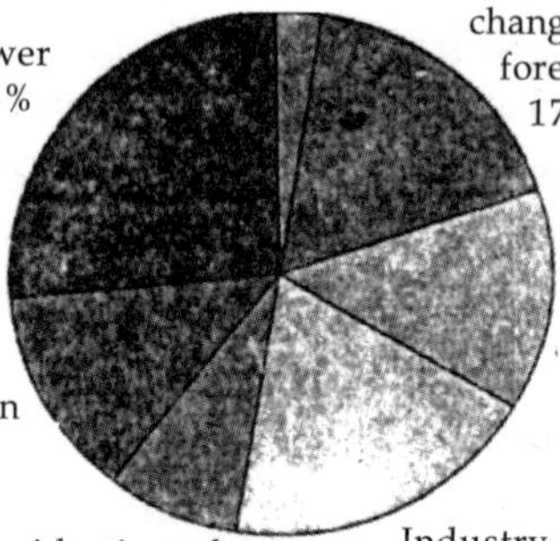

Fig. 1: The ccontribution of different sectors to global warming of CO_2

Plants, through the process of photosynthesis, utilize the energy of sunlight to convert water from the soil ad carbon dixioxide from the air into sugar, starches and cellulose. CO_2 enters a plant through its leaves. Greater atmospheric concentrations tend to increase the difference in partial pressure between the air outside and inside the plant leaves and as a result, more CO_2 is absorbed and converted to carbohydrates. Crop species vary in their response to CO_2. Wheat, rice and soybeans belong to a physiological class (called C_3 plants) that responds readily to increased CO_2 levels. Corn, sorghum, sugarcane, and millets are C_4 plants that follow a different pathway. Higher levels of atmospheric CO_2 also induce plants to close the small leaf openings known as stomata through which CO_2 is absorbed and water vapour is released. Thus under CO_2 enrichment, crops may use less water even while they produce more carbohydrates. This dual effect will likely improve water use efficiency, which is the ratio between crop biomass and the amount of water consumed. The positive impacts of CO_2 enrichment would, to some extent, compensate for the negative impacts of rising mean temperatures (which shorten the growing season of most annual crops and so reduce the yields of current varieties). The possible decline in air quality with increased levels of surface ozone could have serious detrimental effects on crop growth. This positive impact is indicative that agriculture will in many ways help in preserving and combating climate change by adapting to the current stress and help in conserving water. However, the impact is possible in the context of natural farming wherein less carbon intensive agriculture is used. There is greater need to pursue a natural farming with more in situ (or) in farm input resources which would reduce the cost of cultivation as well as help in mitigation.

Biological Evidence of Global Waring

1. *Early blossoming of trees* : Trees are generally a very good bio-indicator of climate change as the flowering in perennial trees takes place as a result of completing the crop-specific required thermal unit/thermal period or degree-days. The very good examples could be the mango trees, which flower according to the thermal regime/period in different parts of the country. The mango trees generally flower in October-November in South India, in December-January in eastern and central India and middle of February-March in north India. But there has been some evidence of flowering of mango trees in December in north India in the year 2004, which was probably due to prevailing higher thermal regime in December. Thus, the flowering behavior of mango, cherry, apple etc. could be used as a very good bio-indicator for climate change.
2. *Appearance of grasses in Antarctica* : Apart from melting the glaciers in the Antarctic region, there has been some evidence of growing grasses in these areas, especially during the year 2004-05, which clearly indicate the possibility of rising atmospheric temperature in these areas.
3. *Changing cropping pattern* : There has been some indication of spatial changes in cropping pattern particularly in the hilly mountain areas of Himachal Pradesh. The traditional regions of apple and other temperate crops which were earlier found suitable for the cultivation of such crops because of prevalence of optimum temperature required for flowering and fruiting of these crops at such places are now gradually becoming warmer due to global warming and may likely to become unsuitable for the cultivation of these temperate crops which require chilling for flowering. Successful cultivation of maize in Bihar during winter season and substituting wheat crop clearly indicates the possibility of global warming and climate change as maize is a C_4 plant and generally performs better under higher temperature.

4. *High latitudinal shifting of temperate crops*: Climate change due to global warming may severely affect the spatial distribution of temperature crops, especially chilling requiring crops viz., apple, apricot, cherry, plum, saffron, cauliflower, cabbage, pea etc., at their traditionally growing places during on as well as off seasons. The areas of temperate crops cultivation in India are mainly confined to higher latitude provinces such as Kashmir, Arunachal Pradesh, Uttarakhand, Sikkim, Himachal Pradesh (Kullu and Manali, Katrayan). The crops grown in these regions may be severely affected in near future due to rise in thermal regime of these regions, as these temperate crops require very narrow range of optimum thermal regime for flowering. There has been some reports on geographical shifting of apple cultivation, as the traditional belts of apple cultivation have become non-conducive for good quality apple production.

Causality in Europe, vulnerability of frog to devastating fungus, killing of large Dimb tree population in Bay Island and northward movement of ticks, bleaching of corals and movement of polar beer, marine fishes and birds are some of other evidences of global warming (Shivay and Rahul, 2008).

Impacts of Global Warming

Biological impacts

1. Direct and indirect effects of rising CO_2
   - C_3 crops are benefitted while C_4 crops are unaffected, but positive effect is manifested only when other factors are not limited.
   - Photosynthetic rate of C_3 crops increased mainly due to reduced photorespiration.
   - Growth and productivity of C_3 crops can be increased directly by rising CO_2 level but may be offset by increased atmospheric temperature and C_3 weeds in C_3 crops.
   - Inputs (fertilizers, irrigated water) and natural resource (radiation, water), use efficiency of C_3 crops is enhanced under rising CO_2 level.
   - C_3 crop plants behave physiologically like C_4 plants without changing their biochemical pathway.

- Biomass, leaf area and yield of C_3 crops are increased significantly without proportional increase in nitrogen uptake, which leads to greater nitrogen use efficiency but lower protein content both in straw and seeds / grains.
- C/N ratio of C_3 plants is likely to increase, thereby depleting nutritional quality especially protein content.
- More beneficial for fodder, sugarcane, potato and root foliage crops where sink is not a limiting factor.

2. Direct and indirect effects of warming
   - Global warming may be beneficial for alpine regions mainly by expanding arable lands and extending crop growth duration, but harmful for temperate, subtropical and tropical regions.
   - High temperature may reduce growth and yield of both C_3 and C_4 crops at their growing sites mainly due to fostering growth and developmental processes leading to shortening of crop growth duration.
   - Global warming reduces growth and productivity of both C_3 and C_4 crops owing to increase in the rate of both dark and photorespiration (respiratory losses).
   - Flowering and fruiting of several temperate crops viz., apple, cabbage, cauliflower, potato, carrot, cherry, plum etc. may be severely affected by global warming.
   - Contrary to CO_2 fertilization, global warming reduces resource utilization efficiency of crops mainly due to low biomass and grain/seed production.
   - High temperature may lead to spatial and temporal loss of crop biodiversity.
   - Chemical integrity and, both physical and nutritional quality, especially sweetness and compaction of protein and starch of variable food and fruit crops may be reduced substantially.
   - Geographical shifting of temperate crops from their traditional areas of cultivation to upper latitude for acquiring optimum thermal regime for flowering and fruiting.

- Spread of existing disease and insect pests and emergence of new biotypes and diseases.
- Greater crop/weed competition for their growth and productivity.

Elavated CO_2 and Photosynthesis

The photosynthetic process sustains life on our planet since, it is the mechanism by which energy from sunlight is used to synthesize bio molecules. Broadly speaking, photosynthesis may be defined as the metabolism by which atmospheric CO_2 is fixed into carbohydrates (photosynthates), with sucrose and starch being the quantitative major end-products of the process. The production of these two metabolites provides organic carbon source for the synthesis of other cellular constituents such as proteins and lipids. In higher plants, photosynthesis takes place in green tissues, mainly in leaves, with the photosynthetic machinery localized within the chloroplast where the newly fixed carbon (namely triose-phosphate or triose-P) can be initially found. Triose P is the immediate source for starch synthesis within the chloroplast; alternatively, this photosynthate can be exported to the cytoplasm and utilized for sucrose production (Fernie *et al.*, 2002). With some exceptions, starch is the principal storage carbohydrate that accumulates in photosynthetic and non photosynthetic tissues of higher plants (Beck and Ziegler, 1989). Long-term storage of starch is confined to plastids (amyloplasts) of cells in specialized non photosynthetic organs (roots, seeds, tubers). Sucrose, meanwhile is the major form by which carbon is moved from photosynthetic to reserve tissues, although some plants may also use it for long-term storage (Edwards and Walker, 1983).

Out of fifteen crops which supply 90% of the world's calories, twelve have the C_3 photosynthetic pathway. C_3 photosynthesis is known to operate at less than optimal CO_2 levels and can show dramatic increase in carbon assimilation, growth and yields. A classical experiment of Kimball (1983) showed biomass increase of 10-143% in several C_3 crops in response to doubling of the ambient CO_2. A literature survey (1994-2009) on the influence of elevated CO_2 among certain C_3, C_4 and crassulacean acid metabolism (CAM) species suggests that most of the C_3 plants showed a significant positive response to photosynthetic acclimation, sorghum and panicum (two C_4 plants) exhibited negative response, whereas Ananas, Agave and Kalanchoe

(CAM plants) showed positive responses to increased CO_2 concentration during growth (Reddy *et al.*, 2010).

The advantage of efficient CO_2 assimilation in C_3 plants has been related to the availability of increased substrate in the atmosphere and in fact that, they do not have to bear the metabolic costs of CO_2 concentrating mechanism at the site of carboxylation (Lang *et al.*, 2004). Photosynthesis in C_3 plants is usually influenced by RuBP (ribose bisposphate) carboxylase-oxygenase (rubisco) (EC 4.1.1.39) and by the accumulation of carbohydrates during carbon assimilation. This activity of the enzyme would cause the combination of CO_2 with RuBP followed by dismutation into two molecules of 3-PGA, which is known as the first committed step in the Calvin-Benson-Bassham cycle (Upretty and Reddy, 2008). As rubisco is substrate-limited by the current atmospheric CO_2 levels, this enzyme has the potential to respond to increases in CO_2 concentration; and have a metabolic control to alter the CO_2 flux during carbon assimilation. Elevated CO_2 is known to be advantageous to the kinetic characteristics of rubisco as it increases the velocity of carboxylation and at the same time competitively inhibits the oxygenase reaction (Ogren *et al.*, 2003). Most of the studies on pot-grown C_3 plants under elevated CO_2 have indicated photosynthetic acclimation, which might be due to soil and nutrient limitation associated with reduced root volume. However, experiments conducted in open top chambers (OTCs) and free atmospheric CO_2 enrichment (FACE) environment showed significant increases in light saturated rates of photosynthesis in several C_3 plants grown at elevated CO_2 (Bosac *et al.*, 1995). The marked increase in net assimilation rates has been explained to be due to increased intercellular CO_2 concentrations.

As implied above, elevated CO_2 atmosphere increases the carboxylation efficiency relative to oxygenation, resulting in reduced photorespiration. Strong reduction in photosynthetic rates under elevated CO_2 conditions has been associated with reduction in the initial slope of A/Ci (A, photosynthetic rate and Ci, internal CO_2 concentration) response curve due to reduced rubisco activity (Lang *et al.*, 2004).

The activity of carbonic anhydrase (CA) (EC 4.2.1.1) was also thought to be crucial in photosynthetic acclimation. CA activities were predicted to enhance the rate of photosynthesis by catalyzing the rapid equilibrium of inorganic carbon and thus increasing the supply of CO_2

across the stroma in the chloroplast (Woodrow, 2009). CA was low in most of the plants exposed to elevated CO_2 (Majean and Coleman *et al.*, 1996), but enhanced CA activities were noticed in Arabidopsis and *Zea mays* (maize or corn), grown at elevated CO_2, indicating difficulties in the interpretation of the role of CA in photosynthetic acclimation (Cervigni *et al.*, 1971). However, research on the response of different isoforms of CA and their polyfunctionality in concentrating CO_2 near the carboxylation site should provide useful evidence for the positive role of CA as a regulator for photosynthetic acclimation.

Changes in photosynthetic rates and acclamatory responses in C_3 crops grown under elevated CO_2 concentration could also be attributed to the feedback metabolic control wherein large accumulation of foliar starch and other carbohydrates could inhibit CO_2 assimilation rates, whereas the plants with potential sinks for carbohydrate translocation and accumulation may not show any down-regulation of photosynthetic capacity suggesting that imbalance in source-sink could be attributed to the variations in the photosynthetic acclimation in different plants (Lang *et al.*, 2004). The relationship between carbohydrate accumulation rates and concomitant increase in respiration in plants under enriched CO_2 is still a matter of controversy. Higher dark respiration rates were recorded in several C_3 plants grown in high CO_2 environment whereas certain C_4 plants did not show any changes in foliar respiration (Bowes, 1993). Further, the reallocation of resources away from the non-limiting processes including rubisco into limiting ones might also result in the acclimation of the photosynthetic apparatus resulting in down-regulation of carbon assimilation rates under elevated CO_2 growth regimes (Maroco *et al.*, 1999).

Table 4: Phenotype characteristics and biomass yields as influenced by CO_2 (ambient–360 ìmol mol^{-1}, elevated–400 ìmol mol^{-1}) in Gmelina arborea recorded at the end of two growth seasons.

Characters	Ambient CO_2	Elevated CO_2
Plant height (cm)	209.45 ± 2.12	359.92 ± 2.78***
Basal diameter (cm)	13.21± 0.59	28.40 + 0.80***
No. of branches	26.20 ± 0.72	44.20 1.19***
Total shoot length (m)	30.73 ± 1.05	59.62 1.43**
No. of leaves / longest shoot	52.70 ± 2.00	108.60 3.12***
Leaf length (cm)	28.10 ± 0.98	37.62 1.12**

Relative plant growth rate	2.97 $\pm$ 0.45	4.08 0.72**
Leaf size expansion rate	3.89 $\pm$ 0.57	9.75 1.02***
Root weight (kg)	3.96 $\pm$ 0.89	5.97 0.85**
Leaf weight (kg)	10.81 $\pm$1.03	15.54 2.12***
Stem weight (kg)	14.86 $\pm$ 0.75	22.13 3.12***
Arial biomass (kg)	25.67 $\pm$ 2.32	37.67 2.98**
Plant biomass (kg)	29.63 $\pm$ 1.67	43.64 3.12***

(Reddy et al., 2010)

A two-season (spring and summer) experiment conducted in the experimental field at the University of Hyderabad (Hyderabad, India) for three consecutive years (2006-2008), using a tree species *Gmelina arborea* Roxb (Verbenaceae) under CO_2-enriched atmosphere in open top chambers (4 x 4 x 4 m), demonstrated a significant up-regulation of photosynthesis throughout the growing season (Table 5). Plants were grown under high CO_2 concentration (460 ìmol^{-1}). After the harvest during all seasons, the biomass yields were markedly higher (48%) in the plants grown under elevated CO_2. Unlike many other reported plant species, growth of Gmelina in elevated CO_2 resulted in increased root volume, stem diameter, altered branching pattern and significant increase in plant height. We attribute the positive correlation between photosynthesis and the morphological characteristics of Gmelina to be due to potential sink capacity which is crucial to the understanding of the physiological, biochemical, genetic and environmental limitations for the productivity of plants grown in CO_2-enriched atmosphere. These potential changes in the growth and development of Gmelina under elevated CO_2 may also be ascribed to increased cell division, cell expansion, cell differentiation and organogenesis, stimulated by increased carbon and more efficient water use. They believe that optimal utilization of resources and well-balanced source-sink activity might enhance carbon gain in plants grown under elevated CO_2. However, the ability of exploiting the extra carbon by any plant species might largely be a function of its inherent structural and physiological attributes, integrated with the plasticity of morphological and anatomical characteristics.

Other factors which can influence plant responses to elevated CO_2 are the growth environment, soil nutrition and the genetic organization of the plant species. The direct effects of rising CO_2 on plant growth and metabolism are a modulation of stomatal conductance, changes in carboxylation capacity, and accumulation of photo-assimilates.

These three regulatory mechanisms will have a wide range of indirect effects on growth and development of plants. Davey *et al.* (2006) postulated that fast growing perennial species would have a greater advantage of having a better sink strength which could result in the up-regulation of carbon metabolism unlike the annual species wherein photosynthetic acclimation has been frequently recorded due to less efficient sink capacity. Different experiments on the effects of elevated CO_2 on photosynthetic capacity in C_3 plants indicate either up- (or) down-regulation, which varies with genetic and interactive environmental factors.

Response of Crops to Elevated CO_2

Wheat

Elevated CO_2 increases the rate of photosynthesis and decreases photorespiration in C_3 plants like wheat. The enzyme responsible for the reaction of CO_2 with ribulose biophosphate is ribulose biophosphate carboxylase-oxygenase (Rubisco). This reaction is not saturated at current CO_2 levels: also oxygen reacts with RUBP, producing phosphoglycolate, which is metabolized, ultimately releasing CO_2 in photorespiration. Doubling CO_2 saturates Rubisco and greatly decreases photorespiration, thus increasing net photosynthesis by 30-40%. Although elevated CO_2 decreases the stomatal conductance with potentially large reductions in water loss, the increased atmospheric CO_2 is more than sufficient to compensate it, so that the internal CO_2 concentration is greatly increased. Wheat exhibits these typical responses.

The stomatal component of photosynthesis is directly affected by elevated CO_2. Samarakoon *et al.* (1995) observed 30-40% of reduction in stomatal conductance in wheat by doubling CO_2. The effects of CO_2 and temperature interaction on wheat biomass and grain yield have been studied in several experiments in UK. Early field studies indicated responses up to 30% increase in yield for a doubling of CO_2. Mitchell *et al.* (1999) using a single genotype of spring wheat cv. Minaret, observed that flag leaf photosynthesis, the major source of assimilate for the grains, increased by 50% and the grain yield by 35% due to elevated CO_2. Similarly, doubling of CO_2 caused 15-27% increase in biomass in experiments stimulating the field temperature (Batts *et al.*, 1998) and a 6-34 % increase when CO_2 was doubled in temperature gradients tunnel. Rawson (1995) obtained increase in biomass ranging from 7 to 36% depending on temperature.

Wheat cultivars vary in their response to both increased temperature (Batts *et al.*, 1998) and to elevated CO_2 (Rawson, 1995; Batts *et al.*, 1998). The variation in the effect of CO_2 may be an indirect consequence of differences in phenology and sink size. Such varietal differences may be important for breeding for climate change. Miglietta and Porter (1992) suggested that slow developing varieties would be better adapted to a warmer climate than faster developing ones and would avoid late spring frosts and onset of dry season. This would not hold if drought began earlier, since faster developing varieties are generally selected in regions prone to late drought (Evans, 1993). Upreti *et al.* (2007) while studying the response of diploid, tetraploid and hexaploid wheats to elevated CO_2 observed that hexaploid responded more than diploid and tetraploid wheats. He attributed it to the greater sink ratio, which was not observed in diploid and tetraploid wheats.

Rice

Rice, being a C_3 plant responds significantly to increased CO_2. Several long and short term CO_2 enrichment studies showed that doubling the current ambient CO_2 increased leaf photosynthesis rate by 30 to 70%. The magnitude of this increase depended on the particular rice cultivars, growth stage and environment (Morrison and Gifford, 1983). Ziska *et al.* (1996) observed that the effect of CO_2 enrichment can be mainly attributed to the responses at unit leaf area level. CO_2 enrichment causes partial stomatal closure, increased stomatal resistance (Uprety *et al.*, 2002), reduced leaf, whole canopy temperature.

Uprety *et al.* (2002) demonstrated that the stomata of lower epidermis were more sensitive to CO_2 enrichment than those of the upper epidermis. They have attributed the reduction in stomatal conductance to balance the resource use and to maintain ci : ca ratio under elevated CO_2. Such changes occurred in rice cultivars through its effect on stomatal size and number.

Uprety *et al.* (2002) also indicated that CO_2 induced increase in leaf size affected the stomatal density by reducing the penetration of radiation and spreading the stomata differentially. The reduction in stomatal density may be a consequence of increase in size of epidermal cells although epidermal cell density decreased indicating greater effect on cell expansion than on cell division. This increase in cell expansion has been attributed to increased osmotic potential of leaf cells with their high saccharide contents, which cause the cells to absorb more water and thus enlarge (Delucia *et al.*,1985). These changes in the stomatal parameters may help rice cultivars to regulate the gas exchange processes and adjust under elevated CO_2 conditions.

Rice displays higher water use efficiency under elevated CO_2 primarily due to increased biomass production and partly due to reduced transpiration. Morrison and Gifford (1983) found CO_2 induced increase in WUE by 53 to 63% in potted rice plants.

Elevated CO_2 increases phonological development rate and hence reduces the number of days to heading (Nakagawa *et al.*, 1993). This CO_2 effect on rice development rate has been shown to be temperature dependent. The CO_2-induced accelerated development rate may be due to increase in stomatal resistance, decreasing the transpiration cooling which increases the tissue temperature and or due to high C:N ratio (Ziska *et al.*, 1996). Baker *et al.* (1990) observed reduction in final main stem leaf numbers with elevated CO_2 due to acceleration in the phenological development. Sakai *et al.* (2001) showed that increase in rice biomass by elevated CO_2 results more from the increase in carbon gain at early than at later stages of growth.

Horie (1993) analyzed published data on long-term CO_2 enrichment studies on biomass production and concluded that under field conditions, CO_2 enrichment increases biomass production by 24% across a wide range of air temperature changes. Generally, the pot-grown plants showed a strong CO_2 and temperature interaction for biomass production; may be that, pot grown plants continue to produce tillers for a longer time than does field-grown rice. This is possibly due to reduced mutual shading in potted plants compared with field grown plants, and is similar to the light interception effects on dry-matter production and their development. Further studies also showed that the relative enhancement of rice biomass production with CO_2 depends strongly on plant nitrogen content. For pot-grown rice plants, CO_2 enrichment, under drought stress conditions resulted in increased relative biomass production compared with drought stress ambient controls (Rogers and Dahlman, 1983). This effect was attributed to higher leaf water potential for the CO_2 enrichment plants under drought conditions.

Several models have been developed for simulating global climate change effects on rice to predict the effects of CO_2, temperature and other environmental factors on rice growth and yield formation processes, and to develop strategies for production technologies adapted to climate change. Model developed by Bannayan *et al.* (2005)

was successful in simulating the CO_2 induced increase in rice yield but failed to explain the interaction with N in the yield response to elevated CO_2.

Soybean

Soybean leaf photosynthetic rates at elevated CO_2 were probably maintained higher because of an increase of one layer of palisade cells in the leaves. This modification of leaf structure provided more surface area of cells within the leaf with a similar amount of Rubisco per unit leaf area, as well as possibly providing better light distribution within the leaf. Campbell *et al.* (1988) found that soluble protein was relatively stable in soybean leaves, and that Rubisco protein was a constant 55% of leaf soluble protein. These two reports showed that Rubisco activity per unit protein was not down-regulated in activity or amount in soybean grown in elevated CO_2.

It is well known that elevated CO_2 decreases photorespiration or the photo-oxidative pathway of C_3 plants by increasing the reactivity of Rubisco for CO_2. The respiration rate per unit mass of plants appears to decrease with elevated CO_2, probably because of the greater relative accumulation of both structural and non-structural carbohydrates and the lesser relative accumulation of proteins. Jones *et al.* (1985) observed that plants grown under double CO_2 and subjected to severe drought yielded about 20% more than non stressed plants grown under ambient CO_2.

The CO_2 influences water use in two ways : firstly, through control in stomatal conductance by decreasing it and also by promoting growth, producing more leaves that will increase the transpiring area per unit of land area. The seasonal cumulative water use decreased by 12% for doubled CO_2 (Jones *et al.*, 1985). This percentage decrease in transpiration is camparable to that simulated by whole crop energy balance models for closed canopies (Boote *et al.*, 1997). The WUE of soybean is always increased by elevated CO_2 at both the leaf and canopy levels. Allen *et al.* (1985) showed that CO_2 enrichment increased WUE mainly by an increase of CER and secondarily by a decrease of transpiration rate.

Maize

The unique combination of the photosynthetic metabolism and leaf anatomy of this C_4 plant is the elevation of CO_2 concentration at the site of Rubisco in bundle. This elevated concentration of CO_2 has two effects. Firstly, competitive inhibition of the oxygenase reaction of rubisco eliminates most of the $C-_2$ photosynthetic oxidative or photo respiratory pathway activity and its expenditure of energy associated with photorespiration. Secondly, it allows rubisco to approach its maximum rate of catalysis despite its low affinity for CO_2. The PEP carboxylase, from C_4 photosynthetic tissues has a high affinity for CO_2, and photosynthesis is saturated at low CO_2 concentration (Lang, 1999).

Increased growth under elevated CO_2 has been reported for a number of C_4 species including maize and sorghum (Wong, 1979). Samarakoon and Gifford (1996) observed 35% more leaf area and 50% more dry matter due to the interactive effect of elevated CO_2 and soil drought in maize.

Reported enhanced growth under elevated CO_2 in C_4 plants like maize has generally been attributed to increased water use efficiency. However, there are reports of photosynthetic stimulation of C_4 plants under elevated CO_2, although this enhancement may be limited to conditions where elevated CO_2 ameliorates the mild drought stress effect (Samarakoon and Gifford, 1996).

Sorghum

Conley *et al.* (2001) observed that elevated CO_2 reduced the cumulative evapo transpiration by 10 % when plants were given ample water and by 4% under severe drought stress. Water use efficiency based on grain yield increased by 9 and 19 %, whereas, based on total biomass, it increased by 16 and 17 % in wet and dry plots, respectively, because of CO_2 enrichment. These observations indicate that in future at high CO_2 environment, water requirements of irrigated sorghum will be lower than at present, while dry land productivity will increase, provided global warming is minimal. Wall *et al.* (2001) even suggested that a rise in atmospheric CO_2 will minimize the deleterious effects of drought on physiological functions and growth, and thereby expand the range where a viable sorghum crop can be raised, especially in more marginal production areas.

In a study on the quantification of the effect of elevated CO_2 on the energy balance and evapo transpiration of sorghum, Triggs *et al.* (2004) reported that WUE in sorghum increased by about 128 % at elevated CO_2 under wet conditions due to saving of water for growth, whereas under dry conditions, it increased by about 16% due to greater relative growth with slightly higher amount of water. Thus, even if further climate change results in less water availability for agriculture, higher atmospheric CO_2 concentration may benefit some C_4 crops.

Cotton

Elevated CO_2 generally enhances the leaf and canopy photosynthesis, growth and productivity of cotton (Kimball, 1983). It often reduces stomatal conductance (Morison, 1987) and increases its water use efficiency.

Cotton canopies grown continuously in elevated CO_2 responded to increasing rates of photosynthesis. Both the initial slopes of the light response curves were greater in elevated CO_2. Additional CO_2 resulted in higher rates of photosynthesis due to the efficiency of the carboxylase fixation sites and diffusion of the gas to fixation sites (Reddy *et al.*, 1995). Radin *et al.* (1987) measured at 52% increase in net leaf photosynthesis under elevated CO_2 and drought conditions.

Stem growth rates increased by 20 %, leaf area by 25 %, while roots and fruits were 30 % and 26 % greater, respectively, due to doubling of CO_2. The harvest index did not change (Reddy *et al.*, 1998).

More carbon was fixed in plants grown in high CO_2 at all levels of water, nutrients and range of temperature. The average temperature above 30°C caused young bolls to abscise. Doubling CO_2 did not ameliorate the adverse effects of high temperature on cotton fruit retention (Reddy *et al.*, 1995). Further, water and nutrients are scarce resources that will be required in greater quantity to utilize high CO_2 environment fully for cotton production. Resource rich economics will benefit more from greater yield potential resulting from increased CO_2 because they can more adequately provide these yield limiting variables.

Black gram

The response of black gram (*Vigna mungo*) to two levels of elevated carbon dioxide (550 and 700 ppm) in terms of growth and yield was investigated and compared with ambient CO_2 level (365 ppm) using open-top chambers. The growth parameters viz., length and weight of root and shoot, root: shoot ratio, leaf area and weight were significantly increased at 700 ppm CO_2 when compared with 550 ppm CO_2 which were 65.4% and 39%, respectively compared to the ambient (chamber) control. The increase in total seed yield at 700 ppm (129%)

was due to an increase in number of pods per plant and 100 seed weight, whereas at 550 ppm (88.7%), it was due to an increased number of pods plant^{-1} and seeds pod^{-1}. The indication of higher responses for root and leaf at initial growth stages at the higher elevated level of CO_2 (700 ppm), which leads to better root establishment, achieving early photosynthetic efficiency and also better biomass production, and its improved partitioning can be reckoned as a positive aspect of increasing concentrations of CO_2 in atmosphere. The harvest index increased significantly to 35.7 and 38.4 % at 550 and 700 ppm, respectively which is a very important phenomenon in pulses for breaking the yield barrier (Table 5). (Vanaja *et al.*, 2007).

Table 5: Yield parameters of black gram under elevated CO_2 at 550 and 700 ppm over the control (365 ppm).

	Treatments			Increase (%)		
	Control	550 ppm	700 ppm	550 ppm vs control	700 ppm vs control	700 ppm vs 550 ppm
Total biomass	61.1	84.9	101.1	39.0	65.4	19.08
No. of pods	158	187	239	18.4	51.3	27.81
Pod weight (g)	35.6	66.5	78.2	86.8	119.7	17.59
Seed weight (g)	17.4	32.9	39.9	88.7	128.9	21.28
100 seed weight (g)	2.59	2.65	3.91	2.3	51.0	47.55
Harvest Index	28.5	38.7	39.5	35.7	38.4	2.07

(Vanaja et al., 2007).

Brassica species

Uprety *et al.* (1995) studied the interactive effect of elevated CO_2 and moisture stress on the morpho-physiologicl processes of Brassica species, *viz., B. campestris, B. juncea, B. carinata* and *B. nigra.* It was observed that the rate of photosynthesis, stomatal resistance and plant growth increased considerably under elevated CO_2. This effect was highly significant under moisture stress conditions mitigating the adverse effect of these components in plants. However, relatively drought susceptible species *B. campestris* and *B. nigra* responded to CO_2 markedly compared to less sensitive *B. carinata* and *B. juncea.* Morison and Gifford (1984) reported that in oil seed rape (*B. napus*), high CO_2 increased the leaf area by 40%, dry weight by 48% and WUE by 50%.

Uprety *et al.* (1997) also demonstrated the sustenance of the stimulatory effect of elevated CO_2 up to seed formation stage in Brassica. The sequestration of carbon in plant parts leads to the accumulation of greater carbohydrate components in Brassica seeds. The maximum increase was in starch content followed by reducing and non reducing sugars. The stress induced decrease was ameliorated by CO_2 enrichment from 19.9 to 1.1 % for starch, from 24.9 to 1.4 % for reducing sugars, from 17.3 to 3.5 % for non reducing sugars, and from 20.12 to 2.76 % for total sugars, respectively. Such a marked recovery among carbohydrate components may be due to greater photosynthate transfer to seeds. Another carbon component, which was significantly boosted by increased CO_2 in Brassica, was its oil content.

Tuber crops

Scientists have repeatedly recognized that species with large below ground sinks for carbon (Farrar, 1996) and with apoplalstic mechanisms of phloem loading (Komor *et al.*, 1996) are likely to be the best responsive to rising atmospheric CO_2 levels. It has been demonstrated that growth or storage sink limitations are possibly major factors constraining responses of plants to elevated CO_2.

Elevated CO_2 has a positive effect due to more efficient energy utilization by suppression of photorespiration. At high PPED, the positive effect is enhanced because the carboxylation rate, linked with the availability of CO_2, is rate limiting. Schapendonk *et al.* (1995) reported 24 to 40 % increase in tuber yield due to doubling CO_2.

The presence of larger sinks for assimilates in tuber and root crops makes these crops more responsive to the elevated CO_2 for large growth and yield (Kimbal, 1983). Late cultivars were more responsive to the elevated CO_2 for tuber yield than a greater tuber mass. Wheeler *et al.* (1994) showed that potato yield response to CO_2 continuously increased upto 1,000 ìmol^{-1}. In this experiment, CO_2 enrichment increased tuber yield and total biomass dry weight by 39 and 34 % respectively, under short day and low light conditions, by 27and 19 % under short day and high light conditions, by 9 and 9 % under long day and lowlight conditions, and dry weight decreased by 9 and 9 % under long day and high light conditions. In other studies on root crops, yield of carrot and kohlrabi (Mortensen, 1994; Wheeler *et al.*, 1994) increased with CO_2 doubling, despite significant interactions between the availability of phosphorus and the CO_2.

Vegetable crops

In most crops, increased CO_2 improves WUE because of decline in stomatal conductance (Rogers and Dahlman, 1993), potentially decreasing drought susceptibility and reducing irrigation requirements. However, the effect of decreased transpiration on vegetable crop yields is unlikely to be large since vegetables are irrigated in most production areas. Physiological disorders, associated with excessive transpiration, may be reduced (Wien, 1997). Respiration of leaves and roots in the dark lowers at high CO_2. The direct, short term and readily reversible effect of CO_2 on respiration has been noted in tomatoe, lettuce, pepper, peas and maize. It is apparently caused by inhibited respiration rather than stimulated carboxylase activity. However root : shoot ratio caused by elevated CO_2 may help in the establishment of vegetative propagated material and in transplants. Bhattacharya *et al.* (1985) observed that elevated CO_2 during propagation increased root number and length in sweet potato. Since propagation and establishment are critical phases for many vegetable crops, especially in terms of increasing uniformity and early yield, increases in these processes from elevated CO_2 would be particularly valuable.

Carbon dioxide enrichment to the extent of about 1,000-1,200 μmol^{-1} mol^{-1} has been used commercially to increase yields of greenhouse grown lettuce (Wittwer, 1986). Hunt *et al.* (1984) indicated about 50% increase in total dry weight in these lettuce plants. Poorter (1993) observed 110 % increase in carrot yield due to doubling of CO_2.

Tree crops

Forest ecosystems cover little more than one-quarter of the earth's land surface (3.5 billion ha; (FAO, 1995). However, approximately 85% of plant carbon and 35% of soil C is found in forests (Kirschbaum and Fischlin, 1996). Forest ecosystems constitute a major terrestrial carbon reservoir (60% stored carbon), which may exert significant impact on atmospheric CO_2 concentration as a sink. However, forests also act as source to increase atmospheric CO_2 through changes in land use such as deforestation and creation of pasture and range land.

Global production of woody forest products is about 4 billion m^3 and that of paper and pulp is 0.5 million tonnes. Over 33 million ha of the world's terrestrial surface is covered with fruit plantation (Goudrianan, 1986). All trees are photosynthesizing C_3 plants (Ceulemans and Saugier, 1991). The predicted rapid changes in CO_2 and increases in temperature will significantly affect currently growing and future forests. Long lived plants have more time to acclimate to changing environmental conditions than short-lived organisms.

In the majority of short term CO_2 enrichment studies on trees, photosynthesis increased, magnitude of response might vary due to inter and intra species differences, variation in growth conditions or because of duration of CO_2 enrichment (Gunderson and Wullschleger, 1994).

Although the major role of a leaf is to be a source, the production of new leaves may be considered as a sink. The CO_2 induced increase in leaf area may be due to enhanced new leaf production and increased individual and total tree leaf area (Guehl *et al.*, 1994). The increased leaf area may affect tree structure and architecture. The larger leaf area and / or altered crown architecture will result in earlier canopy closure, and thus enhanced competition. A reduction in stomatal conductance under elevated CO_2 might have a significant effect on water transport in trees. Hydraulic conductivity decreased with elevated CO_2 (Tognetti *et al.*, 1996).

Fruits represent a major C sink in tree crops. When horticultural productivity is the goal, allocation of resources towards reproductive processes must be maximized. However, tree must also preserve its growth potential for future years; thus a delicate C balance is to be maintained between vegetative and reproductive needs. Among all tree crops, cultivated fruit trees are most adequately supplied with water and nutrients and would not have much constraints to a positive CO_2 response. Idso and Kimball (1997) studied the response of orange trees under elevated CO_2 and observed increase in vegetative tree growth and number of fruits.

Ozone enters the leaf through the stomata and diffuses within the apoplast, where it rapidly decomposes to reactive oxygen species. The impact of ozone on growth is primarily caused by changes in photosynthetic physiology, decreased stomatal conductance and accelerated senescence. Elevated atmospheric CO_2 reduces some of these negative effects through stomatal closure. It also seems unlikely that UV-B radiations would substantially alter the CO_2 responses of trees. In general, trees respond positively to elevated CO_2 and negatively to enhanced UV-B radiation.

CAM. crops

Zhu *et al.* (1997) observed 23 % more dry mass of *Ananas comosus* with 730 ìmol mol^{-1} CO_2 treatment for 4 months. Its nocturnal increase in leaf acidity was 42 % higher at higher CO_2. Graham and Nobel (1996) reported 49 % increase in photosynthesis in *Agave deserti* plants grown under elevated CO_2 and more leaves were produced, with increase in total leaf area, net CO_2 uptake per unit leaf area, and enhanced dry mass accumulation by 88 %.

Summary

Elevated CO_2 cause increase in crop growth and yield due to increased photosynthesis, decreased photorespiration and decreased stomatal conductance. Increase in CO_2 negates negative effect of rise in temperature up to certain degree. A 2°C increase in temperature may cause 10-20 % decrease in grain yield of both rice and wheat due to shorter duration of crop growth. The protein content of soybean grains decrease with increased CO_2 concentration, whereas the total quantity of mineral nutrients and crude protein accumulated in grains per hectare still increased due to increase in crop yield. Elevated CO_2 increases the availability of soil nitrogen and phosphorus because of increased mineralization and activity of phosphatase enzyme in the rhizosphere. C_3 plants are likely to compete even more than now against C_4 crops and vice-versa.

Future Research Priorities

There are several research needs to improve the existing CO_2 enrichment technologies as field experimental tools. The fluid dynamics and heat exchange properties particularly the turbulence and altered flow path of wind within the canopy should be studied to determine the potential effects of increasing CO_2 on plants grown in the field.

Increasing CO_2 may present opportunities to breeders to select for efficient translocation of assimilate from leaves towards roots. Physiological mechanisms involved in down-regulation of rubisco and genetic control on these processes need to be studied. Impacts of interactive effect of elevated CO_2, ozone, temperature, drought and

UV-B radiation need further investigations especially for impact assessment analysis for yield and total biomass production. With the recent advancement in biotechnology, genetic manipulation of plant characters have become possible to result in over- and under-expression of individual genes and introduction of novel genes through the changes in basic metabolism and growth as affected by higher atmospheric CO_2. This might help in constructing plants for future high CO_2 environment. These genetically manipulated plants may turn a powerful tool in studying plant responses to elevated CO_2 and identifying the plant type for future.

References

Allen, L.H. Jr, P. Jones and J.W. Jones. 1985. Rising atmospheric CO_2 and evapo-transpiration. In: *Advances in Evapo Transpiration*. Proceedings of National Conference. ASAF Publication No.14-85. American Society of Agricultural Engineers, St. Joseph, Michigan, pp.13-27.

Baker, J. T. and I. Allen, H. Jr. Boote, K.J., P. Jones and J.W. Jones. 1990. Developmental responses of rice to photoperiod and carbon dioxide concentration. *Agiculture and Forest Meteorology,* 50 : 201-10.

Bannayan, M., K. Kobayashi, H. Kim, M. Lieftering, M. Okada and S. Miura. 2005. Modeling the interactive effects of atmospheric CO_2 and N on rice growth and yield. *Field Crop Research,* 93: 237-51.

Batts, G.R., R.H. Ellis, J. I. L. Morison, P.N. Nkemka, P.J. Gregory and P. Hadley. 1998. Yield and partitioning in crops of contrasting cultivars of winter wheat in response to CO_2 and temperature in field studies using temperature gradient tunnels. *Journal of Agricultural Sciences,* Cambridge, 130 : 17-27.

Beck, E. and P. Ziegler.1989. Biosynthesis and degradation of starch in higher plants. *Annual Review of Plant Physiology and Plant Molecular Biology,* 40: 95-117.

Bhattacharya, S., N.C. Bhattacharya and B. R. Strain. 1985. Rooting of sweet potato stem cuttings under CO_2 enriched environment and with IAA treatment. *Horticultural Science,* 20: 1109-10.

Boasc, C., S. D. L. Gardner, G. Taylor and D. Wilkins. 1995. Elevated CO_2 and hybrid poplar: a detailed investigation of root and shoot growth in physiology of *Populus euramericana,* Primo. *Forest and Ecological Management,* 74:103-106.

Boger, E. 1980. *Biochemical and Photosynthetic Aspects of Energy Production,* pp, 175-180. Pietro A.S. (ed). Acdemic Press, New York.

Boote, K.J., N.B. Pickering and Allen, L.H. Jr. 1997. *Plant modelling : advances and gaps in our capability to predict future crop growth and yield.* In: Allen, L.H. Jr,

Kirkham, M.B., Olsyzk, D.M. and Whitman, C.E. (eds). Advances in Carbon Dioxide Effect Research, ASA Special Publication No.61, ASA-CSSA-SSSA, Madison, Wisconsin, pp.179-28.

Bowes, G. 1993. Facing the inevitable: Plants and increasing atmospheric CO_2 levels. *Annual Review of Plant Physiology and Plant Molecular Biology,* 44: 309-332.

Bowes, G. 1996. Photosynthetic responses to changing atmospheric carbon dioxide concentration. In : Baker N.R. (ed.) *Photosynthesis and the Environment,* pp.367-407. Kluwer Academic Publishers, Dordrecht, Netherlands.

Campbell, W.J., Allen, L.Hr. Jr. and G. Bowes. 1988. Effects of CO_2 concentration on rubisco activity, amount and photosynthesis in soybean leaves. *Plant Physiology,* 88: 1310-16.

Cervigni, T., F. Teofani and C. Bassanelli. 1971. Effect of CO_2 on carbonic anhydrase in *Avena sativa* and *Zea mays. Photochemistry,* 10: 2991-2994.

Ceulemans, R. and B. Saugier. 1991. Photosynthesis. In: Raghavendra, A.S. (ed.) *Physiology of Trees,* pp.21-49. John Wiley & Sons, New York.

Chaudhury, K.B. 2010. Climatic Change and Indian Agriculture. *Kisan Ki Awaaz (August),* pp.4-9.

Ciais, P., P.P. Trans, M. Trolier, J. W. C. White and R. J. Francey. 1995. A large northern hemisphere terrestrial CO_2 and indicated by the C^{13} and C^{12} ratio of atmospheric CO_2. *Science,* 269: 1098-102.

Cole, C.V., C. Cerri, K. Minami, A. Mosier, N. Rosenberg and D. Sauerbeck. 1996. Agricultural options for mitigation of greenhouse gas emissions. In: Watson, R., Zinyowera, M.C. and Moss, R. (eds). *Climate change 1995. Impacts, Adaptations and Mitigation of Climate Change: Scientific and Technical Analyses,* IPCC Working Group II, Cambridge University Press, Cambridge, U.K, pp. 745-71.

Collete, E. 1995. Will plant profit from high CO_2. *Science,* 268: 645-55.

Conley, M.M., B. A. Kimball, T. J. Brooks, Printer, P.J. Jr., D. J. Hunsaker, G. W. Wall, N. R. Adam, R. L. LaMorte, A. D. Matthias, T. L. Thompson, S. W. Leavitt, M. J. Ottman, A. B. Cousins and J.M. Triggs. 2001. CO_2 enrichment increases water-use efficiency in sorghum. *New Phytologist,* 151: 407-12.

Davey, P.A., H. Olcer, O. Zakhleniuk, C.J. Bernacchi, C. Calfapietra, S.P. Long and C. A. Raines. 2006. Can fast growing trees escape biochemical down-regulation of photosynthesis when grown throughout their complete production cycle in the open air under elevated carbon dioxide. *Plant Cell and Environment,* 29: 1235-1244.

Delucia, E.H., T.W. Sasek and B.R. Strain. 1985. Photosynthetic inhibition after long term exposure to elevated levels of atmospheric carbon dioxide. *Photosynthesis Research,* 7: 175-84.

Dugas, W.A., S. A. Prior and H. H. Rogers. 1997. Transpiration from sorghum and soybean growing under ambient and elevated CO_2 concentrations, *Agriculture and Forest Meteorology*, 83 : 37-48.

Edwards, G. E. and D. A. Walker. 1983. *C_3, C_4: Mechanisms of Cellular and Environmental Regulation of Photosynthesis*. Berkeley, C.A: University of California Press.

Eswaran, H., Van De Berg and P. Reich. 1993. Orange carbon in soils of the world. *Journal of Soil Science Society of America*, 57: 192-94.

Evans, L.T. 1993. *Crop Evolution, Adaptation and Yield*. 500 pp. Cambrdge University Press, Cambridge, U.K.

FAO. 1995. *Production Year Book*, 49: 70-71.

Farrar, J.F. 1996. Sinks, integral parts of whole plant. *Journal of Experimental Botany* 47: 1273-80.

Fernie, A.R., L. Willimitzer and N. Trethewey. 2002. Sucrose to starch: a transition in molecular plant physiology, *Trends Plant Science*, 7: 35-41.

Graham, E.A. and P. S. Nobel. 1996. Long term effects of a doubled atmospheric CO_2 concentration on the CAM species *Agave deserti*. *Journal of Experimental Botany*, 47: 61-69.

Guehl, J. M., C. Picon, G. Aussenac and P. Gross. 1994. Interactive effects of elevated CO_2 and soil drought on growth and transpiration efficiency and its determinants in two European forest tree species. *Tree Physiology*, 14: 707-24.

Gunderson, C.A. and S. D. Wullschleger. 1994. Photosynthetic acclimation of forest trees to a doubling of atmospheric CO_2: a broader perspective. *Photosynthesis Research*, 39: 369-88.

Hall, A.E. and Allen, H.Jr. 1993. *Designing Cultivars for the Climatic Conditions of the next century*. Buxton, D.R., Shibles, R., Forsberg, R.A., Blad, B.L., Asay, K.H., Paulson, G,.M. and Wilson, R.F. (eds), International Crop Science Society of America, Madison, Wisconsin. Pp. 291-97.

Horie, T. 1993. Predicting the effect of climate variation and elevated CO_2 on rice yield in Japan. *Journal of Agricultural Meteorology*, 48: 567-74.

Houghton, J.T., L.G. MeiraFilho, B. A. Callander, N. Harris, A. Kattenberg and K. Maskell. (eds.). (1996). *Climate Change 1995 : The Science of Climate Change* IPCC. 572 pp. Cambridge University Press, Cambridge, U.K.

Hunt, R., J. W. Wilson, D. W. Hand and D. G. Sweeney. 1984. Integrated analysis of growth and light interception in winter lettuce. I. Analytical methods and environmental influences. *Annals of Botany*, 54: 743-57.

Idso, S.B. 1995. CO_2 and Biosphere. The Incredible Legacy of the Industrial Revolution. Special Publication, University Minnesota, St. Paul, USA.

Islam, K.R., CL. Mulchi and A. A. Ali. 1999. Tropospheric carbon dioxide or ozone enrichments and moisture effects on soil organic carbon quality. *Journal of Environmental Quality*, 28: 1629-36.

International Panel on Climate Change, Report, 2001.

Jones, P., J. W. Jones and Allen, L. H. Hr. 1985. Seasonal carbon and water balances of soybean growth under CO_2 and water stress treatments in sunlit chambers. *Transaction of the American Society of Agriculture Engineers* 28: 2021-28.

Kaur, R. and Rajni. 2010. Impact of increasing CO_2 on crop productivity. *Indian Farming (June)*, 23-27.

Keeling, C.D., M. Whorf, T. P. Wahlen, and J. Van der Plicht. 1995. Inter annual extremes isn't the rate of rise of atmospheric carbon dioxide since 1980. *Nature*, 375: 666-70.

Kimball, B.A. 1983. Carbon dioxide and agricultural yield: An assemblage and analysis of 430 prior observations. *Agron. J.*, 75: 779-788.

Kirschbaum, M.U.F. and A. Fischlin. 1996. Climate change impacts on forests. In :Watson, R.T. Zinyowera, M.C. and Ross, R.H. (eds.). *Climate Change 1995 Impacts, Adaptations and Mitigation of Climate Change: Scientific Technical Analysis*. Pp. 99-129, Cambridge University Press, Cambridge, U.K.

Komor, E., G. Orlich, A. Weig and W. Kockenberger. 1996. Phloem loading is not metaphysical, only complex; towards a unified model of phloem loading. *Journal of Environmental Botany*, 47: 1155-64.

Lal, R. 1997. Residue management conservation tillage and soil restoration for mitigating greenhouse effect by CO_2 enrichment. *Soil Tillage Research*, 43 : 81-107.

Lal, R., J. M. Kimble, R.F. Foller and V. Cole. 1998. *Potential of US crop land for Carbon Sequestration and Greenhouse Effect Mitigation*. USDA - NRCS, Washington, DC. Ann. Arbor Press, Chelsea, Michigan, 128 pp.

Lang, S.P., E. A. Ainsworth, A. Rogers and D. R. Ort. 2004. Rising atmospheric carbon dioxide. Plants FACE the future. *Annual Review of Plant Biology*, 55: 591-628.

Majean, N. and J. R. Coleman. 1996. Effect of CO_2 on carbonic anhydrase and ribulose-1,5, bisphosphate carboxylase / oxygenase expression in pea. *Plant Physiology*, 112: 569-574.

Maroco, J.P., G.E.I. Edwards and M.S.B. Ku. 1999. Photosynthetic acclimation of maize to growth under elevated levels of carbon dioxide. *Planta*, 210: 115-125.

Miglietta, F. and J. R. Porter. 1992. The effect of climate change on development in wheat: analysis and modeling. *Journal of Environmental Botany,* 43: 1147-58.

Mitchell, R.A.C., D. W. Lawlor, V. J. Mitchell, C. L. Gribbard, E. M. White and J. R. Porter. 1999. Effects of elevated CO_2 concentration and increased temperature on winter wheat-test on ARCWHEATI simulation model. *Plant, Cell and Environment,* 18: 736-48.

Morison, J.I.L. 1987. Plant growth and CO_2 history. *Nature,* 327 : 560.

Morrison, J.I.L. and R. M. Gifford. 1983. Stomatal activity to carbon dioxide and humidity. A comparison of two C_3 and C_4 grass species. *Plant Physiology,* 71: 789-96.

Mortensen, L.M. 1994. Effect of elevated CO_2 concentrations on growth and yield of 8 vegetable species in a cool climate. *Scientia Horticulture,* 58: 177.

Nakagawa, H., T. Horie, J. Nakano, H. Y. Kim, K. Wada and M. Kobayashi 1993. Effect of elevated CO_2 and high temperature on growth and development on rice. *Journal of Agricultural Meteorology,* 48: 799-802.

Oren, W.L. 2003. Affixing the O_2 to rubisco: Discovering the source of photorespiratory glycolate and its regulation. *Photosynthetic Research,* 53-63.

Paustian, K., O. Andren H. H. Janzen, R. Lal, P. Smith, G. Tian, H. Tiessen, M. VanNoordwijk and P. L. Woomer. 1997. Agricultural soils as a sink to mitigate CO_2 missions. *Soil Use and Management,* 13: 230-44.

Poorter, H. 1993. Inter specific variation in the growth response of plants to an elevated ambient CO_2 concentration. *Vegetation,* 104/105: 77-97.

Radin, J.W., B. A. Kimball, D. L. Hendrix and J. R. Mauny. 1987. Photosynthesis of cotton plants exposed to elevated level of carbon dioxide in the field. *Photosynthesis Research,* 12: 191-203.

Rawson, H.M. 1995. Yield responses of two wheat genotypes to carbon dioxide and temperature in field studies using temperature gradient tunnels. *Australian Journal of Plant Physiology,* 22: 23-32.

Reddy, A.R., G. K. Rasineni and A. S. Raghavendra. 2010. The impact of global elevated CO_2 concentration on photosynthesis and plant productivity. *Current Science,* 99: 46-57.

Reddy, K.R., H. F. Hodges and J. M. McKinion. 1995. Carbon dioxide and temperature effects on Pima cotton development. *Agron. J.,* 87: 820-26.

Reddy, K.R., R. R. Robana, H. F. Hodges, X, J. Liu and J. M. McKinion. 1988. Influence of atmospheric CO2 and temperature on cotton growth and leaf characteristics. *Environmental and Experimental Botany,* 39: 117-29.

Rogers, H.H and R. C. Dahlman. 1993. Crop responses to CO_2 enrichment. *Vegetation,* 104: 117-312.

Sakai, H., K. Yagi, K. Kobayashi and S. Kawashima. 2001. Rice carbon balance under elevated CO_2. *New Phytologist,* 150: 241-49.

Samarakoon, A.B. and R. M. Gifford. 1996. Elevated CO_2 effects on water use and growth of maize in wet and drying. *Australian Journal of Plant Physiology,* 23: 53-62.

Samarakoon, A.B., W. J. Muller and R. M. Gifford. 1995. Transpiration and leaf area under elevated CO_2 effects of soil water status and genotype in wheat. *Australian Journal of Plant Physiology,* 22: 33-44.

Schapendonk, A.H.C.M., P. Dijkstrs, J. Groenwold, C. S. Pot and Geijen, S.C., van de. 1995. Carbon balance and water use efficiency of frequently cut *Lolium perenne* L. Swards at elevated carbon dioxide. *Global Change Biology,* 3: 207-16.

Shivay, Y.S. and A. Rahul. 2008. Effect of global warming on crop productivity. *Kurukshetra (July),* pp.13-33.

Tognetti, R., A. Giovannelli, A. Longobucco, F. Miglietta and A. Raschi. 1996. Water relations of oak species growing in the natural CO_2 spring of Rapolano (central Italy). *Annales des Sciences Forestieres,* 53: 475-85.

Torbert, H.A., S. A. Prior, H. H. Rogers and C. W. Wood. 2000. Review of elevated atmospheric CO_2 effects on agro-ecosystems: residue deposition processes and soil C storage. *Plant and Soil,* 224: 59-73.

Trans, P.P. and P. S. Bakwin. 1995. Climate change and carbon dioxide for ever. *Ambio,* 24: 367-78.

Triggs, J.M., B. A. Kimball, P. J. Jr. Pinter, G. W. Wall, M. M. Conley, T.J. Brooks, R. L. La Morte, N.R. Adam, M. J. Ottmam, A. D. Matthias, S. W. Leavitt and R. S. Cerveny. 2004. Free air CO_2 3enrichment effects on the energy balance and evapotranspiration of sorghum. *Agriculture and Forestry Meteorology* 124: 63-79.

Upretty, D.C., N. Dwivedi and R. Mohan. 1977. Interactive effects of rising CO_2 and drought on mineral nutrient concentrations of Brasica. *Science and Culture,* 63: 291-292.

Upretty, D.C., B. S. Bisht, N. Dwivedi, D. C, Saxena, R. Mohan, A. Raj, G. Paswan, A. P. Mitra, S. C. Garg, M. K. Tiwari, H. K. Maini and D. Singh. 2007. Comparison between Open Top Chamber (OTC) and Free Air CO_2 Enrichment (FACE) Technologies to study the response of rice cultivars to elevated CO_2. *Physiology and Molecular Biology of Plants,* 13: 259-66.

Upretty, D.C. and V. R. Reddy. 2008. *Rising atmospheric carbon dioxide and crops.* ICAR, New Delhi, p. 114.

Uprety, D.C., N. I. Dwivedi, V. Jain and R. Mohan. 2002. Effect of elevated CO_2 on the stomatal parameters of rice cultivars. *Photosynthetica,* 40: 315-19.

Uprety, D.C., R. S. Mishra and Y. P. Abrol. 1995. Effect of elevated CO_2 and moisture stress on photosynthesis and water relations in Brassica species. *Journal of Crop Science and Agronomy,* 175: 231-37.

Vanaja, M., P. Raghuram Reddy, N. Jyothi Lakshmi, M. Maheswari, P. Vagheera, P. Ratnakumar, M. Jyothi, S.K. Yadav and B. Venkateswarlu . 2007. Effect of elevated atmospheric CO2 concentrations on growth and yield of black gram (*Vigna mungo* L. Hepper) – a rainfed pulse crop. *Plant soil environ,* 53 (2): 81– 88.

Wall, G.W., T. J. Brooks, N. R. Adam, A. B. Copusins, B. A. Kimball, P. J. Pinter, Jr. LaMore, R.L., J. Triggs, M. J. Ottman, S. W. Leavitt, A. D. Matthias, D. G. Williams and A. N. Webber. 2001. Elevated atmospheric CO_2 improved sorghum plant water status by ameliorating the adverse effects of drought. *New Phytologist,* 152: 231-48.

Wheeler, R.M., C. L. Ackowiak, J. C. Sager and W. M. Knott. 1994. Growth of soybean and potato at high CO_2 partial pressure. *Advances in Space Research,* 14: 251-55.

Wien, H.C. 1997. Papper. In: *The Physiology of Vegetable Crops,* pp.259-93. CAB International, Wallingford, U.K.

Wittwer, S.H. 1986. Worldwide status and history of CO_2 enrichment: an overview. In: Enoch, H.Z. and Kimball, B.A. (eds). *Carbon dioxide Enrichment of Greenhouse Crops,* Vol. I, pp.3-15. CRC Press, Boca Raton, Florida.

Wong, S.C. 1979. Elevated atmospheric partial pressure of CO_2 and plant growth. *Oecologia,* 44: 68-74.

Woodrow, I.E. 2009. *Flux control analysis of the rate of photosynthetic assimilation. In Photosynthesis in silico* (eds. Laiska, A., Nedbal, L. and Govindjee), Springer, The Netherlands. pp. 349-360.

Zhu, J., D. P. Bartholomew and G. Goldstein. 1997. Effect of elevated carbon dioxide on the growth and physiological responses of pineapple, a species with crassulacean acid metabolism. *Journal of American Society of Horticulture Science,* 122: 233-37.

Ziska, L.H., R. C. Sicher and J. A. Bunce. 1999. The impact of elevated carbon dioxide on the growth and gas exchange of three C_4 species differing in CO_2 leak rates. *Physiologia Plantarum,* 105: 74-80.

Ziska, L.H., W. Weerakoon, O. S. Namuco and R. Pamplona. 1996. The influence of nitrogen on the elevated CO_2 response in field-grown rice. *Australian Journal of Plant Physiology,* 23: 45-52.

□□□

Chapter 3

Castor with Relation to Abiotic Stress and Recent Molecular Advances on Downstream Plant Responses

M.K.KALARANI, P.KALAISELVI, S.R.VENKATACHALAM, A.SENTHIL AND V.PALANISAMY
Tapioca and Castor Research Station, Tamil Nadu Agricultural University, Yethapur, P.G. Palayam-636 119, Salem, Tamil Nadu, India email : kalatnau@yahoo.co.in

Castor plant (*Ricinus communis* L.) belonging to Euphorbiaceae family is one of the ancient oilseed crops of the world. Castor oil is important industrial oil and it is the only source of hydroxylated ricinoleic acid. India is one of the primary centres of origin of castor and dominates world castor production accounting for 53 % of area and 70 % of production. Productivity of castor in India is higher (1392 kg/ha) than the world average (1037 kg/ha).Castor productivity in India is more than the world average and it ranks first among the major castor producing countries viz., China and Brazil.

India has the distinction of being the only country to commercially exploit hybrid technology in castor since 1970s.The phenomenal five fold increase in productivity and eight fold increase in production with an impressive 4% compound annual growth rate in productivity are good indicators of growth in research and development efforts in the crop. India exports large amounts of castor oil and few of its products to different countries which was worth off nearly Rs 2253 crores (2008-09) and has established a virtual monoplant grip on the international market (Hegde, 1998).

Castor belongs to the monospecific genus *Ricinus* which is believed to have polyphyletic origin with four centers of diversity of which India is the one. India is the first country to commercially exploit heterosis by two line hybrid breeding based on indigenous stable 'S' type pistillate line VP1. At present, 14 hybrids and more than 18 varieties are released either at the central or state level. In Tamil Nadu, two hybrids are released and two are in pipeline. The second green revolution is expected to emerge from rainfed or dryland agriculture where castor is an important component. It is note worthy to mention that castor registered highest growth rate in terms of production and productivity during last two decades as compared to other agricultural crops. The spectacular progress in terms of area, production and productivity in castor is due to the development of hybrid castor. All the modern day castor hybrids are semi dwarf and they are highly productive under irrigated ecosystem. These hybrids are however not that productive under rainfed ecosystem as they lack drought tolerance. Abiotic stress is a major production constraint for castor under rainfed condition. Since physiological knowledge on this hardy crop is very much limited, understanding of physiological responses to its environment will be useful in improving the productivity of this crop. Concerted efforts are required to tackle these problems for enhancing productivity, so as to remain competitive in the world market. Some of the recent research highlights in castor to manage the abiotic stress tolerance are discussed below.

Recent Research Highlights of Abiotioc Stress Responses in Castor

1. Water use efficiency (WUE) with relation to root traits for drought tolerance

Efficient use of stored water, increased biomass productivity per unit water use and highest conservation of vegetative biomass to economic yield are the ultimate goal of any drought management research. Among the many traits associated with drought tolerance, root characters and water use efficiency (WUE) are considered to be the most relevant. Good root system is essential for crop plants to grow under moisture deficit conditions. Specific Leaf area (SLA), SPAD chlorophyll meter reading (SCMR), carbon isotope discrimination (^{13}C) are the traits used as surrogates for measuring WUE. Improvement of these traits has been shown to increase the productivity in several species (Sheshashayee *et al.,* 2003).

Lakshmamma *et al.* (2010) conducted experiments with sixty four genotypes and were sown in two specially constructed structures to identify castor germplasm and breeding lines for better water use efficiency (WUE) and root traits. In each structure, 32 lines were sown in two replications. Observations on various growth parameters, WUE traits, root characters and total dry matter (TDM) were recorded at the time of harvest (90 days after sowing). Growth characters were presented in (Table 1). Among the different root characters, root volume and root dry weight showed strong positive correlation ($r>0.80$) compared to root length ($r = 0.32$) with TDM. Leaf area index (LAI) and stem girth also showed strong positive correlation of >0.80 with TDM. Among the 64 lines screened, thirteen lines (RG 1450, RG1611, RG1661, RG1826, RG2122, RG2149, RG2714, RG2797, RG2826, RG2850, RG3063, GCH 5 and 48-1) showed root length >230cm, root volume of >250ml, root dry weight of >40g, LAI >3.0 and TDM >350g/ plant.

2. *Osmotic adjustment with relation to drought tolerance and yield*

A study was conducted by Babita *et.al* (2010) to analyse the association of osmotic adjustment (OA) with drought tolerance and yield in castor (*Ricinus communis* L.). Hybrids (GCH 4, DCH 32 and DCH 177) and their respective parents (VP 1, 48 1, LRES 17, DCS 5, DPC 9 and DCS 9) were assessed for their osmotic adjustment, leaf water relations, accumulation of compatible solutes in relation to seed yield in response to moisture stress at primary spike development stage. Osmotic adjustment (OA) increased with increasing stress period up to 33 days and the increase was more rapid in the high OA (HOA) genotypes. HOA genotypes also had higher leaf relative Water Content (RWC) and maintained higher leaf water potentials ($Ø_l$) compared to Low OA (LOA) genotypes underwater deficit. Genotypes with HOA accumulated greater levels of proline, total soluble sugars (TSS), total free amino acids (FAA) and potassium than those with LOA in response to water deficit. Contribution of TSS was the maximum (61%), compared to FAA (17%), proline (12%) and potassium (2.8%) to the ψ_s at 33 days after imposing water deficit indicating that sugars were the major contributors towards OA in castor leaves. A positive relationship existed between OA of expanded leaf of 33 days after imposing stress ($r = 0.8539$) and total seed yield under water -limited conditions. In various castor genotypes tested, HOA genotypes had higher total seed yield than genotypes with LOA. Genotypic variability exists for OA and it is a heritable trait in castor. Hybrids followed their superior parents in terms of OA. HOA genotypes of castor produced significantly higher seed yield than LOA genotypes. Accumulation of TSS contributed largely to the OA in castor.

Table 1: Growth characters of selected castor genotypes

Germplasm lines	Pl.ht (cm)	Leaf no	Node no	Stem girth (cm)	SCMR	SLA	LAI	^{13}C	TDM	Root characters		
										Length (cm)	Volume (ml)	Dry weight (g)
RG 1450	168.5	27	20	10.0	46.9	2.57	4.9	-27.736	422	264	311	56
RG 1611	72.3	32	13	9.3	48.9	2.91	4.6	-28.824	395	232	247	44
RG 1661	32.8	32	8	10.0	49.5	1.79	2.9	-28.476	357	256	377	42
RG 1826	68.8	28	14	9.2	48.1	2.23	4.3	-28.453	427	219	269	52
RG 2122	80.5	33	14	9.9	48.2	1.92	4.1	-28.271	453	260	328	55
RG 2149	100.8	40	16	11.2	49.6	2.06	5.5	-28.866	638	232	391	69
RG 2714	87.0	41	19	10.0	51.1	1.61	4.0	-29.222	415	239	375	54
RG 2797	188.3	11	23	10.5	48.3	2.09	2.7	-28.254	312	258	263	48
RG 2826	132.3	37	19	9.9	53.3	1.99	2.6	-27.500	317	263	254	47
RG 2850	130.3	19	17	10.2	54.2	1.97	3.3	-27.291	379	214	263	50
RG 3063	196.8	32	23	10.8	48.7	2.13	4.9	-27.370	430	250	353	62
GCH-5	133.8	28	19	10.5	45.9	1.67	3.6	-28.866	438	258	336	55
48-1	107.0	30	17	10.7	44.2	2.16	2.5	-28.744	396	235	315	54
SEm ±	14.8	3.14	1.32	0.52	1.50	0.29	0.49		27.4	20.4	23.3	6.6
CD (P=0.05)	42.4	9.0	3.8	1.5	4.3	0.37	1.41		78.9	24.4	66.9	16.2

(*Source:* Lakshmamma *et al.*, 2010)

3. Scope of Vesicular Arbuscular Mycorrhiza (VAM)

Phosphorus (P) is essential for root growth, seed and fruit development and for early ripening. It is a constituent of several organic compounds including oils and amino acids. Phosphorus deficiency retards plant growth, root growth and delay in flowering. The P recommended for rainfed castor varies from 15 to 40 kg P_2O_5/ha depending upon the fertility and moisture retention status of the soils. The red soils where castor is cultivated in Andhra Pradesh are deficient in available N and available P and zinc. Productivity of these soils is low due to low

Water holding capacity, soil erosion, nutrient runoff losses, organic matter depletion, poor native fertility, deficiencies of micro nutrients and sulphur, soil crusting and hardening.

Aziz *et al.* (2009) was found that there exists the association between VAM and castor crop. It was noted that *Glomus faciculatum, G.constrictum* and *G.mosseae* were the predominant VAM species to be associated on castor crop in chalka (red sandy loams) soils of Andhra Pradesh. In normal to alkaline soils, the predominant VAM species found in the castor rhizosphere were *Glomus mosseae, Glomus microcarpum* and *Glomus macrocarpum.* The present review shows the potential possibility of using VAM for improving the castor productivity in marginally fertile soils.

Many researchers have shown that productivity of castor could be improved by various agronomic manipulations in the red soils *viz.,* applying balanced fertilizers and manures (Bheemaiah *et al.,* 1998 and Hegde, 1998) by improving moisture conservation (Subba Reddy *et al.,* 1996), and by using phosphate solublizers (Arangarasan *et al.* 1999 and Baby and Bapi Reddy, 1998).

The research literature pertaining to the use of vesicular-arbuscular mycorrizas (VAM) in improving the P nutrition of castor crop grown in india is hardly available. However, it is known that occurrence of VAM is ubiquitous and they are known to increase the P and moisture uptake in all agricultural crops. Plants infected with mycorrhizal fungi exhibits significant increase in the yield as compared with non-mycorrhizal plants.

The available literature suggests the positive association of vesicular arbuscular mycorrhiza with castor crop, the scope of using VAM in improving the P nutrition of castor crop grown in P deficient soils and for improving its productivity. It has been highlighted that *Glomus fasciculatum* and *Glomus constrictum* were predominant VAM species found in chalka soil (red soil of AP) and *Glomus mosseae, Glomus microcarpum* and *Glomus macrocarpum* species were found to dominate in normal to alkaline soil conditions when castor was grown as host plant. The available information provides ample scope for research in exploiting VAM in castor.

4. Agronomic management practices for nutrient and drought management

In Telangana districts of Andhra Pradesh, castor is mostly grown as rainfed crop during *kharif* season in sequence with sorghum and pigeonpea. Primary racemes of dryland castor develops during later part of rainy season. As the crop duration extends beyond rainy season, the productivity phase, particularly the development of secondary and tertiary racemes are dependent on residual soil moisture and unseasonal rainfall. The productivity of the rainfed castor is low as it suffers from inadequate application of nutrients and also from moisture stress caused by intermittent dry spells. Resource poor farmers do not use much inputs and as such the productivity is low, and fluctuating.

Venkateswarlu (2009) carried out an experiment to evolve simple, low cost moisture conservation and integrated nutrient management (INM) practices to improve productivity and minimize drought effects at Hayathnagar Research Farm of Central Research Institute for Dry Land Agriculture (CRIDA), Hyderabad. The trial consisted of 8 treatments with and without application of farm yard manure (FYM) as main plots and four sub-plots as application of supplemental N (25%) through gliricidia making conservation furrows between castor rows, application of additional dose of 10 kg N/ha and 100% N through inorganic fertilizers (50-30-0).The results of the experiment showed that the bean yield in rainfed castor with and without FYM (1095 and 967 kg/ha) were highest under application of 25% N through gliricidia and 75% N through inorganic fertilizers. Next best treatment was conservation furrow between castor rows (1040 and 909kg/ha) followed by top dressing of additional dose of 10 kg N/ha (1001 and 850 kg/ha). The lowest bean yield (883 and 735 kg/ha) was obtained with 100 % of N through inorganic fertilizers (50-30-0). Application of FYM @ 10 t/ha improved bean yield and filled capsule number compared to no FYM application. The bean yield enhancement under different INM treatments was about 13 to 20% with FYM application over no application.

5. *Transport, synthesis and catabolism of abscisic acid (ABA) under phosphate deficiency and moderate salinity*

The flows of abscisic acid (ABA) were investigated by Dieter jeschke *et al.* (1997) in whole plants of castor bean grown in sand culture under either phosphate deficiency or moderate salinity. Xylem transport of ABA in P-deficient plants was stimulated by a factor of 6 whereas phloem transport was affected only very slightly. ABA deposition into leaves of P-deficient plants was not appreciably different from the controls (Fig. 1) because of strong net degradation in leaves. Since conjugation of ABA was strongly reduced in all the organs of P deficient plants, ABA was presumably metabolized mainly to phaseic acid and dihydrophaseic acid. The increased import of ABA occurred predominantly into fully differentiated but not senescent leaves and showed a good correlation with the inhibition of leaf conductance under P deficiency (Fig. 2). As with low-P-plants, salt stress increased ABA synthesis in roots and associated transport in the xylem. However, salinity caused a distinctly greater accumulation of ABA in the leaves, stem segments and the apex than in P-deficient plants. As opposed to P deficiency, ABA export in the phloem from the leaves was stimulated by salinity (Fig. 3). Modelling of ABA flows within an individual leaf over its life cycle showed that young growing leaves imported ABA from both phloem and xylem, whereas the adult non-senescent leaves were a source of ABA and thus provided a potential shoot to root stress signal as well as an acceptor for reciprocal signals from root to shoot. In senescing leaves, ABA flows and accumulation were somewhat retarded and ABA was lost in net terms by export from the leaf.

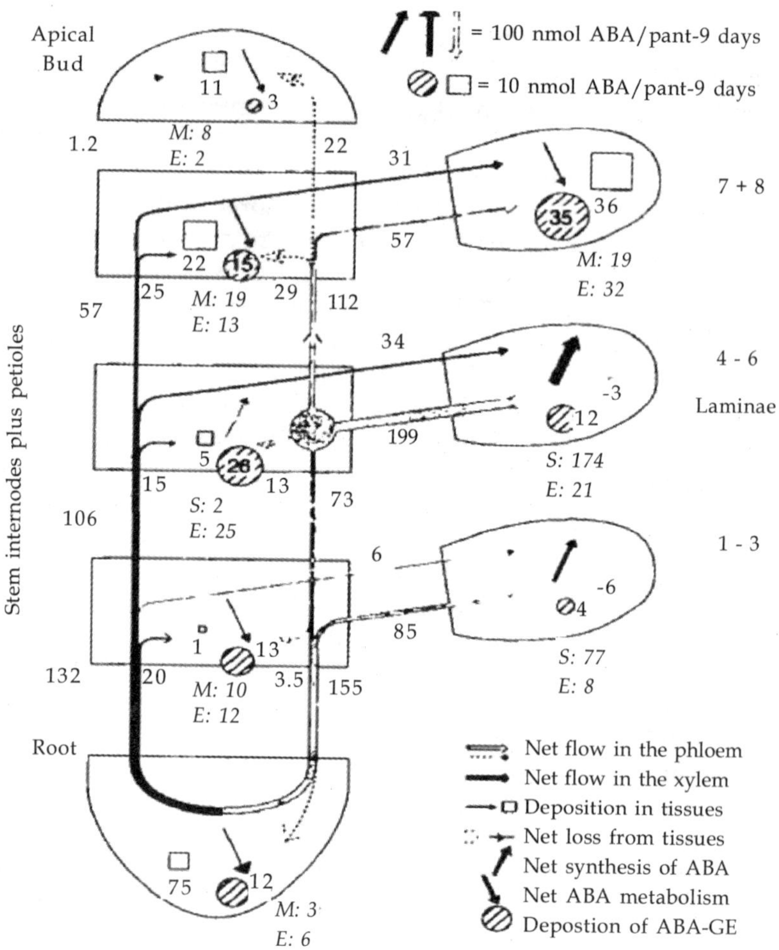

(Source: Dieter eschke et al. (1997)

Fig. 1: Flow profiles for transport, utilization, synthesis or metabolism and esterification of ABA in control plants of Ricmus communis L. over the study period 44-53 DAS. The plants were fed with 0.5 mol m^{-3} $H2PO4^{-}$. Dotted or hatched arrows indicate net flows of total ABA (free and conjugated ABA) in the phloem, black ones net flows in the xylem. The width of arrows (flows in xylem and phloem) and the area of squares (incorporation) and circles (esterification) are drawn in proportion to rates of flow, deposition or esterification of ABA (for the scale see the key at the top). The numbers indicate net rates in nmoles ABA $plant^{-1}$ $(9\ d)^{-1}$. The net rates of ABA synthesis (S), metabolism (M) or esterification (E) are given in the same units. The numbers under each stem segment denote uptake from the phloem (right) or the xylem (left). For the sake of clarity flows into the young, growing leaves 7 and 8 into or out of the maturing and mature leaves 4-6 and the oldest leaves 1-3 and the subtending stem segments plus petioles have been combined.

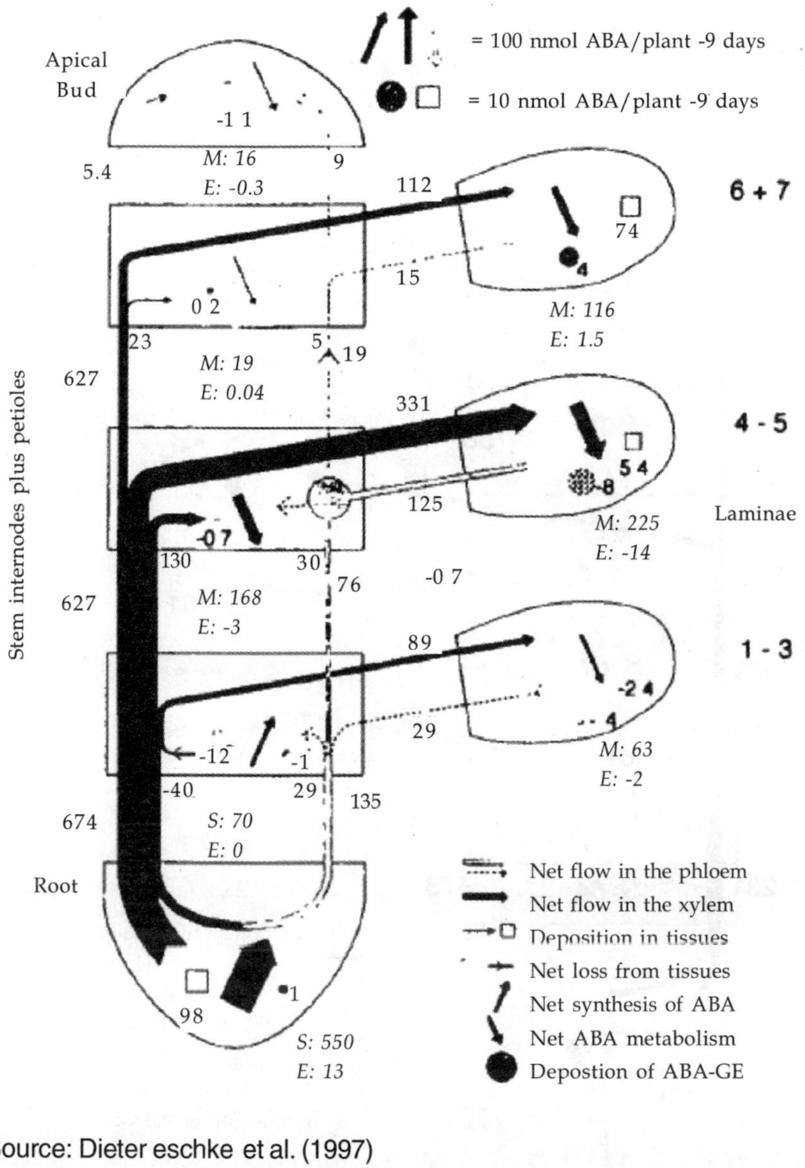

(Source: Dieter eschke et al. (1997)

Fig. 2: Flow profiles for transport, utilization, synthesis or metabolism and esterification of ABA in P-deficient plants of Ricinus communis L over the study period 44-53 DAS. Phosphate had been withdrawn at 29 DAS and 5μM $H_2PO_4^-$ was then given at 41 DAS. As in Fig 1, the arrows indicate net flows of total (free and conjugated) ABA in phloem (dotted) and xylem (black). Note that the scale for arrows and squares is the same as in Fig. 1, highlighting the substantial increase in xylem transport of ABA. In this model the young leaves 6 and 7 were bulked in the calculations. Other details as in Fig. 1.

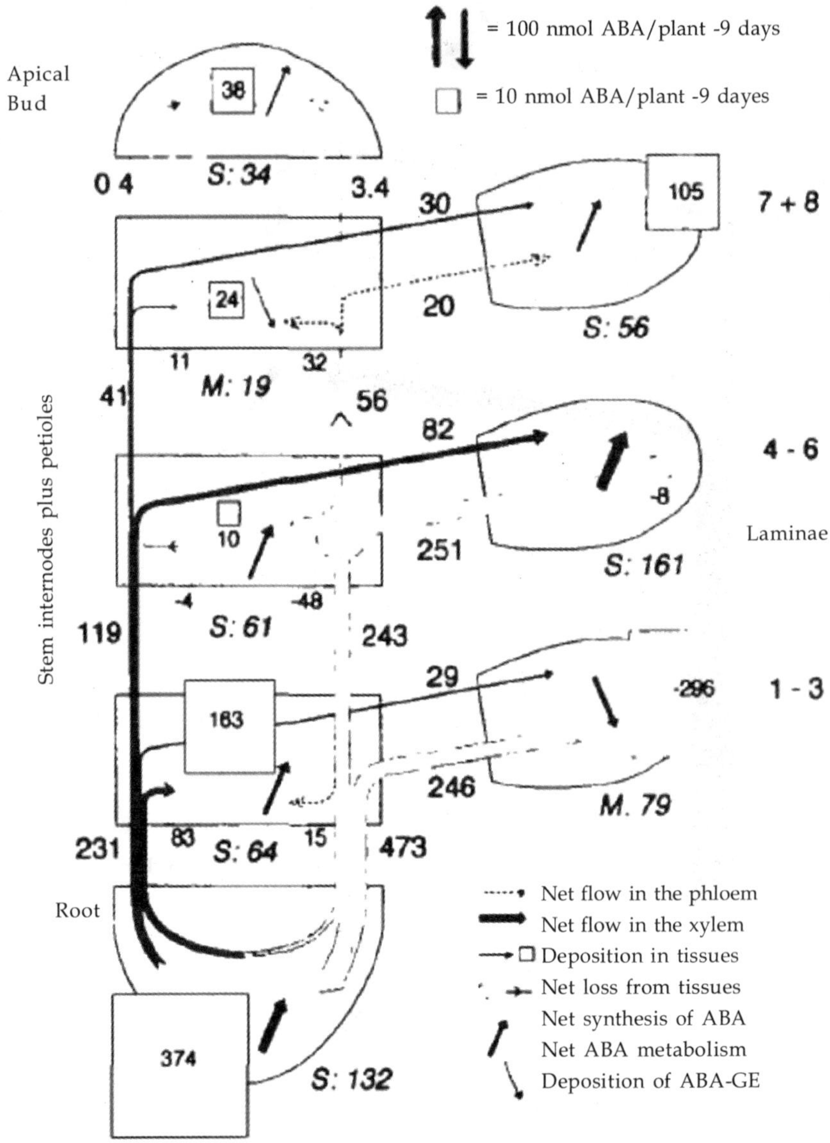

(Source: Dieter eschke et al. (1997)

Fig. 3: Flow profiles for transport, utilization and synthesis or metabolism of ABA in salt-treated plants of Ricinus communis L. over the study period 44-53 DAS. The plants were exposed to a mean salinity of 128 mM NaCl starting at 35 DAS. Details as in Fig. 1. However, since conjugated ABA had not been measured, there are no data for the formation of glucose esters in this flow scheme. Same scale for arrows and squares as in Figs 1 and 2.

6. Flows of elements, ions and abscisic acid and site of nitrate reduction under potassium limitation

A pot culture experiment with *Ricinus communis* was conducted by Andreas *et al.* (2002) The plants were cultivated in quartz sand and supplied daily with a nutrient as the nitrogen source and either with full strength potassium (1.3 mol m^{-3}, control) or 8% potassium (0.1 mol m^{-3} , K limitation). Although the final fresh weight of the root -shoot ratio was not affected by K^+ limitation, the root-shoot ratio was increased due to a relatively increased root growth and inhibited development of younger shoot parts. Owing to K^+ limitation, photosynthesis was slightly decreased, while dark respiration of the shoot markedly decreased and root respiration was nearly doubled. The transport of carbon in the pholoem and to some extent in the xylem was greater and the root was favoured in the partitioning of carbon. This was also true for nitrogen and potassium which were taken up at lower rates, particularly potassium. In these two cases, a high remobilization and recycling from the old part of the shoot was observed. By contrast, uptake of sodium was 2,4- fold higher under K limitation and this resulted in increased flows in the plants which was discussed generally as a means for charge balance (in combination with a slight increase in uptake of magnesium and calcium). Nitrate reduction took place in the same portion in the root and shoot. This was a shift to the root compared to the control and points to an inhibition of xylem transport caused by limitation of K^+ as an easily permeating counteraction. Low K^+ supply also resulted in an increased biosynthesis of ABA in the roots (265%). This caused a slightly increased deposition of ABA in the roots (193 %) and a 4 and 6 fold higher root-to-shoot and a doubled shoot to root ABA signal in the xylem or phloem respectively. The high degration of ABA in the shoots prevented ABA accumulation.

7. Highly variable sex expression in castor

Castor is a sexually polymorphic species occurs in nature as mostly monoecious. The basic sex forms are monoecious, pistillate, pistillate with interspersed staminate flowers (ISF) and sex revertants (Stam and Podkuichenko, 1970). Sex expression in castor is highly influenced by environmental condition (Shiffriss, 1960). Winter, low temperature (<30°C), young plants, high nutrition promote female flowers and shift the balance towards femaleness on a spike. Summer or rainy season, high temperatures (>32°C), old plants, late order spikes, low nutrition promote male flowers on a sapike and incline towards maleness

(Lavanya, 2002). The role of exogenous and endogenous growth harmones like gibberellic acid, silver nitrate and ethylene in shifting the female and male tendency has been well documented (Ramesh *et al.*, 2000)

Temperature is probably the main environmental component affecting sex expression. Moderate temperature promotes female tendency. High temperature promotes male tendency. Production of F_1 hybrids castor seed is dependent on this tendency to predominant femaleness (Shifriss, 1956). High temperature at flowering i.e. 41°C (105°F) or above for even a relatively short period results in blasting of flowers and poor seed set. This effect is more marked if high temperatures coincide with period of water stress. Prolonged cloudy weather with high temperatures during flowering also reduces the amount of seed set. Regulation of sex especially improving pistillate flower production by exogenous application of hormones will be of great importance in order to obtain higher yield (Ramesh *et al.*, 2000).

Investigation was made to study the effect of exogenous application of chemicals on pistillate production efficiency, seed setting percent and seed yield by Kalarani *et al.* (2010) in Tapioca and Castor Research Station, Yethapur, Salem, Tamil Nadu Agricultural University,. This investigation was carried out with castor varieties of TMV 5 and TMV 6 and hybrids of TMVCH 1 and YRCH 1. Chemicals viz., Gibberellic acid- GA_3 (100ppm), Naphthaline acetic acid - NAA (40ppm), Salicylic acid (100 ppm), Ethrel (50 ppm), Brassinosteroid (0.1 ppm) and Plant Growth Regulator consortia (250 ml/ha) were sprayed ten days before initiation of flowering. Among the chemicals tried, irrespective of the genotypes, PGR consortia induced earliness (10 days), increased more number of pistillate flowers (7 to 45 flowers) and developed compact raceme with 92 per cent of seed set. Among the hybrids, YRCH 1 recorded more yield (1744 kg/ha) followed by TMVCH 1 (1388 kg/ha) in PGR consortia sprayed plot. Among the varieties, TMV 6 recorded higher yield (1215kg/ha) followed by TMV 5 (1128 kg/ha) in PGR consortia treatment. PGR consortia developed in India increased the seed yield when sprayed at 10 days before appearance of flowering in castor genotypes. This exogenous application of hormones change the endogenous level of the naturally occurring hormones and allow for desired change. PGR consortia could maintain hormonal balance in the crop plants and initiate flowering, induce femaleness, improve seed set and increase seed yield. Hence it is concluded that the response of plant hormones vary with variety of plants. Even a single variety responds differently depending on its age, environmental conditions, physiological stage of development and nutritional status. As for as this PGR concerned, the growing conditions did not affect its action but its effectiveness mainly depend on the variety of crop plant and stage of treatment in relation to plant growth.

Crop Improvement Strategies in Castor

Abiotic stress resistance

The screening techniques have been developed for drought and salinity resistance and some genotypes resistant to these stresses have been identified. however, systematic programme for screening the available germplasm, breeding material to identify the sources of resistance and their use in specific breeding programme is advocated to evolve high yielding varieties and hybrids suitable for rainfed and salinity affected area has been well documented (Pathak, 2009; Lavanya and Solanki, 2010).

Drought tolerance/resistance

Very limited work has been done on castor drought resistance in the world as it is a drought tolerant crop grown on marginal and poor soils. Among the different genotypes studied, GCH 4 and its male parent, 48-1 showed drought resistance due to their quick response of stomata coupled with leaf reflectance and transmittance properties (Rao *et al.*,1998). GAUCH 1 also showed better adaptability to drought. A deep root system and regenerating capacity of the shoot enabled it to extend its growth period. Deep root system and high root volume enabled GCH 5 drought tolerance. Among 260 lines screened during 2000-02, using temperature induction response technique at Directorate of Oilseed Research, Hyderabad, 43 germplasm lines showed acquired tolerance to stress with >80% recovery with induction. Among these,14 lines were found highly tolerant to stress. These lines include RG 122, 226, 232, 235, 242, 247, 1096,1117,1701,1713,1729,1826,111 and 2732 (Lakshmamma *et al.*, 2004).

Castor is predominantly a drought resistant crop, capable of surviving even adverse weather conditions, hardy soils and during prolonged drought. The wide disparities in productivity levels of rainfed and irrigated situation ranging from 1500 to 2500 kg/ha emphasize the need to idendify the promising parental lines and germplasm accessions to strengthen the drought resistance breeding programme. Most of the released hybrids and varieties of castor are not bred specially for drought conditions though some of them do well under stress situations.

Several disease resistant germplasm accessions and breeding lines were identified as drought tolerant using standard drought screening techniques like Temperature Induction Response Technique (TIR) in lab and in field by imposing stress during *rabi* or *summer*. Drought resistant traits linked with root characteristics and WUE are being used for identification of drought tolerant genotypes for further use in breeding programmes (Lakshmamma *et al.*, 2003). Screening procedure for root traits related to drought resistance like root length, volume, penetrability, thickness, density were standardized at Directorate of Oilseeds Research by growing the genotypes in specially constructed raised structures on the ground. Screening for shoot traits like Spad Chlorophyll Meter Reading (SCMR), Specific Leaf Area (SLA), Leaf Area Index (LAI), Total Dry Matter (TDM), plant height, leaf number, stem girth and root traits like length, volume, weight, traits related to WUE like ^{13}C and ^{18}O were used to develop an index for identifying the drought resistant genotypes and later confirmed in field screening (Lakshmi and Lakshmamma, 2009). Several genotype were identified as drought resistant through different screening techniques like TIR, Field study and root structure (Table. 2).

Table 2: Promising drought resistant genotypes identified

Screening technique	Drought resistant genotypes
Field	'RG17, RG 18, RG 22, RG 25, RG 43, RG 52, RG 68, RG 72, and RG 90; low DSI-RG 72 (11%), RG 17 (22%) and RG 52 (27%);0.5DSI-RG89, 111,122,293,297, 298,332,707,1117,1815,1449;<0.5 DSI-RG 122,214,298,332,724,900.
TIR	RG 122, RG 226, RG 232, RG235, RG242, RG 247, RG 1096 and RG1117.
Root structure	51 germplasm, GCH-5, 48-1, DCS-99*

DSI-Drought susceptibility index; *to be confirmed in field

(*Source:* Lakshmamma *et al.*, 2003; 2004; Lakshmamma and Lakshmi Prayaga, 2006a; 2006b; 2010)

Among 11 germplasm accessions screened in field under stress, spad chlorophyll meter reading (SCMR) increased under stress, specific leaf area (SLA) did not vary with stress and SCMR is negatively correlated with SLA (>-0.90) both in control and stress treatments. Genotype RG 297 showed high SCMR, low SLA, and high TDM both in control and stress treatments. This genotype also showed higher seed yield and HI in stress with less percent reduction in seed yield compared to control and with low DSI values.

The epicuticular wax coating called bloom in drought resistance has a specific role in reducing the transpiration losses. A study on 12 germplasm lines (tolerant *to Fusarum* wilt and differing in bloom/ epicuticular wax load) under field conditions indicated RG 2317, RG 923 with <30% yield reduction and <0.5 DSI as drought resistant. A significant and substantial increase in bloom 17-18% compared to control was reported due to imposition of stress in two crop growth stages.

Resistance to drought is polygenically controlled in several crops. The availability of standard screening procedures coupled with confirmed sources of drought resistant indicate the scope for further intensifying the research in collaboration with molecular biologist for QTL mapping as in the case of other crops.

Salt resistance/tolerance

Castor can tolerate alkalinity up to pH 8.5 and acidity up to pH 5-4.5. As castor is grown mostly under rainfed situation and as a predominant crop under semi-arid tropical conditions where inland salinity is often encountered, limiting crop yields to variable degrees, identification of salt resistant genotypes forms the low cost technology for the resource-poor farmers. Across salinity levels, the genotypes 48-1, GC 2, GCH 4, DCS 5 and Kranti maintained lower NA/K ratios in their plant tissue than other genotypes indicating their tolerance to salinity. Raghavaiah *et al.* (2006) evaluated 20 genotypes under 4 levels of Salinity, 2,4,6,8 dS/m in terms opf growth parameters such as length of root and shoot, dry matter and vigour and found DCH-32,48-1,DCS-9, Kranti as most promising for saline condition. During evaluation of 6 pre-release genotypes of castor including 4 hybrids at 0,6,8 and 10 dS/m, it was observed that genotypes having tolerance to salinity showed a mechanism of adaptation to excessive soluble salts through higher germination and higher K, Na/k ratio (Table. 3).

Table 3: Screening of salt tolerant genotypes with various screening techniques and criteria

Screening technique	Salt tolerant genotypes	Selection criteria
Laboratory (10 dS/m)	TMV 5, DCH 151, DPC 9, DCH 32, DCH 149.	Better germination, shoot and root growth of seedlings
Field	GCH 5, DCH 32, 48-1, DCP 9 and GC 2.	Higher seed yield, high test weight
Pot culture	48-1, DCH 151, DCH 149, DCH 32, DCS 86, TMV 6, DPC 9, DCS 82, GC 2.	Better germination, shoot and root growth of seedlings, lower Na/K, higher Ca/N in leaf tissue

(Source: Raghavaiah et al., 2002; 2003; 2006)

Laboratory screening procedures for salt tolerance were standardized and several genotypes were screened using a gradient of salt concentration (0,4,6,8,10 dS/m) for their response to salinity in terms of germination, shoot and root length, shoot and root dry weight of the seedlings. Field screening for salt tolerance was conducted in natural sodic conditions of farmer fields (Raghavaiah *et al.*, 2002; 2003 2006).

The sources of resistance identified in parental lines like DPC 9, TMV 5, 48-1, DCS 82, DCS 86 and GC 2 can further be explored for the manifestation of heterosis. Genetics of resistance to salinity, identification of molecular markers for salinity resistance and the mechanism of salt tolerance in resistance sources is worth exploring keeping in view the potential danger of increasing soil salinity.

Tissue culture and crop improvement

Tissue culture techniques *viz.*, somaclonal variation, haploid breeding and *in vitro* selection can be very effectively utilized for creation of variability and selection of desirable traits. Suitable and effective protocols required are to be developed and standardized in these areas, so that these techniques are utilized as an aid in crop improvement programme.

Genetic improvement of any crop species through genetic engineering technique requires an efficient *in vitro* regeneration system which is rapid, reproducible and applicable to broad range of genotypes. The shoot proliferation system using meristematic ex-plants has been developed and genotypes especially the parents of released and promising hybrids for their genetic transformation for required traits *viz.*, disease, insect pest, drought and salinity resistance are to be identified.

Transgenic for abiotic stresses

The first-generation genetically modified crops harbouring single gene that allowed them to tolerate herbicide application or insect attack have been grown in a number of countries since 1990's. The second-generation transgenic crops with stacked genes or traits that allow crops to tolerate environmental stresses such as drought , cold, salt, heat, or flood are now being tested in confined fields trials around the world. In case of castor, transgenic development is at infancy and there are no commercially released products till date. With regard to abiotic stresses, an Israli biotechnology company-Rahan Meristem (http://www.rahan.co.il) is developing transformation protocols for castor to develop transgenic with resistant to salinity and drought. The company has isolated several genes from unicellular algae that confer resistance to drought and salinity which are expected to improve the performance of castor in rainfall climates.

Recent Molecular Advances on Down-Stream Plant Responses to Abiotic Stress

Global effects on desertification, soil salinization, atmospheric carbon di oxide enrichment, nutrient imbalances (including mineral toxicities and deficiencies), and effects of other pollutants are predicted to cause dramatic changes in the environment of agricultural lands. In a world where population growth exceeds food supply, researchers should focus efforts to find solutions that may help plants overcome stress caused by increasingly challenging environmental conditions (Alcázar *et al.*, 2006).

Plants respond and adapt to continuous environmental fluctuations by appropriate physiological, developmental and biochemical changes to cope with these stress conditions. The stress in plants is an induced physiological situation when there is severe or constant change in the environment or when normal conditions are aggressive, altering the physiological and adaptive pattern of plants

(Savio Pinho dos Reis *et al.*, 2012). As an example of the changes that induce abiotic stress in plants, we can mention the variations of temperature, moisture, aqueous saline, soil pH, radiation, and pollutants, such as heavy metals and mechanical damage (Ahuja *et al.*, 2010). All of these environmental modifications produce physiological reactions in their cells of genetic origin (Bhatnagar-Mathur *et al.*, 2008). These abiotic stress conditions also cause extensive losses to agricultural production worldwide, because they affect negative plant development and productivity.

It is estimated that less than 10% of the world's arable lands may be free of major environmental stresses. Up to 45% of the world's agricultural lands are subject to continuous or frequent drought and 19.5% of irrigated agricultural lands are considered saline (Flowers *et al.*, 1995). Also, crops and other plants are routinely subjected to a combination of different abiotic stresses (Mittler, 2006). In drought areas, for example, many crops encounter a combination of drought and other stresses, such as heat or salinity. Together, these environmental stresses reduce average yield of major crop plants by 50% to 70%.

Heat-shock proteins (Hsps)

Temperatures above the normal optimum cause heat stress at different levels in all living organisms. Heat stress often associated with salinity and drought stress disturbs cellular homeostasis and causes denaturation and dysfunction in many proteins leading to severe retardation in growth, development and even death. Worldwide, extensive agricultural losses are attributed to heat often in combination with drought or other stresses.

Heat shock proteins are assumed to play a central role in the heat stress response and in tolerance to high temperatures in all plants and other organisms (Vinocur and Altman, 2005). The heat stress response is a highly conserved reaction caused by the exposure of an organism tissue or cells to sudden high temperature stress and it is characterized by rapid induction and transient expression of Hsps. As many molecular chaperones are stress proteins and many of them were originally identified as heat-shock proteins, the names of these molecular chaperones follow their early nomenclatures and are referred here as Hsps (Wang *et al.*, 2004). So, these proteins primarily function as molecular chaperones to control the proper folding and conformation of both structural (cell membrane) and functional (enzyme) proteins ensuring the correct function of many cellular proteins under conditions of elevated temperature. The primary protein structure for Hsps is well conserved in organisms ranging from bacteria and other prokaryotes to eukaryotes such as higher animals and plants. This conservation ensures a close involvement in the protection of the organism against heat shock and the maintenance of homeostasis (Vinocur and Altman, 2005)

Hsps are located in both the cytoplasm and organelles such as the nucleus, mitochondria, chloroplasts and endoplasmatic reticulum. Five major families of Hsps are conservatively recognized, with some of them designated by their approximate molecular weights: the Hsp70 (DnaK) family; the chaperonins (GroEL and Hsp60); the Hsp90 family; the Hsp100 (Clp) family; and the small Hsp (sHsp) family, with proteins with molecular weight between 15 and 42 kDa. The best known families are the chaperonins and the Hsp70 family (Wang *et al.*, 2004). The transcription of Hsp encoding genes is controlled by regulatory proteins called heat stress transcription factors (Hsfs) which exist as inactive proteins mostly found in the cytoplasm (Baniwal et al., 2004). Thus, the heat stress response is controlled by Hsfs, which are activated by the presence of Hsps acting by binding to the highly conserved heat shock elements in the promoters of target genes (Hsps) activating expression of them (Flowers *et al.,* 1995). Heat shock transcription factors and the promoter heat shock elements are among the most highly conserved transcriptional regulatory elements in nature (Hahn *et al.*, 2004). In addition to mediating a relatively large part of the defense response of eukaryotes to heat stress, Hsfs are also thought to be involved in different pathological conditions, cellular responses to oxidative stress, heavy metals, amino acid analogs and metabolic inhibitors and certain developmental and differentiation processes (Miller and Mittler, 2006; Hahn *et al.*, 2004; Scharf *et al.*, 2012).

However, heat is not the only stress treatment that leads to elevated expression of many Hsps. Some are also induced by cold, salt, drought and other non-heat stresses (Vierling, 1991). A study with *Arabidopsis* revealed there are some Hsps induced by different stresses with changes in expression under a number of environmental conditions (Swindell *et al.*, 2007). Also, an enhanced salt tolerance was obtained in *Escherichia coli* transformed with the cytosolic chaperonin CCP-1a from *Bruguiera sexangula,* resulting in a significant osmoprotective effect (Yamada *et al.*, 2002). Salt stress induced the expression of six different

Hsps in rice (Ye *et al.*, 2012). Hsp100 was up-regulated under salt and drought stress but not under heat stress (Hu *et al.*, 2009). In rice, there was an increased expression of Hsp70 under different kinds of stress (Goswami *et al.*, 2010). Overexpression of Hsp17.6A improves heat and osmotic tolerance in *Arabidopsis* (Hu et al., 2009). In *Ginkgo biloba,* all three GbHSPs of the study were up-regulated under cold stress, whereas extreme heat stress only caused up-regulation of one of them (Cao *et al.*, 2012). Heavy metal stress tolerance was also reported. For instance, Hsp90.3 was up-regulated in *Arabidopsis* under cadmium and arsenic stresses (Song *et al.*, 2012). It is clear that the main role of the majority of Hsps/Hsfs is to increase heat tolerance. But there are a number of proteins in plants involved in many other abiotic stresses individually or combined that help plants respond to different degrees of environmental changes.

Osmoprotectants

High soil salinity is one of the important environmental factors that limits distribution and productivity of major crops causing yield losses. It reduces the ability of plants to take up water thus leading to reduction in growth rate due to hormonal signal generated by the roots (Munns, 2002). Although salinity is a major problem for agriculture in arid and semi-arid regions of the world, this environmental adversary can be observed in other areas as well, since it affects approximately 40% of irrigated land to various degrees (Sahi *et al.*, 2006; Ashraf and Akram, 2009).

However, to combat this problem, a major strategy for the utilization of salt-affected lands is to grow salt-tolerant crops or cultivars on such soils. This strategy is considered the most efficient and economical means of utilizing salt-affected soils and hence a permanent solution to the problem. The most efficient way for achieving this is through a transgenic approach (Ashraf and Akram, 2009). As different stress-regulated genes have roles in salt tolerance, transgenic research promotes the alteration of expression levels of native genes or by incorporating exogenous genes for a desired trait (Munns, 2002).

Salt stress results in a wide variety of physiological and biochemical changes in plants, such as the activation of salt-inducible genes such as transcription factors (Ji *et al.*, 2012), bZIP (Yang *et al.*, 2009), LEA (Costa *et al.*, 2011), RING zinc-finger (Reis *et al.*, 2012) and the large scale production and accumulation of osmolytes. Plants accumulate derivatives of these low molecular weight solutes to mitigate the detrimental effects of salt stress by lowering the water potential of cells or by protecting various cellular structures and proteins during stress. The accumulation of compatible osmolytes involved in osmoregulation allows additional water to be taken from the environment (Kishor *et al.*, 2005). On the basis of this understanding, enzymes that catalyse the biosynthesis of these compatible osmolytes are considered to be examples of salt-stress-tolerance effectors (Munns, 2002; Sahi *et al.*, 2006).

Proline

Proline is an amino acid known to occur widely in higher plants and in response to environmental stresses (especially salt or osmotic stress) and normally accumulates in large quantities. Under salt or osmotic conditions, it contributes to the stabilization of proteins, membranes and subcellular structures in cytosol and protecting cellular functions by scavenging reactive oxygen species. Furthermore, it is known to induce expression of salt stress responsive genes which possess proline-responsive elements (Kishor *et al.*, 2005; Ashraf and Foolad, 2011). Accumulation of proline could be due to *de novo* synthesis or decreased degradation, or both and it is synthesized from glutamate and ornithine. In plants, the main pathway is from glutamate which is converted to proline by two successive reductions catalyzed by pyrroline-5-carboxylate synthases (P5CS) and pyrroline-5-carboxylate reductases (P5CR), respectively. Ornithine is the alternative precursor for proline which can be transaminated to P5C by Orn-d-aminotransferase (OAT) a mitochondrial located enzyme (Kishor *et al.*, 2005; Verbruggen and Hermans, 2008)

The down-regulation of proline dehydrogenase (ProDH) also increases proline levels. The transcription of antisensing ProDH improved the production of proline in tobacco (Stein *et al.*, 2011). In the mulberry, the activities of proline dehydrogenase were reduced with progressive increase in water stress (Chaitanya *et al.*, 2009). Also, there are studies about the proline alternative pathway. The activities of ornithine transaminase were increased in *Morus alba* with low leaf water potentials (Chaitanya et al., 2009). In cashew plants, it was reported that under salt-induced stress the levels of ornithine were increased (Rocha *et al.*, 2012). In *Hibiscus tiliaceus* a contradictory result was reported. Formate dehydrogenase negatively associated with the accumulation of proline in response to osmotic stress was found highly up-egulated during salt stress (Yang *et al.*, 2011).

Glycine Betaine

Glycine betaine (GB) is a fully *N*-methyl-substituted quaternary ammonium derivative of glycine that is found in bacteria, hemophilic archaebacteria, marine invertebrates, mammals and plants (Chen and Murata, 2011). Among the many compounds known in plants, GB is accumulated at high levels in response to abiotic stress mainly to salt or osmotic stress. Levels of accumulated GB are generally correlated with the extent of stress tolerance and vary considerably among plant species and organs (Chen and Murata, 2008). GB is abundant mainly in chloroplast where it plays a vital role in the adjustment and protection of the thylakoid membrane, thereby maintaining photosynthetic efficiency and its highly ordered state at non-physiological salt concentrations via ROS (Relative Oxygen Species) scavenging. Also, at lower concentrations, GB effectively stabilizes the quaternary structures of enzymes and complex proteins (Chen *et al.*, 2008; Peleg *et al.*, 2011).

In higher plants, GB is synthesized and accumulated in chloroplasts from serine via ethanolamine, choline and betaine aldehyde. Choline is converted to GB through a two-step pathway: first to betaine aldehyde by choline monooxygenase or an enzyme with similar function which is then converted to GB by betaine aldehyde dehydrogenase (BADH). Although other pathways such as direct N-methylation of glycine is also known the pathway from choline to glycine betaine has been identified in all GB-accumulating plant species.

Many researches corroborate the hypothesis that GB plays an important role in tolerance to abiotic stress showing natural accumulators such as spinach, maize, sugar beet, and barley that synthesize GB during exposure to salt, drought, and low temperature stresses (Khan *et al.*, 2009). Furthermore, both the exogenous application of GB and the introduction of the GB-biosynthetic pathway genes to create lines of GB-accumulating transgenic plants increased the tolerance of such plants to osmotic stress.

Trehalose

Trehalose, a non-reducing disaccharide sugar is a compatible solute composed of two molecules of glucose bound by an α, α (1-1) linkage. It is well known as an abiotic stress protectant in a wide range of organisms including bacteria, fungi, invertebrates and plants, working in the stabilization of membranes and lipid assemblies and in the stabilization of biological macromolecules in the folded state forming a molecular complex. These protective properties make trehalose an ideal stress protectant (Luo *et al.*, 2008). In plants, its biosynthesis is catalyzed by two key enzymes trehalose-6-phosphate synthase (TPS) and trehalose- 6-phosphate phosphatase (TPP) (Li *et al* , 2011). It is not thought to accumulate to detectable levels in most plants with the exception of the desiccation-tolerant "resurrection plants". Indeed for years trehalose in plants was thought to be restricted to these resurrection plants. Later, it was detected in other plants with the application of inductors, such as validamycin A, an inhibitor of trehalase, the trehalose degrading enzyme (Fernandez *et al.*, 2010).

The exogenous application or induction of trehalose without genetic engineering to obtain salt or osmotic tolerance is also common. Li *et al.* 2010 has reported thermal tolerance with the addition of trehalose in wine yeast. The induction with validamycin A improved the salt tolerance of *Medicago truncatula* (López *et al.*, 2009). Exogenous trehalose recovered rice seedlings from salt stress (Nounjan *et al.*, 2012). Finally, trehalose treatment reduced the adverse effects of salinity stress on the metabolic activity of maize seedlings (Zeid, 2009).

Other sugar alcohols

Mannitol is a widely distributed sugar alcohol found in many plants and other organisms. It is synthesized in mature leaves from mannose- 6-phosphate through the combined action of a NADPH-dependent mannose-6-phosphate reductase (M6PR) and a mannitol-6-phosphate phosphatase. It is transported to sink tissues where it can be either stored or oxidized to mannose by an NAD-dependent mannitol-1phosphate dehydrogenase (mtlD) (Gupta and Kaur, 2005; Conde *et al.*, 2007). In addition to its role as a primary photosynthetic product and translocated carbohydrate in many species mannitol production may confer several potential advantages, functioning in osmotic and salt stress tolerance by serving, when accumulated as a compatible solute or osmoprotectant (Tang *et al.*, 2005)

Sorbitol, a six-carbon sugar alcohol and another important representative of osmoprotectants, is a primary photosynthate, a major translocated form of sugars and the preferential storage carbohydrate of fruit trees, mainly woody members of the Rosaceae family (Tari *et al.*, 2010; Li *et al.*, 2012). Sorbitol may be involved in providing tolerances against abiotic and biotic stresses. It plays a key role in osmotic adjustment under drought, cold and salinity conditions. Its synthesis shares a common hexose phosphate pool with sucrose production in cytosol. In source leaves, glucose-6-phosphate is converted to sorbitol-6-phosphate by sorbitol-6-phosphate dehydrogenase (S6PDH). Afterward, sorbitol-6-phosphate is used to form sorbitol via sorbitol-6-phosphatase by dephosphorylation. The reaction catalyzed by S6PDH is the key regulatory step in sorbitol synthesis (Kanamaru *et al.*, 2004; Liang *et al.*, 2012).

LEA Protein

Late embryogenesis abundant (LEA) proteins were first described almost 30 years ago in cotton seeds, where they are specifically produced and accumulated during late embryo development (Dure et al., 1981). LEA protein expression correlates closely with the acquisition of tolerance against drought, freezing, and salinity stresses in many plants (Bray, 1993; Shih *et al.*, 2008). The term "late embryogenesis abundant" has been used for more than 20 years to describe genes or proteins. However, numerous LEA genes were found to be not expressed during seed development and expressed only in vegetative tissues. In addition, considering the animal LEA proteins, the term "late embryogenesis abundant" cannot be applied to all situations. Dure *et al.* (1989) suggested a substitute term "water stress protein", which is not acknowledged widely, possibly because it refers to proteins induced only during water stress. In the past decade, LEA proteins were also found in animals (Hand *et al.*, 2011) and also in nematodes (Browne *et al.*, 2002; Goyal *et al.*, 2005). LEA proteins are mainly small molecular weight proteins ranging from 10 to 30 kDa (He and Fu, 1996); they have been grouped into various LEA families based on the occurrence of amino acid motifs (Dure *et al.*, 1989; Wise, 2003) and can confer protection against different stress conditions; in general, groups 1, 2 and 3 are considered as the three major groups containing most members of the protein family. Group 1 LEA proteins (Pfam 00477) are mostly present in plants and they contain at least one copy of a 20 amino acid motif. For example, group 1 LEA protein of wheat confers tolerance to osmotic stress in yeast (Swire-Clark and Marcotte, 1999). Group 2 LEA proteins or dehydrins (Pfam 00257) are also found in algae and share a common K-segment present in one or several copies; many dehydrins also contain an S-segment (polyserine stretch) that can undergo extensive phosphorylation and a Y-domain (DEYGNP), similar to the nucleotide-binding site of plant and bacterial chaperones. Group 3 LEA proteins (Pfam 02987), also found in nematodes and prokaryotes contain at least one copy of a 11 amino acid motif. Also, the group 3 LEA protein of rice has been found to induce resistance to drought when over-expressed transgenically (Xiao *et al.*, 2007).

References

Ahuja, I., R.C.H. de Vos, A.M. Bones and R.D. Hall. 2010. Plant molecular stress responses face climate change. *Trends Plant Sci.*, 15: 664–674.

Alcázar, R., F. Marco, J.C. Cuevas, M. Patron, A. Ferrando, P. Carrasco, A.F. Tiburcio and T. Altabella. 2006. Involvement of polyamines in plant response to abiotic stress. *Biotechnol. Lett.*, 28: 1867–1876.

Andreas, D. P., W. Dieter Jeschke and W. Hartung. 2002. Flows of elements, ions and abscisic acid in Ricinus communis and site of nitrate reduction under potassium limitation, *J. Exp. Bot.*, 53 (368): 241-250.

Arangarasan, V., S.P. Palaniappan and S. Chellaiah. 1999. Response of castor to Azospirilum and phosphobacteria inoculation. *J. Oilseeds Res.*, 16(1): 121-122.

Ashraf, M. and N.A. Akram. 2009. Improving salinity tolerance of plants through conventional breeding and genetic engineering: An analytical comparison. *Biotechnol. Adv.*, 27: 744–752.

Ashraf, M. and M.R. Foolad. 2007. Roles of glycine betaine and proline in improving plant abiotic stress resistance. *Environ. Exp. Bot.*, 59: 206–216.

Aziz Qureshi.A, R.D. Prasad and G. Suresh. 2009. Scope of vesicular arbuscular mycorrhiza (VAM) in castor , *Ricinus communis* L, *J. Oil seeds. Res.*, 27:189-191.

Babita, M., M. Maheswari, L.M. Rao, A.K. Shankar and D. Gangadhar Rao. 2010. Osmotic adjustment, drought tolerance and yield in castor (*Ricinus communis* L.) hydrolysis. *Environ. Exp. Bot.*, 69 (3): 243-249.

Baby, A. and Bapi Reddy. T. 1997. Effect of moisture conservation practices and time of nitrogen application on yield of castor . *J. Oil seeds. Res.*, 14(2): 225-228.

Baby, A. and Bapi Reddy. T.1998. Integrated management in rainfed castor. *J. Oil seeds. Res.*, 15(1): 115-117.

Baniwal, S.K., K. Bharti, K.Y. Chan, M. Fauth, A. Ganguli, S. Kotak, S.K. Mishra, L. Nover, M. Port and K.D. Scharf. 2004. Heat stress response in plants: A complex game with chaperones and more than twenty heat stress transcription factors. *J. Biosci.*, 29: 471–487.

Bhatnagar-Mathur, P., V. Vadez and K.K. Sharma. 2008. Transgenic approaches for abiotic stress tolerance in plants: Retrospect and prospects. *Plant Cell Rep.*, 27: 411–424.

Bheemaiah,G., T. Madhusudhan, M.V.R. Subramaniyam and S. Ismail. 1998. Effect of green leaf manuring and nitrogen application on growth and yield of rainfed castor alley cropped with white popanic (*Leucaena leucocephala*). *Indian J. Agrl. Sci.*, 68 (11): 722-725.

Bray, E.A. 1993. Molecular responses to water deficit. *Plant Physiol.*, 103: 1035–1040.

Browne, J., A. Tunnacliffe and A. Burnell. 2002. Anhydrobiosis: Plant desiccation gene found in a nematode. *Nature,* 416: 1038-1041.

Cao, F., H. Cheng, S. Cheng, L. Li, F. Xu, W. Yu and H. Yuan. 2012. Expression of selected ginkgo biloba heat shock protein genes after cold treatment could be induced by other abiotic stress. *Int. J. Mol. Sci.* 13: 5768–5788.

Chaitanya, K.V., G.K. Rasineni and A.R. Reddy. 2009. Biochemical responses to drought stress in mulberry (*Morus alba* L.): Evaluation of proline, glycine betaine and abscisic acid accumulation in five cultivars. *Acta Physiol. Plant.,* 31: 437–443.

Chen, T.H.H. and N. Murata. 2011. Glycinebetaine protects plants against abiotic stress: Mechanisms and biotechnological applications. *Plant Cell Environ.* 34: 1–20.

Chen, T.H.H. and N. Murata. 2008. Glycinebetaine: An effective protectant against abiotic stress in plants. *Trends Plant Sci.*139: 499–505.

Conde, C.; P. Silva, A. Agasse, R. Lemoine, S. Delrot, R. Tavares and H. Geros. 2007. Utilization and transport of mannitol in *Olea europaea* and implications for salt stress tolerance. *Plant Cell Physiol.*, 48: 42–53.

Costa, C.N.M., A.B. Santa-Brígida, B.N. Borges, M.A. Menezes-Neto, L.J.C.B. Carvalho, C.R.B. de Souza. 2011. Levels of MeLEA3, a cDNA sequence coding for an atypical late embryogenesis abundant protein in cassava, increase under *in vitro* salt stress treatment. *Plant Mol. Biol. Rep.*, 29: 997–1005.

Dieter Jeschke, D. P. Andreas. S. P. John and Wolfram Hartung. 2012. Transport, synthesis and catabolism of abscisic acid (ABA) nin intact plants of castor bean (*Ricinus communis* L.) under phosphate deficiency and moderate salinity. *J. Exp. Bot.*, 48(314): 1737-1747.

Dure, L., M. Crouch, J. Harada, T.H.D. Ho, J. Mundy, R.S. Quatrano, T. Thomas and Z.R. Sung, 1989. Common amino acid sequence domains among the LEA proteins of higher plants. *Plant Mol. Biol.,,* 12: 475–486.

Dure, L., S.C.Greenway and G.A.Galau. 1981. Developmental biochemistry of cottonseed embryogenesis and germination: Changing messenger ribonucleic acid populations as shown by *in vitro* and in vivo protein synthesis. *Biochemistry,* 20: 4162–4168.

Flowers, T.J. and A.R. Yeo. 1995. Breeding for salinity resistance in crop plants: Where next? *Aust. J. Plant Physiol.,* 22: 875–884.

Goswami, A.,R. Banerjee and S. Raha. 2010. Mechanisms of plant adaptation/ memory in rice seedlings under arsenic and heat stress: Expression of heat-shock protein gene HSP70. *AoB Plants,* 2010, doi:10.1093/aobpla/ plq023.

Goyal, K., C. Pinelli, S.L. Maslen, R.K. Rastogi, E. Stephens and A. Tunnacliffe. 2005. Dehydration-regulated processing of late embryogenesis abundant protein in a desiccationtolerant nematode. *FEBS Lett.,* 579: 4093–4098.

Gupta, A.K. and N. Kaur. 2005. Sugar signalling and gene expression in relation to carbohydrate metabolism under abiotic stresses in plants. *J. Biosci.,* 30: 761–776.

Hahn, J.S., Z. Hu, D.J. Thiele and V.R. Iyer. 2004. Genome-wide analysis of the biology of stress responses through heat shock transcription factor. *Mol. Cell. Biol.,* 24: 5249–5256.

Hand, C.S., M.A. Menze, M. Toner, L. Boswell, D. Moore. 2011. LEA proteins during water stress: Not just for plants anymore. *Annu. Rev. Physiol.,* 73: 115–134.

He, J.X. and J.R. Fu. 1996. Research progress in Lea proteins of seeds. *Plant Physiol. Commun.,* 32: 241–246.

Hedge, D.M. 1998. integrated nutrient management for production of sustainability of oilseeds-a review. *J. Oilseeds Res.,* 15(1): 1-17.

Hu, W., G. Hu and B. Han. 2009. Genome-wide survey and expression profiling of heat shock proteins and heat shock factors revealed overlapped and stress specific response under abiotic stresses in rice. *Plant. Sci.,* 176: 583–590.

Jha, Y., R.B. Subramanian and S. Patel. 2011. Combination of endophytic and rhizospheric plant growth promoting rhizobacteria in *Oryza sativa* shows higher accumulation of osmoprotectant against saline stress. *Acta Physiol. Plant.,* 33: 797–802.

Ji, X., Y. Wang and Liu, G. 2012. Expression analysis of MYC genes from *Tamarix hispida* in response to different abiotic stresses. *Int. J. Mol. Sci.,* 13: 1300–1313.

Kalarani, M. K., V. Palanisamy, S. Venkatesan and S.R. Venkatachalam. 2010. Increasing productivity by improving pistillate efficiency and seed set in castor, *Ricinus communis L J. Oil seeds Res.*, 27: 62-64.

Khan, M.S., X. Yu, A. Kikuchi, M. Asahina and K.N. Watanabe. 2009. Genetic engineering of glycine betaine biosynthesis to enhance abiotic stress tolerance in plants. *Plant Biotechnol.*, 26: 125–134.

Kishor, P.B.K., S. Sangam, R.N. Amrutha, P.S. Laxmi, K.R. Naidu, K.R.S.S. Rao, S. Rao, K.J. Reddy, P. Theriappan and N. Sreenivasulu. 2005. Regulation of proline biosynthesis, degradation, uptake and transport in higher plants: Its implications in plant growth and abiotic stress tolerance. *Curr. Sci.*, 88: 424–438.

Lakshmamma .P, P. Lakshmi, K. Anjani and C. Lavanya. 2010. Identification of castor genotypes for water use efficiency (WUE) and root traits, *J. Oilseeds Res.*, 27: 187-189.

Lakshmamma, P. and P. Lakshmi. 2006a. Identified the sources of tolerance for drought in castor, *Ricinus communis L. J. Oilseeds Res.*, 23(2):348-352.

Lakshmamma, P. and P. Lakshmi. 2006b. Screening castor *Ricinus communis L.* germplasm lines for thermal tolerance. *J. Oilseeds Res.*, 23 (2): 353-355.

Lakshmamma, P. and P. Lakshmi. 2010. Variability for water use efficiency traits and drought tolerance in castor (*Ricinus communis* L.) germplasm lines. *J. Oilseeds Res.*, (In press).

Lakshmamma, P., P. Lakshmi, and Anjani. K. 2004. Screening of castor germplasm accession for drought tolerance. *Indian Journal of Plant Physiol.*, 9(4): 447-450.

Lakshmamma, P., P. Lakshmi, and P. Padmavathi. 2003. Morpho-physiological plants traits related to drought tolerance in castor (*Ricinus communis* L). *Agrl. Sci. Digest*, 23(1): 1-5.

Lakshmi P. and P. Lakshmamma. 2009. New technique for root studies. *DOR Newsletter*, 14(4): 1-2.

Lavanya, C. 2002. Sensitivity and variation of sex expression in response to environmental changes in castor. *Indian J. Genetics Plt. Breeding*, 62(3): 232-237

Lavanya. C and S.S Solanki. 2010. Crop Improvement of Castor: The Challenges Ahead, Hegde, D.M. (Ed.). 2010. In: Research and development in Castor: Present Status and Future Strategies. Indian Society of Oilseeds Research , Hyderabad, pp.36-54.

Li, F., H. Lei, X. Zhao, R. Tian and T. Li.2012. Characterization of three sorbitol transporter genes in micropropagated apple plants grown under drought stress. *Plant Mol. Biol. Rep.*, 30 :123–130.

Li, H., H. L Wang, J. Du, G. Du, J.C. Zhan and W.D. Huang. 2010. Trehalose protects wine yeast against oxidation under thermal stress. *World J. Microbiol. Biotechnol.,* 26: 969–976.

Liang, D., M. Cui, S. Wu, F.W. Ma. 2012. Genomic structure, sub-cellular localization, and promoter analysis of the gene encoding sorbitol-6-phosphate dehydrogenase from apple. *Plant Mol. Biol. Rep.,* 30: 904–914.

López, M., N.A. Tejera and C. Lluch. 2009. ValidamycinA improves the response of *Medicago truncatula* plants to salt stress by inducing trehalose accumulation in the root nodules. *J. Plant Physiol.,* 166: 1218–1222.

Luo, Y., W.M. Li and W. Wang. 2008. Trehalose: Protector of antioxidant enzymes or reactive oxygen species scavenger under heat stress? *Environ. Exp Bot.,* 63: 378–384.

Miller, G. and R. Mittler. 2006. Could heat shock transcription factors function as hydrogen peroxide sensors in plants? *Ann. Bot.,* 98: 279–288.

Mittler, R. 2006. Abiotic stress, the field environment and stress combination. *Trends Plant Sci.,* 11: 15–19.

Munns, R. 2002. Comparative physiology of salt and water stress. *Plant Cell Environ.,* 25: 239–250.

Nounjan, N., P.T. Nghia and P. Theerakulpisut. 2012. Exogenous proline and trehalose promote recovery of rice seedlings from salt-stress and differentially modulate antioxidant enzymes and expression of related genes. *J. Plant Physiol.,* 169: 596–604.

Pathak, H.C. 2009. Crop improvement Strategies in Castor, Hegde, D.M. (Ed.). 2009. Vegetable oils scenario: Approaches to meet the growing demands. Indian Society of Oilseeds Research, Hyderabad, pp. 82-94.

Peleg, Z., M.P. Apse and E. Blumwald. 2011. Engineering salinity and water-stress tolerance in crop plants: Getting closer to the field. *Adv. Bot. Res.,* 57: 405–443.

Raghavaiah, C.V., C. Lavanya, Shaktikumaran and T. Jeevan Royal, 2006. Screening castor *(Ricinus communis)* genotypes for salinity tolerance in terms of germination, growth and plant ion composition. *Indian J.Agrl. Sci.,* 76(3): 196-199.

Raghavaiah, C.V., Y. Muralidharadu, T. Joeseph Jeevan Royal, P. Ammaji, C. Lavanya, and P. Lakshmamma. 2002. Influence of salinity stress on germination and early growth of castor *(Ricinus communis L)* genotypes. *Indian J. Agrl Sci.,* 72(10):601-603.

Raghavaiah, C.V., Y. Muralidharadu, T. Joeseph jeevan Royal, P. Ammaji, C. Lavanya, and P. Lakshmamma. 2003. Castor, *Ricinus communis L.,* production in relation to certain agronomic measures under sodic vertisol conditions. *J. Oilseeds Res.,* 20(2): 287-291.

Ramesh, T., D.V.V. Reddy, T. Venkat Reddy, and M.M.K. Durga Prasad. 2000. Effect of ethrel on sex expression and yield of castor (*Ricinus communis L). J. Oilseeds Res.,* 17(2): 279-281.

Rao, D.G., M. Vanaja, K.C. Lakkincni and P.R. Reddy.1998. Suitability of excised leaf water retention capacity (ELWRC) technique for screening castor genotypes for yield. *J. Oilseeds Res.,*15: 280-287.

Reis, S.P., L.S.C. Tavares, C.N.M. Costa, A.B. Santa-Brígida and C.R.B de Souza. 2012. Molecular cloning and characterization of a novel RING zinc-finger protein gene up-regulated under *in vitro* salt stress in cassava. *Mol. Biol. Rep.,* 39: 6513–6519.

Rocha, I.M.A., V.A. Vitorello, J.S. Silva, S.L. Ferreira-Silva, R.A. Viégas, E.N. Silva and J.A.G. Silveira. 2012. Exogenous ornithine is an effective precursor and the ä-ornithine amino transferase pathway contributes to proline accumulation under high N recycling in salt-stressed cashew leaves. *J. Plant Physiol.,* 169: 41–49.

Sahi, C., A. Singh, E. Blumwald and Grover, A. 2006. Beyond osmolytes and transporters: Novel plant salt-stress tolerance-related genes from transcriptional profiling data. *Physiol. Plant.,* 127: 1–9.

Savio Pinho dos Reis, Aline Medeiros Lima and Claudia Regina Batista de Souza, 2012. Recent molecular advances on downstram plant response to abiotic stress. *Int J. Mol. Sci.,* 13: 8628-8647.

Scharf, K. D., T. Berberich, I. Ebersberger and L. Nover. 2012. The plant heat stress transcription factor (Hsf) family: Structure, function and evolution. *Biochim. Biophys. Acta,* 1819: 104–119.

Sheshshayee, M.S., H. Bindumadhava, A.G. Shankar, T.G. Prasad and M. Udayakumar. 2003. Breeding strategies to exploit WUE for crop improvement. *J. Plt Bio.,* 30 (2):253-268.

Shiffries, O. 1960. Conventional and unconventional system controlling sex variation in *Ricinus. Journal of Genetics,* 57: 573-578.

Shih, M.D., F.A. Hoekstra and Y.I.C. Hsing. 2008. Late embryogenesis abundant proteins. *Adv. Bot. Res.,* 48: 211–255.

Stam, Y.M. and G.V. Podukuichenko.1970. In: Castor .Ed.Moshkin, V.A., Oxonian Press Pvt. Ltd.New Delhi. pp.48

Stein, H., A. Honig, , G. Miller, O. Erster, H. Eilenberg, L.N. Csonka, L. Szabados, C. Koncz and A. Zilberstein. 2011. Elevation of free proline and proline-rich protein levels by simultaneous manipulations of proline biosynthesis and degradation in plants. *Plant Sci.,* 181:140–150.

Subba Reddy, G., D. Gangadhara Rao, S. Venkateswarlu and V. Maruthi.1996. Drought management option for rainfed castor in Alfisols. *J. Oilseeds Res.,* 13(2): 200-207.

Swindell, W.R., M. Huebner and A.P. Weber. 2007. Transcriptional profiling of *Arabidopsis* heat shock proteins and transcription factors reveals extensive overlap between heat and non-heat stress response pathways. *BMC Genomics,* 8:1–15.

Swire-Clark, G.A. and W.R. Marcotte. 1999. The wheat LEA protein Em functions as an osmoprotective molecule in *Saccharomyces cerevisiae. Plant Mol. Biol.,* 39: 117–128.

Tang, W., X. Peng, and R.J. Newton 2005. Enhanced tolerance to salt stress in transgenic loblolly pine simultaneously expressing two genes encoding mannitol-1-phosphatedehydrogenase and glucitol-6-phosphate dehydrogenase. *Plant Physiol. Biochem.,* 43: 139–146.

Tari, I., G. Kiss, A.K. Deér, J. Csiszár, L. Erdei, A. Gallé, K. Gémes, F. Horváth, P. Poór and Á. Szepesi. 2010. Salicylic acid increased aldose reductase activity and sorbitol accumulation in tomato plants under salt stress. *Biol. Plant.,* 54: 677–683.

Venkateswarlu, S. 2009. Evaluation of FYM and certain management practices for nutrient and drought management in rainfed castor. *J. Oilseeds Res.,* 27: 394-396.

Verbruggen, N. and C. Hermans. 2008. Proline accumulation in plants: A review: *Amino Acids,* 35: 753–759.

Vierling, E. 1991. The roles of heat shock proteins in plants. *Annu. Rev. Plant Physiol. Plant Mol. Biol.,* 42: 579–620.

Vinocur, B. and A. Altman, 2005. Recent advances in engineering plant tolerance to abiotic stress: Achievements and limitations. *Curr. Opin. Biotechnol.,* 16: 123–132.

Wang, W., B. Vinocur, O. Shoseyov and A. Altman. 2004. Role of plant heat shock proteins and molecular chaperones in the abiotic stress response. *Trends Plant Sci.,* 9: 244–252.

Wise, M.J. 2003. LEAping to conclusions: A computational reanalysis of late embryogenesis abundant proteins and their possible roles. *BMC Bioinform.,* 4:1–19.

Xiao, B., Y. Huang, N. Tang and L. Xiong. 2007. Over-expression of a LEA gene in rice improves drought resistance under the field conditions. *Theor. Appl. Genet.,* 115: 35–46.

Yamada, A., M. Sekiguchi, T. Mimura and Y. Ozeki. 2002. The role of plant CCT in salt- and osmotic-stress tolerance. *Plant Cell Physiol.,* 43: 1043–1048.

Yang, G., R. Zhou, T. Tang, X. Chen, J. Ouyang, L. He, W. Li, S. Chen, M. Guo, X. Li, C. Zhong and S. Shi. 2011. Gene Expression Profiles in Response to Salt Stress in *Hibiscus tiliaceus*. *Plant. Mol. Biol. Rep.*, 29: 609–617.

Yang, O., O.V. Popova, U. Süthoff, I. Lüking, K.J. Dietz and D. Golldac. 2009. The *Arabidopsis* basic leucine zipper transcription factor AtbZIP24 regulates complex transcriptional networks involved in abiotic stress resistance. *Gene*, 436:45–55.

Ye, S.F., S.W. Yu, L.B. Shu, J.H. Wu, A.Z. Wu, L.J. Luo. 2012. Expression profile analysis of 9 heat shock protein genes throughout the life cycle and under abiotic stress in rice. *Chin. Sci. Bull.*, 57: 336–343.

Zeid, I.M. 2009. Trehalose as osmoprotectant for maize under salinity-induced stress. *Res. J. Agric. Biol. Sci.*, 5: 613–622.

❑❑❑

Chapter 4

Breeding of Mungbean (*Vigna radiata* L Wilczek) for Environmental Stress Tolerance Under Nature Saline Conditions

B. SUNIL KUMAR, M. PRAKASH AND B. MUTHUSELVAN
Department of Genetics and Plant Breeding, Faculty of Agriculture, Annamalai University, Annamalai nagar 608 002, India.

Abstract

A study on line x tester analysis in green gram (*Vigna radiata* (L)Wilczek) was conducted between 2010-2012 under natural saline environment at the Experimental Farm, Department of Genetics and Plant Breeding, Faculty of Agriculture, Annamalai University, Chidambaram with an objective to assess the combining ability of the parents and to know the nature of gene action in respect of seed yield and its ten component traits *viz.*, plant height, days to 50 per cent flowering, days to maturity, number of pod bearing clusters per plant, number of pods per cluster, pod length, 100 seed weight, total dry matter at harvest and harvest index involving ten varieties as lines and two varieties as testers and their 20 hybrid combinations. The genetic analysis revealed that the most of the traits were under control of non-additive gene effects. The male parents CGG-09-22, RM8-661 and Thuraiyur Local were found as good general combiners for most of the traits. Among the female parents, CO6 was relatively good general combiner for most of the traits than ADT 3. The *per se* performance of testers reflected that TARM-2 recorded highest seed yield of 5.57 g plant^{-1} and followed by AU 1 (4.93 g plant^{-1}) and RM8-661 (4.67 g plant^{-1}). Two hybrids *viz.*, CO6 x CDM Local and CO 6 x Thuraiyur Local exhibited significant *sca* effects with good *per se* performance for seed yield per plant under natural saline environment.

Introduction

Mungbean (*Vigna radiata* / Wilczek) is considered as one of the important legume crops characterized by a relative high content of proteins rich in leucine, phenylalanine, lysine, valine, isoleucine and certain vitamins (Oplinger *et al.*, 1990). Mungbean is also an important pulse crop with exploited medicinal properties. During their growth and development plants areexposed to abiotic (high and low temperature, salinity, drought, radiation etc.) and biotic (pathogen, fungus etc.) stress factors, which decrease their yield and affect the quality of their products (Habashy *et al.*, 2008). Moreover, over 6% of the world's total land area is affected by salinity and sodicity (Munns, 2005). Most crops are salt sensitive or hypersensitive plants (glycophytes) in contrast to halophytes which are the native flora of saline environments. Some halophytes have the capacity to accommodate extreme salinity because of very special anatomical and morphological adaptations or avoidance mechanisms (Flowers *et al.*, 1986). The tolerance of salt stressed plants is dependent on their ability to accumulate inorganicsolutes from external media in their tissues. Water stress is one of the major environmental constraints on plant growth and combined with high temperature and soil physiochemical deficiencies restricts crop production all over the world (Agarwal *et al.*, 1999 and Padder *et al.*, 2012).

Salinity decreases soil water potential and causes osmotic stress (Karlberg *et al.*, 2006). Osmotic unadjustment imposes on the plants the subsequent secondary drought stress and reduces their ability to obtain nutrients from the root medium (Sairam *et al.*, 2002). Furthermore, salinity directly affects nutrient uptake. It causes not only high sodium (Na^{+}) and chloride (Cl^{-}) accumulation in plants, but also antagonistically affects the uptake of essential nutrient elements such as potassium (K^{+}), calcium (Ca^{2+}) and magnesium (Mg^{2+}) in competition with Na+and also nitrate (NO^{3-}) in contrast with Cl^{-}(Sairam *et al.*, 2002; El-Hendawy *et al.*, 2005). Moreover, salt stress increases the saturation degree of membrane fatty acids and leads to conformational changes and loss of function of membrane proteins. Changes in the membrane structure/ composition in turn enhance plasma membrane permeability (Azizpour *et al.*, 2010).

The nature and magnitude of gene action involved in expression of quantitative traits is important for successfuldevelopment of crop varieties. Also, the correct choice of parents for hybridization is crucial for developmentof cultivars. Combining ability analysis provides such information so as to frame the breeding programme effectively. This analysis helps in identification of parents with high general combining (*gca*) and parental combinations with high specific combining (*sca*). The estimation of additive and non-additive gene action through this technique may be useful in determining the possibility of commercial exploitation of heterosis and isolation of pure line among the progenies of the goodhybrids.

Materials and Methods

Soil type

The experimental site soil type was deep alluvial, weakly cracking vertisol. The mean soil status and condition during the experimental season is given below.

Soil Properties	Value
Mechanical Analysis	
Course sand (%)	2.44
Fine sand (%)	5.82
Silt (%)	35.20
Clay (%)	54.20
Organic carbon (%)	2.23
Total nitrogen (%)	0.11
C/N ratio	20.13
Soil texture	Clay
Chemical Analysis	
E.C.	2.53 dsm^{-1}
pH	7.76
Irrigation Water Quality	
Soluble Cations (Cmol (+) Kg^{-1})	
Na +	1.89 ml l^{-1}
K +	0.33 ml l^{-1}
Ca ++	2.72 ml l^{-1}
Mg ++	2.84 ml l^{-1}
Soluble Anions	
HCO_3^-	7.13 ml l^{-1}
Cl^-	14.61 ml l^{-1}
SO_4-	4.27 ml l^{-1}
CO_3	Nil

Plant materials

The experimental materials comprised of twelve genotypes of which, ten genotypes (AU 1, RM8-661, TARM-2, Thuraiyur Local, Trichy Local, CDM Local, LGG 460, CGG-09-04, RM3-665 and CGG-09-22) were used as females (designated as lines) and two genotypes (ADT3 and CO6) designated as testers were used as males. These parents were crossed to produce 20 F_1 hybrids according to linex tester mating design (Kempthorne, 1957). This study was conducted from 2010–2012 at Experimental Farm, Department of Genetics and plant breeding, Faculty of Agriculture, Annamalai University, Annamalai Nagar, Tamilnadu. Single seedlings of each entry were planted at a spacing of 30 cm x 10 cm. A uniform population of 25 plants in a row per replication was maintained with a randomized complete block design with three replications.

The analysis of variance was carried out for 11 characters *viz.*, plant height, days to 50 per cent flowering, days to maturity, number of clusters per plant, number of pods per cluster, number of pods per plant, pod length, hundred seed weight, total dry matter at harvest, harvest index and seed yield per plant were evaluated based on standard evaluation system.

Statistical Analysis

Data were recorded on ten randomly selected plants from parents and F_1's plant samples. Combining ability analysis was done using line x tester method (Kempthorne, 1957). The variances for general combining ability and specific combiningability were tested against their respective error variances derived from ANOVA reduced to mean level. Significance test for *gca* and *sca* effects were performed using t-test.

Results and Discussion

The analysis of variance revealed highly significant differences among the crosses, genotypes, parents, parent vs. crosses, crosses and line x tester interaction for all the characters studied (Table 1), whereas for females, number of pods per cluster, number of pods per plant and pod length were found significant. The significant difference of mean squares between parents and crosses for all the traits indicates that they are suitable for combining ability studies. They also found significant differences among parents vs. crosses. The significant differences between line x tester interaction for these traits, indicated that specific combining ability attributed heavily in the expression of these traits and provide the importance of dominance or non-additive variances for all the traits. Several researchers have reported the predominance of dominant gene action for a majority of the yield traits. The significant mean squares of lines and testers also revealed the prevalence of additive variances for the traits studied.

Major role of non-additive gene effects in the manifestation of all the traits was observed by higher value of specific combining ability variance (σ^2*sca*) than the general combining ability variance (σ^2*gca*), ratio of σ2*gca*/ó2*sca* being less than one, and degree of dominance $(\sigma^2A/\sigma^2D)^{1/2}$ being greater than one. When the ratio between variance due to *gca* and *sca* was studied, the predominance of non- additive type of gene action was observed for many traits *viz.*, plant height, days to 50 per cent flowering, days to maturity, number of pods per cluster, number of pods per plant, seed yield per plant, total dry matter at harvest and harvest index. The present results indicate the preponderance of non-additive gene action in the expression of all the traits studied and a very good prospect for the exploitation of non-additive genetic variation for traits through hybrid breeding. Similar results of non-additive gene action was reported by many authors, Naidu and Satyanarayana (1993a), Ayyangouda Patil *et al.* (2003), Singh *et al.* (2007), Khan *et al.* (2007), Gawande *et al.* (2001), Barad *et al.* (2008), Gopi *et al.* (2006) and Sunil Kumar and Prakash (2011) (Table 2).

The close examination of results revealed that none of the parents showed significant *gca* effects simultaneously in the desired direction for all the traits studied. However, to find the good parent for subsequent development, variation in *gca* effects was estimated among lines and testers for all the traits. Among female parents, CO 6 exhibited significant *gca* effects for number of clusters per plant, number of pods per plant, seed yield per plant, and total dry matter at harvest, while ADT3 was found as good general combiner for days to maturity, number of pod bearing clusters per plant and pod length.

Of the 10 males, CGG-09-22 was found as a good general combiner for days to 50 per cent flowering, days to maturity and seed yield per plant where as AU 1 was also found as a good general combiner for days to 50 percent flowering. For plant height, none of the lines and testers were found as good general combiners. Furthermore the parents

Table 1: Analysis of variance for various traits in green gram (*Vigna radiata* L. Wilczek).

Source of variation	df	Plant height	Days to 50% floweri-ng	Days to maturity	No. of clusters per plant	No. of pods per cluster	No. of pods per plant	Pod length	100-seed weight	Total dry matter at harvest	Harvest index	Seed yield per plant
Replication	2	148.13**	7.71	19.90	0.29	3.75	0.22	0.12	0.19	6.54	4.88	0.22
Parents	11	54.65*	7.12**	132.20**	9.56**	153.60**	0.66**	2.29**	0.21	112.74**	67.04**	2.57**
Females	1	35.04	1.50	4.00	0.16	15.68*	0.59*	6.93**	0.16	14.72	50.22	0.20
Males	9	52.45*	7.73**	144.16**	6.15 **	120.80**	0.70**	1.96**	0.24	69.26**	36.07*	2.09**
Females Vs Males	1	94.12*	7.20	102.75**	49.71**	586.80**	0.36	0.64	0.03	602.07**	362.55**	9.29**
Hybrids	19	55.60**	13.85**	48.70**	12.58**	120.47**	2.42**	2.11**	0.70**	87.43**	332.69**	15.19**
Parents Vs hybrids	1	144.90*	50.62**	6.66	1.39**	70.22**	2.02**	16.19**	2.21**	5.98	752.49**	36.81**
Error	62	21.39	2.67	14.08	0.15	3.24	0.14	0.61	0.25	4.33	17.53	0.33
S.E±		2.67	0.94	1.16	0.22	1.03	0.21	0.45	0.29	1.20	2.41	0.33
C.D at5%		7.47	2.64	3.26	0.64	2.91	0.60	1.26	0.81	3.36	6.76	0.94
C.D at1%		9.93	3.51	4.34	0.85	3.86	0.80	1.67	1.08	4.47	8.99	1.25

Table 2: Analysis of variance of combining ability of various traits in green gram (Vigna radiata L. Wilczek).

Source of variation	df	Plant height	Days to 50% floweri-ng	Days to maturity	No. of clusters per plant	No. of pods per cluster	No. of pods per plant	Pod length	100-seed weight (g)	Total dry matter at harvest	Harvest index	Seed yield per plant
Replication	2	438.38**	7.35	9.81	0.78*	0.35	1.39	0.21	0.23	5.72	4.9	0.13
Females	1	672.01**	5.40	77.06**	36.50**	3.73**	72.18**	27.11**	6.01**	263.21**	15.26	11.68**
Males	9	19.36	15.77**	57.52**	9.66**	2.81**	185.23**	0.79	0.55	74.98**	93.29**	180.81**
Females Vs Males	9	23.35	12.88**	36.73**	12.84**	1.89**	51.07**	0.67	0.25	80.35**	207.361 **	7080.71 **
Error	38	20.15	2.73	3.62	0.17	0.17	4.03	0.62	0.31	5.75	23.18	93.7079
S.E+	-	0.72	0.02	0.27	0.06	0.01	1.56	0.03	0.01	0.16	2.83	120.087
σ^2 *gca*	-	0.67	0.006	0.024	0.014	0.018	0.099	0.01	0.01	0.006	0.046	0.52
σ^2 *sca*	-	1.06	3.38	11.03	4.22	0.57	15.68	3.000	1.000	24.86	61.39	232.00
σ^2 *gca*/σ^2 *sca*	2	438.38**	7.35	9.81	0.78*	0.35	1.39	0.21	0.23	5.72	4.9	0.13

AU 1, Thuraiyur Local and CGG-09-04 were the top performing and identified as desirable parents for early maturity. CGG-09-22 was observed to have *gca* in desirable direction for the traits viz., days to 50 per cent flowering, days to maturity and seed yield per plant; Thuraiyur Local and RM8-661 for number of clusters per plant and number of pods per cluster; RM3-665 for harvest Index, 100 seed weight and seed yield per plant (Table 3).

The parents with good general combining ability also exhibited good *per se* performance. This is true with the parent RM3-665 for seed yield per plant and other traits too. For hundred seed weight and harvest index, RM3-665 was a good combiner compared to other genotypes (Table 4).

The crosses CO6 x CDM Local and CO6 x Thuraiyur Local exhibited significant *sca* effects for number of pod bearing clusters per plant, number of pods per plant, total dry matter at harvest, harvest index and seed yield per plant except for pods per cluster. The cross ADT 3 x RM8-661 exhibited significant *sca* effects. CGG-09-04 was also a found to be good combiner for seed yield per plant indicating its uniqueness in transmitting additive genes to their progenies. These parents mentioned above can be of greater practical utility in breeding programme, to evolve varieties with maximum yield potential.

The *sca* effects of hybrids revealed that none of the hybrids showed significant positive specific combining ability effects for all the characters. The cross combinations ADT3 x CDM Local exhibited significant *sca* effect in desirable direction for days to 50 per cent flowering, and this cross has also showed significant *sca* effects for number of clusters per plant and number of pods per plant (g). The cross combination, ADT3 x RM8-661 had significant *sca* effects in a desirable direction and this cross was followed by CO6 x CDM Local and CO6 x CGG-09-04 for days to maturity. Top ranking hybrids with significant *sca* effect along with mean value is presented in (Table 5).

Among the twenty crosses, the cross CO6 x RM8-661 showed highly significant *sca* effects with good per se performance which had the parents of positive *gca* effects for number of pods per plant. CO6 x RM8-661 was also found as a good cross having significant *sca* effect with high per se performance. Whereas ADT3 x Trichy Local and ADT3 x CDM Local also exhibited significant *sca* effect but the mean values do not correlate with that of *sca* effect. It may be due to the

Table 3: General Combining Ability (gca) effects of lines and testers for various traits in green gram (Vigna radiata L. Wilczek).

Parents	Plant height	Days to 50% floweri-ng	Days to maturity	No. of clusters per plant	No. of pods per cluster	No. of pods per plant	Pod length	100-seed weight	Total dry matter at harvest	Harvest index	Seed yield per plant
LINES											
AU 1	1.86	-2.67**	4.77**	1.29**	0.22	2.93**	-0.62	-0.05	0.97	1.96	0.27
RM8-661	-0.60	-1.33	-1.10	1.47**	1.01**	8.12**	-0.51	-0.13	4.22**	-1.33	0.43
TARM-2	2.20	1.67*	0.40	0.99**	-0.38*	0.67	0.58	-0.38	1.32**	-5.09	-0.60*
Thuraiyur Local	2.21	0.50	1.23	1.94**	0.11	6.48**	-0.08	-0.22	4.69**	-3.43	0.23
Trichy Local	-2.10	2.00**	2.57**	-0.83**	0.93**	2.42**	0.18	-0.22	-4.84**	-2.35	-1.62**
CDM Local	-1.29	1.33	3.57**	-0.58**	0.36*	0.90	0.16	0.45	0.20	-3.66	-0.60*
LGG 460	-2.22	0.50	0.07	-1.54**	-1.16**	-10.48**	0.29	0.03	-4.71**	-4.69*	-1.93**
CGG-09-04	1.60	0.50	2.57**	-0.86**	0.31	-0.48	0.19	0.20	0.84	0.84	-0.87**
RM3-665	-0.30	-0.17	1.23	-0.93**	-0.71**	-6.56**	-0.14	0.53*	-0.58	4.81**	4.98**
CGG-09-22	-1.35	-2.33**	-5.77**	-0.96**	-0.25	-4.01	-0.05	-0.22	-4.01**	1.32	0.48**
TESTERS											
CO 6	-3.35	0.30	1.13 **	-3.35**	-0.25**	1.10**	-0.67**	0.32	2.09**	-0.50	0.44**
ADT 3	3.35	-0.30	-1.13 **	3.35**	0.25**	-1.10**	0.67**	-0.32	-2.09**	0.50	-0.44**
SE (gca for Lines)	1.83	0.67	0.77	0.17	0.17	0.81	0.32	0.22	2.85	1.44	0.53
SE (gca for Testers)	0.81	0.33	0.34	0.07	0.07	0.36	0.14	0.10	0.97	0.87	0.12
Number of parents with desirable gca	0	2	2	5	4	5	1	1	4	1	3

Table 4: Best parents identified on the basis of gca and per se performance for different traits in green gram.

Trait	Parents	gca	Parents	per se performance
Plant height	-	-	CO 6	21.44
			CGG-09-22	21.98
			CGG-09-04	23.20
			CDM Local	23.42
Days to 50 per cent flowering	AU 1	-2.67	CGG-09-22	36.80
	CGG-09-22	-2.33	Trichy Local	38.40
			ATD 3	38.40
Days to maturity	CGG-09-22	-5.77	AU 1	57.06
	ADT 3	-1.13	Thuraiyur Local	55.20
			CGG-09-04	58.14
Number of clusters per plant	Thuraiyur Local	1.94	Trichy Local	4.45
	RM8-661	1.47	CDM Local	4.35
	CO 6	1.13	RM8-661	4.11
	AU 1	1.29		
	TARM-2	0.99		
Number of pods per cluster	RM8-661	1.01	TARM-2	3.20
	Trichy Local	0.93	Thuraiyur Local	2.74
	CDM Local	0.36		
			Trichy Local	2.62
	CO 6	0.25	AU 1	2.62
Number of pods per plant	RM8-661	8.12	TARM-2	8.28
	Thuraiyur Local	6.48	Trichy Local	7.83
	AU 1	2.93	RM8-661	6.76
	Trichy Local	2.42	AU 1	6.62
	CO 6	1.10		

Pod length	ADT 3	0.67	RM3-665	6.93
			LGG 460	6.80
			AU 1	6.31
100 seed weight	RM3-665	0.53	CGG-09-04	3.20
			Trichy Local	2.66
Total dry matter at harvest	Thuraiyur Local	4.69	AU 1	14.50
	RM8-661	4.22	Trichy Local	12.46
	CO 6	2.09	TARM-2	11.69
	TARM-2	1.32	CDM Local	11.07
Harvest index	RM3-665	4.81	TARM-2	19.59
			RM8-661	18.98
			Thuraiyur Local	17.61
Seed yield per plant	RM3-665	4.98	AU 1	4.93
	CGG-09-22	0.48	RM8-661	4.67
	CO 6	0.44	RM3-665	4.61

combination of parent of negative x positive *gca* effects for this trait. Similar results have been reported earlier by Reddy and Sreeramulu (1982) and Aher *et al*. (1999). For number of pods per plant, four crosses exhibited *sca* effects in which CO6 x RM8-661 had highest *sca* effects and good mean performance. Another cross ADT 3 x CDM Local also exhibited significant *sca* effects but mean performance was much lower when compared with that of *sca* effects which may be attributed to non-additive gene action. The crosses CO6 x RM8-661 and ADT 3 x Trichy Local exhibited significant *sca* effects with good per se performance. Such crosses having good per se performance and significant *sca* effects are likely to yield transgressive segregants in segregating population. The cross CO6 x RM3-665 and ADT 3 x Thuraiyur Local exhibited positive significant *sca* effects for harvest index. The cross CO6 x RM8-661 exhibited significant *sca* effects for number of pod bearing clusters per plant, number of pods per plant and total dry matter at harvest (Table 6).

In general, no single cross exhibited significant *sca* effects for all the traits. The cross CO6 x RM3-665 exhibited significant *sca* effect for three traits *viz*., seed yield per plant (g), number of pods per cluster and harvest index.

Table 5: Specific Combining Ability (sca) effects of lines and testers for various different traits in green gram.

Parents	Plant height	Days to 50% flowering	Days to maturity	No. of clusters per plant	No. of pods per cluster	No. of pods per plant	Pod length	100-seed weight	Total dry matter at harvest	Harvest index	Seed yield per plant
CROSSES											
CO 6 X AU 1	0.91	1.63	2.13	0.50*	0.10	1.85	0.60	-0.32	0.49	-1.05	-0.28
CO 6 X RM8-661	0.28	0.37	4.47**	3.07**	-1.17**	4.24**	0.17	0.10	4.68**	-4.42	-0.71
CO 6 X TARM-2	-1.29	-0.63	-0.70	-0.12	-0.10	-1.08	0.09	-0.15	0.12	-0.87	0.18
CO 6 X Thuraiyur Local	-1.10	-1.83	-1.20	1.87**	-0.16	4.53**	-0.57	0.02	6.26**	6.20**	0.84*
CO 6 X Trichy Local	1.21	-1.30	-0.20	-00.26	-0.63**	-3.73	-0.32	0.02	-1.16	-0.22	-0.39
CO 6 X CDM Local	-1.94	-2.73*	-3.53**	-1.18**	0.04	4.25**	0.10	0.18	6.14**	9.67**	2.24**
CO 6 X LGG 460	-0.70	2.20*	1.80	-1.41**	-0.01	-0.96	-0.21	-0.23	0.16	2.00	-0.48
CO 6 X CGG-09-04	0.48	0.20	-3.37**	-0.83**	-0.45	-0.16	0.12	0.27	-4.29**	-3.07	0.16
CO 6 X RM3-665	-2.32	-0.47	1.13	-1.06**	-0.64*	0.35	-0.25	0.18	-0.24	-2.47	0.39
CO 6 X CGG-09-22	4.46	0.70	-0.53	-0.86**	0.35	-0.76	0.25	-0.15	-0.24	-3.87	-0.58
ADT 3 X AU 1	-0.91	-1.63	-2.13	-0.50*	-0.10	-1.85	-0.60	0.32	-0.49	1.05	0.28
ADT 3 X RM8-661	-0.28	-0.37	-4.47**	-3.07**	1.17**	-4.24**	-0.17	-0.10	-4.68**	4.42	0.71
ADT 3 X TARM-2	1.29	0.63	0.70	0.12	0.10	1.08	-0.09	0.15	-0.12	0.87	-0.18
ADT 3 X Thuraiyur Local	1.10	1.83	1.20	-1.87**	0.16	-4.53**	0.57	-0.02	-6.26**	-6.20**	-0.84*
ADT 3 X Trichy Local	-1.21	1.30	0.20	00.26	0.63**	3.73	0.32	-0.02	1.16	0.22	0.39
ADT 3 X CDM Local	1.94	2.73*	3.53**	1.18**	-0.04	-4.25**	-0.10	-0.18	-6.14**	-9.67**	-2.24**
ADT 3 X LGG 460	0.70	-2.20*	-1.80	1.41**	0.01	0.96	0.21	0.23	-0.16	-2.00	0.48
ADT 3 X CGG-09-04	-0.48	-0.20	3.37**	0.83**	0.45	0.16	-0.12	-0.27	4.29**	3.07	-0.16
ADT 3 X RM3-665	2.32	0.47	-1.13	1.06**	0.64*	-0.35	0.25	-0.18	0.24	2.47	-0.39
ADT 3 X CGG-09-22	4.46	-0.70	0.53	0.86**	-0.35	0.76	-0.25	0.15	0.24	3.87	0.58
No of crosses with sirable *sca* effects	0	1	3	7	3	3	0	0	4	2	2

Table 6: Top ranking hybrids identified based on significant sca effects, their mean for different characters along with their performance in green gram.

Trait	Crosses	Mean	sca effects
Plant height	Nil	-	-
Days to 50 % flowering	CO 6 X CDM LOCAL	40.00	-2.73**-
	ADT 3 X LGG 460	40.80	2.20*
Day to maturity	CO 6 X CDM LOCAL	65.06	-3.53**-
	CO 6 X CGG-09-04	58.14	3.37**-
	CO 6 X RM8-661	64.00	4.47**
Number of pod bearing clusters per plant	CO 6 X RM8-661	4.11	3.07**
	CO 6 X Thuraiyur Local	3.32	1.87**
	ADT 3 X CDM LOCAL	4.35	1.18**
	ADT 3 X CGG-09-04	2.55	0.83**
	ADT 3 X LGG 460	2.98	1.14**
Number of pods per cluster	ADT 3 X RM8-661	2.44	1.17**
	ADT 3X RM3-665	2.26	0.64**
	ADT 3 X Trichy Local	2.62	0.63**
Number of pods per plant	CO 6 X Thuraiyur Local	6.12	4.53**
	CO 6 X CDM LOCAL	6.42	4.25**
	CO 6 X RM8-661	6.76	4.324**
	ADT 3 X Trichy Local	7.83	3.78**
Pod length (cm)	NIL	-	-
Hundred seed weight (g)	NIL	-	-
Total dry matter at harvest (g)	CO 6 X Thuraiyur Local	9.00	6.26**
	CO 6 X CDM LOCAL	11.07	6.14**
	CO 6 X RM8-661	10.13	4.68**
	ADT 3 X CGG-09-04	10.41	4.29**
Harvest index (%)	CO 6 X RM3-665	17.27	9.67**
	CO 6 X Thuraiyur Local	17.61	6.20**
Seed yield per plant (g)	CO 6 X RM3-665	4.46	2.24**
	CO 6 X Thuraiyur Local	3.87	0.84*

The present study on combining ability revealed presence of both additive and non-additive variances with preponderance of non-additive variances ($\sigma^2 sca$) for all the traits studied thus indicating the ample scope of hybrid breeding programme under salinity conditions and needs to be exploited following appropriate breeding procedures. Further, based on results of the nature and magnitude of gene action simultaneously with combining ability, we could not identify any parent/cross suitable for all the desirable attributes. Thus hybridization of parents with good *gca* effects has an ample scope of throwing promising transgressive segregants in the segregating generations to be used as commercial varieties. The promising crosses with desirable *sca,* effects for most of the characters and involving good/good or average/good as combinations will be advanced to next generations to derive super segregants for most of the traits having favourable attributes from restored cross combinations.

References

Agarwal, R.M., Pandey Rashmi and Gupta Sunita. 1999. Certain aspects of water stress induced changes andtolerance mechanism in plants. *J. Indian Bot.,* 78: 225-259.

Aher, R.P., Datur D.V. and Sonawane, Y.P. 1999. Combining ability in mungbean. *Crop Res.,* 18 :256-260.

Ayyangouda Patil., Kajjidoni, S.T. and Salimath, P.M. 2003. Genetic analysis of morpho- physiological traits in greengram. *Karnataka J. Agric. Sci.,* 16: 542-547.

Azizpour, K., Shakiba, M.R., Khosh Kholgh Sima N., Alyari, H., Moghaddam, M., Esfandiari, E.and Pessarakli, M. 2010. Physiological response of spring durum wheat genotypes to salinity. *Jl Plant Nutrition.* 33: 859-873.

Barad, H.R., Pithia, M.S. and Vachhani,J.H. 2008. Heterosis and combining ability studies for economic traits in genetically diverse lines of mungbean (*Vigna radiata*(L.) Wilczeck). *Legume Res.,* 31(1):68-71.

El-Hendawy, SE., Hu Y. and Schmidhalter,U. 2005. Growth, ion content, gas exchange, and water relations of wheat genotypes differing in salt tolerances. *Australian Journal of Agricultural Research.,* 56: 123–134.

Flowers, T.J., Hajibagheri, M.A. and Clipson, NJW. 1986. In Halophytes. *The Quart Rev. Biol.,* 61: 313-337.

Gawande, V.L., Patil, J.V., Kute, N.S., Dhole, V.J. and Patil, D.K. 2001.Heterosis studies in mungbean [*Vigna radiata*(L.)Wilczek]. *NewBotanist.,* 28(1/4): 127-134.

Gopi,P., Prakash, M., Sarvanan, K., Kumar, B.S., Karuppiah, P. and Ganesan, J. 2006, Studies on combining ability in greengram (*Vigna radiata* (L.) Wilckzeck). *Res. on Crops.*,7(3):743-747.

Habashy, N.R., Zaki, R.N. and Mahmood A watef A. 2008. Maximizing tomato yield and its quality under salinity stress in a newly reclaimed soil. *Journal of Applied Sciences Research.*, 4(12): 1867-187.

Karlberg, L., Ben-Gal, A., Jansson,P.E. and Shani, U. 2006. Modelling transpiration and growth in salinity-stressed tomato under different climatic conditions. *Ecological Modelling*. 190: 15–40.

Kempthorne, O.1957. An introduction to genetic statistics. The IOWA State University Press (Eds.).John Wiley and Sons Inc., New York, pp.545.

Khan, M. G., Ahmad, W., Khattak, G.S., SirajudDin and Ahmad, H. 2007. Studies on detection of epistasis and estimates of gene effects for secondary yield characters in *Vigna radiata* (L.) Wilczek. *Sarhad J.Agric.*, 23(4):1013-1017.

Munns, R. 2005. Genes and salt tolerance: bringing them together. *New Phytol;* 167: 645-663.

Naidu, N.V. and Satyanarayana, A.1993. Heterosis and combining ability in mungbean. *Indian J. Pulses Res.*, 6:38-44.

Oplinger, E S., Hardman, L.L., Kaminski, A.R., Combs, S.M. and Doll, J. D. 1990. Mungbean. Alternative Field Crops Manual. University of Wisconsin Extension, Cooperative Extension.

Padder, Burhan M., Yadav, Rajni and Agarwal, R. M. 2012. Effect of salinity and water stress in mungbean (*Vigna radiata*) L. Wilczek var. Hum1. *Plant Sciences Feed 2.*, (9): 130-134.

Reddy,P. R. and Shreeramulu, C. 1982. Heterosis and combining ability for yield and yield component in mungbean. *Genet. Agr.*, 36 :297-307.

Sairam, R.K., Rao, K.V. and Srivastava, G.C. 2002. Differential response of wheat genotypes to long term salinity stress in relation to oxidative stress, antioxidant activity and osmolyte concentration. *Plant Science.*, 163: 1037–1046.

Singh, V. K., Tyagi, K., Tomer, A. K., Singh, M. N. and Nandan, R. 2007. Gene action for yield and yield attributing traits in mungbean{*Vigna radiata*(L.)Wilczek}. *Legume Res.*,30 (1):29-32.

Sunil Kumar, B. and Prakash, M., 2011. Heterosis for biometric and biochemical components in mungbean (*Vigna radiata* (L.) Wilczek.). *Karnataka J. Agric. Sci.*, 24 (4): (523-524).

❑❑❑

Chapter 5

Climate Mitigation Strategies in Seed Pre-Sowing Techniques

K.SIVASUBRAMANIAM, K.SUJATHA, K.SELVARANI, K.BALAKRISHNAN & V. VIJAYALAKSHMI
Department of Seed Science and Technology
Agricultural College and Research Institute, Madurai -625 104

In India, nearly 70% of cultivated land is rainfed and accounts for about 42 % of the total quantity of produced food grains. Quality seeds play a major role, along with improved package of practices lead to enhanced productivity. The low productivity under rainfed condition is due to use of poor quality seeds, soil moisture deficit, low and erratic rainfall and improper crop management. The most common impediment faced by an Indian farmer is the failure of monsoon, which in its extreme manifestation is called drought. In some of the regions, erratic rainfall leads to drought during the vegetative phase upsetting the water balance of a plant and as a consequence, the physiological functions contributing to growth and yield are deranged. Prevalence of drought during the initial phase of plant life is torturous leading to insufficient crop establishment (Matarira *et al.*, 2004). Germinating seeds in low moist conditions face hardships like high temperatures, high solute concentration at seed-water interface leading to peak osmotic potential, rapid soil drying and crust formation (Parera and Cantliffe, 1994). Safe guarding seeds during initial stage of germination will give a special impetus for the seed to overcome the moisture stress condition and develop into a vigorous plant. Seed hardening is a practice adopted to alleviate the moisture stress or making the plants resistant to moisture stress (Kaur, 2002; Giri and Schilinger, 2003). A simple method to alter the physiological and biochemical nature of the seed in order to induce the factors responsible for drought resistance is termed as seed hardening (Henckel, 1964).

Seed Hardening

Different physiological activities within the seed occur at different moisture levels (Vertucci and Leopold 1984; Taylor, 1997) and the last physiological activity is the germination process observed as radicle emergence. The initiation of radicle emergence requires high seed water content. By limiting seed water content, all the metabolic steps necessary for germination can occur without the irreversible act of radicle emergence. Prior to radicle emergence, the seed is considered desiccation tolerant, thus the primed seed moisture content can be decreased by drying. After drying, hardened seeds can be stored until time of sowing.

Pre-sowing hardening of seeds is one of the best method that results in modifying the physiological and biochemical nature of seed so as to get the characters that are favourable for drought resistance. Pre-sowing hardening is the result of extensive physiological reorganization induced by dehydration process. It can be done with water or dilute chemical solutions or growth regulating compounds or commonly available natural tonics like coconut water or milk. Hardening induces early germination, better root and seedling growth and enhances the yield potential of the crop varieties.

Steps in Seed Hardening

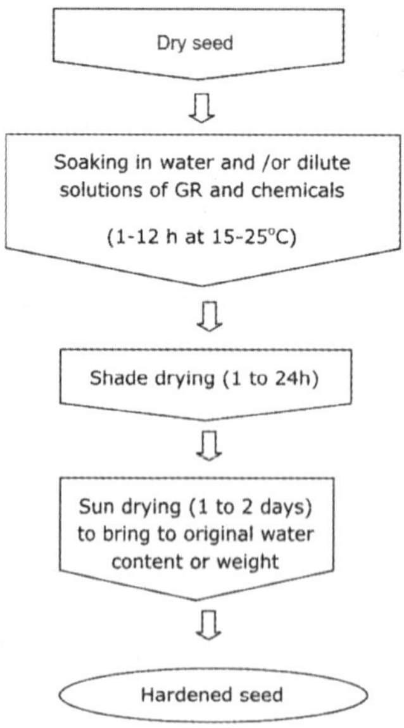

Principle of Seed Hardening

During hardening process, a number of physicochemical changes occur and modifies the protoplasmic characters, increasing the physiological activity of the embryo and associated structures, eventually leading to higher absorption of water, increase in the elasticity of cell and development of a stronger and efficient root system.

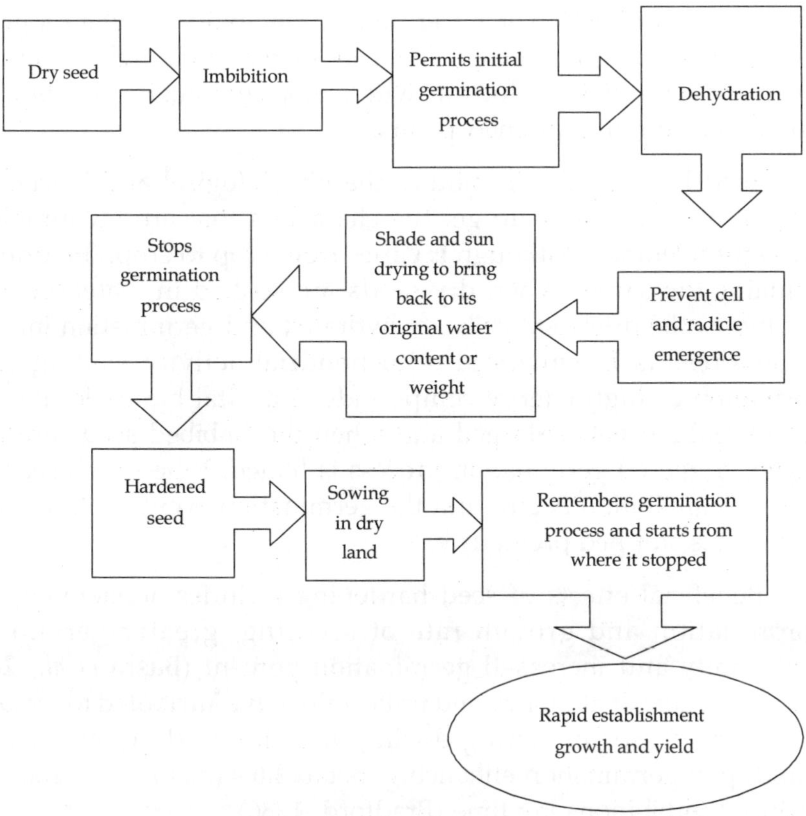

Mode of Action

The basics for pre-sowing seed hardening is the significant increase in hydrophilic property of protoplasmic colloids namely viscosity and elasticity; increased phosphorylation activity of their mitochondria; reduction in solute leakage by improving cell membrane integrity (Simon and Raja Harun, 1972), resumption of rate of protein and RNA synthesis characterized during the initial period of imbibition and shortening of the time of DNA replication in the second hydration period. Simulation of long lived mRNA under moisture stress conditions; simultaneous protein and proline content increase after hardening treatments and quenching of free radicals. Pre-sowing treatments also initiate the formation of vital bio-molecules, stimulate mitochondrial activity and preserve cellular ultra structures which would allow plants to resist adverse edapho-c1imate conditions. The consequences of some of these cellular changes are claimed to include a more xeromorphic structure with higher rate of photosynthesis, lower rate of respiration, lower water deficit, the ability to retain a greater quantity of water, a more efficient root system with higher root shoot ratio and less yield reduction when subjected again to drought as compared to non-hardened plants.

Seed hardening will modify the physiological and biochemical nature of seeds, so as to get the characters that are favourable for drought tolerance. Although it varies from crop to crop, the principle remains the same. When dry seeds are soaked in water/chemical solutions, the quiescent cells get hydrated and germination initiated. It also results in enhanced mitochondrial activity leading to the formation of high energy compounds and vital bio-molecules. The latent embryo gets enlarged and when the imbibed seeds are dried again, triggered germination process is halted. When such seeds are sown, reimbibition begins and the germination event continues from where it is stopped previously.

Beneficial effects of seed hardening includes accelerated rapid germination and growth rate of seedling, greater germination uniformity and increased germination percent (Basra *et al.,* 2005). Increased germination rate and uniformity have attributed to metabolic repair processes occurring during imbibitions (Bray *et al.,* 1995), buildup of germination enhancing metabolites (Basra *et al.,* 2005) and reduced imbibitions lag time (Bradford, 1986).

Process

The hardening resulting from pre-sowing treatments is due to a number of phyto-chemical changes within the cytoplasm including greater hydration of colloids, higher viscosity and elasticity of the protoplasm, increase in hydrophilic and decrease in lipophilic colloids, increase in the temperature required for protein coagulation and increase in bound water content. Pre-sowing treatments also initiate the formation of vital bio-molecules, stimulate mitochondrial activity and preserve cellular u1trastructres which would allow plants to resist adverse edapho-climate conditions. The consequences of some of these cellular changes are claimed to include a more xerormorphic structure with higher rate of photosynthesis, lower rate of respiration, lower water deficit, the ability to retain a greater quantity of water and a more efficient root system with higher root shoot ratio and less yield reduction when subject to again to drought as compared to non-hardened plants (May *et al.*, 1962). Root is the main plant part as far as moisture extraction and nutrient absorption are concerned. Seed hardening with chemicals and simple water were found to increase root growth even at the seedling stage. This will have a favourable influence on dry land and post monsoon season situations. Increase in root dry weight helps in maintaining high moisture status of plant leaf and concomitant increase in productivity. The hardened plants develop a more extensive system, thus enabling them to survive better under drought conditions. It is possible that early radicle emergence and seedling treatment on planting in the field following seed hardening treatments simply gives the plant a better start than non-hardened plants. By emerging early in the growing season, seedlings will be able to compete more efficiently with weeds. Further, the germination will be more synchronized which might ultimately in a uniform crop population (Basra *et al.*, 2005). Thus, a pretreatment or hardened plant might survive adverse environmental stresses more easily because of its advanced state of development.

Basis of Seed Hardening

Henckel (1964) enumerated the following as the basis of pre-sowing seed hardening.

Physiological Changes

1. Significant increase in hydrophilic property of the protoplasmic colloids namely the viscosity and elasticity.
2. Quality of the proteins are changed.
3. Increase in osmotic potential.
4. These overall changes leads to increase in water holding capacity of plants.

Metabolic changes

1. Increased respiration.

2. High level of synthetic reaction even during drought; leaves of hardened plants have more starch.
3. Increased phospharylation activity of their mitochondria.
4. Higher rate of photosynthesis because of increase in the bound water.
5. Higher accumulation of organic phosphorus and nucleo proteins.

Anatomical and Morphological Changes

1. Hardened plants have a more xeromorphic morphology than unhardened ones.
2. More extensive and denser network of veins and ribs.
3. Epidermal and stomata cells are smaller.
4. Number of stomata per unit leaf area is greater.
5. Foliage area is increased.
6. Faster recovery from atmospheric drought.
7. Greater total and absorbing surface in the root system, as well as more number of primary roots.
8. Leaves of hardened plants have more of starch.

Method of Seed Hardening

Seeds are soaked in water and allowed to absorb moisture up to 30-35 per cent of their weight under well aerated condition and kept in swollen condition for 1-12 h depending upon the crop species at 25°C (lower temperature of 15°C is found more beneficial). Then, the seeds are spread in a thin layer under shade for drying for 1 to 2 days. After shade drying, they are sun dried for 1 to 2 days to bring back to the original moisture content or weight. The hardened seeds are used for sowing. The treatment is repeated for more times depending upon the kind and variety of crops. This can also be tried by using various concentrations of number of chemicals. High temperature treatment of seed can also be attempted for inducing resistance to drought.

Chemicals Used for Seed Hardening

In addition to water, the following chemicals are recommended:

Chemicals	Role
Potassium nitrate	Involved in the physiological process of stomatal movement, Osmo-regulation, cell wall & membrane permeability.
Calcium chloride	Used to activate ATPase and amylase enzymes and to participate in starch metabolism.
Ammonium sulphate	Used as a constituent of amino acids like cystine, methionine and cystenine.
Potassium chloride	It is required for activation of enzyme process.
Zinc sulphate	It induces the dehydrogenase, alcohol dehydrogenase and lactic acid dehydrogenase enzyme activity.
Potassium dihydrogen phosphate	Used as a constituent of phospholipids, nucleotides and sugar phosphate.

Growth regulators	Role
Gibberrellic acid	Increase the hydrolytic enzymes.
CCC, Kinetin	Biosynthesis of t RNA and closely related to nucleic acid synthesis.
Ascorbic acid	Involved in several oxidation and reduction reactions.
Riboflavin	Participate in photo oxidation of IAA.
Biotin	Involved in amino acid fatty acid metabolism.

Vitamins	Role
Vit.K3	Involved in catalytic and regulating functions in cell metabolism.
Nicotinic acid	It interacts with IAA in auxin mediated process, involved in hydrogen transfer process.
Pantothenic acid	Used as a component of many co enzymes
Adenine	Increases the rate of DNA replication, rate of RNA and protein biosynthesis.

The concentration of this chemical, duration of soaking and seed to solution ratio varies depending upon the crop species.

Plant Products Used for Seed Hardening

Apart from chemicals which have proven track record in according improved seed physiological status to hardened seeds, experiments using plant parts have been carried out. Such organic products from botanical sources are easily available to the farmer and are cheap alternative to chemical methods. Examples of plant products are - Garlic (*Allium sativum* L.) extract, Coconut (*Cocos nucifera* L.) water, leaf extracts of Prosopis (*Prosopis juliflora* (Sw.)DC.), Moringa (*Moringa oleifera* Lam.), Pungam (*Millettia pinnata* (L.) Panigrahi) and Neem (*Azadirachta indica* L). A list of experiments tried for various crops and their effect has been provided below.

Table 1: List of crops and botanical seed priming treatments tried along with their promotory effects.

S.No	Crop	Chemicals used for seed hardening	Effect on crop	Authors
1	Rice	Leaf extracts of Prosopis, Pungam and Acacia (1%).	Increased seed quality parameters.	Nagaraj, S. (1996).
2	Black gram	Herbal priming with Chicory 20 %, Holy basil 15 %.	Maintained vigour and viability.	Vanitha, C. (2005).
3	Maize	Chicory 20 % Holy basil 15 %.	Improved vigor.	Vanitha, C. (2005).
4	Sunflower	Chicory 20 % Holy basil 15 %.	Higher plant establishment.	Vanitha, C. (2005).
5	Blackgram	0.5 % KNO_3 +2% Prosopis leaf extract.	Improved seed quality and shelf life of seeds.	Surulirajan, E. (2007).
6	Blackgram	0.5 % KNO_3 +2% moringa leaf extract.	Increased germination.	Surulirajan, E. (2007).
7	Cowpea	Occimum 10%, Chicory 20%, Calotropis 15%.	Increased seed quality parameters , growth and yield attributes.	Sujatha, K. and K. Ramamoorthy, (2008).
8	Sesame	Tamarind leaf extract 2%.	Increased seed quality parameters.	Suma, N. (2005).

Seed hardening techniques recommended for crops

S.No	Crop	Salt	Method
1.	Rice	1 % Potassium chloride.	40 g in 4 L of water. 4 kg seeds required for an acre is soaked for 16 hrs and then dried back to original moisture content under shade.
2.	Pearl millet	2 % Potassium chloride.	40 g in 2 L of water. 3.2 kg seeds required for an acre is soaked for 16 hrs and then dried back to original moisture content under shade.
3.	Ragi	0.2 % Sodium chloride.	8.4 g in 4.2 L of water. 6 kg seeds required for an acre is soaked for 16 hrs and then dried back to original moisture content under shade.
4.	Sorghum	2 % Potassium Di hydrogen phosphate.	120g of salt is dissolved in 6 L of water. 6 kg seeds required for an acre is soaked for 6 hrs and then dried back to original moisture content under shade.
5.	Cotton	2 % Potassium chloride / 1 % Pungam leaf extract.	78 of salt is dissolved in 3.9 L of water. 6 kg of seeds required per ac is soaked for 10 hrs or 650g of fresh pungam leaf paste mixed with 6.5L of water.
6.	Red gram	100 ppm Zinc Sulphate.	300mg salt in 3 L of water for soaking. 8 kg of seed/ac for 3 hrs with constant stirring and dried back to original moisture content.
7.	Black gram and Green gram	100 ppm Mangenese sulphate.	300mg salt in 3 L of water. soak 8 kg of seed/ac for 3 hrs with constant stirring and dried back to original moisture content.
8.	Bengal gram	1 % Potassium Di hydrogen phosphate.	24 g in 2.4 L of water. 8 kg of seeds required per ac is soaked for 4 hrs followed by shade drying.
9.	Sunflower	2 % Potassium chloride.	Dissolve 60g of $ZnSO_4$ in 3 L of water. Soak 6kg seeds for 16 hr followed by shade drying.
10.	Groundnut	0.5 % Calcium Chloride.	60 g of $CaCl_2$ in 1.2 L of water. Soak 40kg seeds for 4 hr followed by shade drying.

Advantages

1. Accelerates rapid germination and growth rate of seedling.
2. Plants from the treated seeds recover quickly from wilting when compared to plants from untreated seeds.
3. Flowering is slightly accelerated in treated plants.
4. Induces resistance to drought and salinity.
5. Seeds also withstand higher temperature (80-105°C) for prolonged periods (24-48h) without loss of viability.
6. By emerging early, seedlings will be able to compete more effectively with weeds.
7. Treated plants are generally better in growth and yield.

Caution

- Use wide mouth vessel for soaking seeds during hardening.
- Constant stirring during hardening (every 3-5hrs) is essential for effective hardening.
- Hardened seeds cannot be stored for more than 3 weeks. Hence, hardening should be practiced just prior to arrival of monsoon i.e. after the receipt of first showers. The hardened seeds must be stored in cool, dry place. Treatment using biofertilizers, bio-control agents can be done on the day of sowing.

Seed Hardening for Cereals

1. Rice

Chemical hardening

Excellent review of various chemicals used for seed hardening and their effect on rice crop has been discussed by Solaimalai and Subburamu (2004). They had collected voluminous information regarding presowing seed treatment and can be summarized as follows:

- Pre soaking of seeds with 75 ppm NAA, IAA, 10 ppm GA_3 or triacontanol 100 ppm resulted in higher dry matter production and yield.
- Soaking in succinic acid (0.25%) led to increase in tiller number.

- One molar solution of KH_2PO_4 or 1% solution of $KMnO_4$ for 24 hours or 4% Manganese sulphate or 0.05 or 0.1% copper sulphate solution or 0.5% hydrogen peroxide solution or 0.1% ammonium molybdate and 0.5% sodium molybdate or water, sodium dihydrogen phosphate (350 ppm), aluminum nitrate (200 ppm), KH_2PO_4 (2%), sodium chloride (3%) individually resulted in better stand establishment and resulted in greater production of dry matter.
- Seed treatment with fresh biodigested slurry @ 50% + potassium dihydrogen phosphate 1% + penshibao @ 50 ml/ha resulted in higher germination percentage, root and shoot lengths and yield of rice.
- Combo treatment of rice seeds soaked in 0.54 M brassinolide and 150 mM sodium chloride for 24 h exhibited increased seed germination and plant stand in the main field.

Osmo Hardening

Osmo hardening is a techqnique wherein osmoticants are used for inducing drought hardiness. Ananda and Reddy (2002) reported that seed treatment with 1% calcium chloride solution recorded significantly more grain yield of paddy and also caused for increase in plant height, dry matter accumulation, grain number per ear head and 1000-grain weight as compared to that of untreated control under rainfed condition. Susmitha (2006) revealed that in rice, seed hardening with one per cent KCl for 12 h and coating with polykote @ 3 g kg^{1} seeds significantly increased the seed germination, shoot and root length, dry matter production, vigour index and á-amylase, field emergence, chlorophyll content, protein content, and proline content over control. Farooq *et al.* (2006) concluded that osmopriming and hardening can be successfully integrated for vigor enhancement in coarse rice and osmohardening with KCl performed better than control with 19% more yield (Table 2).

Table 2: Effect of seed hardening on germination and yield in direct seeded coarse rice.

Treatments	Days to 50% emergence	Tillers ($No.m^{-2}$)	Branches/ panicle	1000 kernel weight (g)	Kernel yield ($t.ha^{-1}$)
Hardening	4.39	713.3	24.0	17.00	3.03
Osmo hardening (KCl)	4.00	738.5	22	19.00	3.23
Osmo hardening ($Ca\,Cl_2$)	4.30	716.0	24.3	16.33	3.11
Control	5.35	623.3	21.2	16.33	2.71

Farooq et al. (2006)

Bio-Hardening

Seed treatment with cow urine is known to have beneficial effect on germination, growth components viz., plant height, number of leaves, leaf area and yield components like number of grains, tiller number, grain weight and yield of crops. This has been attributed to the fact that cow urine contains physiologically active substances viz., growth regulators, nutrients (Kamalam Joseph and Rajappan Nair, 1989) and trace elements (Munoz, 1988). Seeds soaked in 10% cow dung extract for 12 hours registered more germination, root and shoot length compared with untreated seeds (Kamalam Joseph and Rajappa Nair, 1989). Cow's urine (5%), cow dung extract (10 %) and nitric acid (0.1 N) treatments resulted in higher root length and vigour index in paddy (Kamalam Joseph and Rajappa Nair, 1989). Salakinkop *et al.* (1996) at Mugad (Karnataka) noticed that the seed treatment with cow dung extract increased the plant height, dry matter production and yield of paddy under rainfed condition as compared to untreated seeds.

Seeds treated with 2.5% neem kernel extract gave more vigorous seedlings than untreated seeds (Kareem *et al.*,1989). In rice, seed treatment of talc based formulation of P. *fluorescens* at the rate of 5 and 10 g kg^{-1} of seed increased the germination, seedling vigour and seedling emergence when compared to control (Kumar *et al.*, 2000). Ramamoorthy *et al.* (2000) observed that seed biofortification with *Azospirillum* increased certain physiological traits *viz.*, amylase activity, speed of germination, seedling length and seedling dry weight in low and high vigour seeds in rice. Mishra and Sinha (2000) found that rice seeds when treated with *Pseudomonas fluorescens* @10 g kg^{-1} increased the germination percentage by 26.3–52.6 per cent and there was similar increase in root length, shoot length and fresh weight of seedlings. Prosopis leaf extract (1%) seed hardening and pelleting using potassium di hydrogen orthophosphate and DAP (1%) was able to improve germination and yield of finger millet (Vigneswari *et al.*, 2005). Seeds soaked in 0.5% sea weed (*Sargassum* sp.) followed by drying were effective only in stored seeds enhancing the speed of germination, seedling length, dry matter production, dehydrogenase activity in rice var. MDU 5 (Ramamoorthy *et al.*, 2005).

2. Maize

Kulkarni and Eshanna (1988) revealed that seed treatment with $CaCl_2$ in the ratio of 1:3 (seed: solution) recorded significantly higher plant height, LAI and total dry matter at different growth stages in maize as compared to untreated control. Pre-sowing seed treatment with growth regulators can be made to promote vegetative growth as well as to increase productivity. Fasui *et al.* (1996) observed increased germination rate, seedling vigour, seedling fresh weight, seedling height and root length of maize when the seeds were soaked in solution of 0.3 per cent $CaCl_2$ and 0.1per cent $ZnSO_4$ either individually or in combination.

Maize seeds soaked in 20 ppm GA_3 for 30 min improved some germination traits, but could not affect grain yield (Subedi and Ma, 2005). Hardened seeds usually would exhibit rapid germination when absorb water under field conditions (Ashraf and Foolad, 2005). *Azospirillum* inoculation was found to improve physiological status of maize plant by ameliorating the harmful effects of water short fall during the flowering period which will help in increase in grain production (Casanovas *et al.*, 2003).

Priming

Soaking seeds in water i.e. hydropriming improved early and vigorous crop establishment in maize (Nagar *et al.*, 1998). The treatment effects of priming on field emergence were more pronounced at early stage (7 DAS) (10.6 to 28.9%) than at later stages of crop growth (14 and 21 DAS). The hydro-primed maize seed produced consistently and significantly longer shoots after 5 days than the untreated control (Matthews and Hosseini, 2007).

3. Wheat

Chikkappaiah (2010) conducted seed hardening experiment with nine treatments viz., pre-sowing seed treatments with water soaking, $CaCl_2$ (2%), CCC (500 and 1000 ppm), KNO_3 (1% and 2%), KCl (1% and 2%) and control and the results are summarized as follows:

- The plant height increased significantly from 30 DAS to harvest due to the seed hardening treatment with 2% $CaCl_2$ as compared to control. The treatment CCC @ 1000 ppm and 500 @ ppm showed decreased plant height.
- Number of green leaves and tillers differed significantly among the treatments. Seed hardening with CCC (1000 ppm) recorded significantly more number of green leaves and tillers while significantly lowest number were recorded in control.
- Dry matter accumulation is increased significantly due to seed hardening with CCC (500 and 1000 ppm) and KCl (1%). Significantly lowest values were recorded by control.
- Leaf area and leaf area index increased significantly due to seed hardening with CCC (1000 ppm), CCC (500 ppm), $CaCl_2$ (2%), KCl (1%), KCl (2%). Significantly lowest values were recorded by control.
- Significant differences were observed between the treatments with respect to dry matter production and its distribution into different plant parts (leaves, stem and panicle). Seed hardening with CCC (1000 ppm) recorded significantly more total dry matter followed by CCC (500 ppm) and significantly lowest in control.
- Growth parameters like AGR, CGR, RGR and NAR differed significantly among the treatments at all the growth stages. The pre-sowing seed hardening technique with CCC 1000 @ ppm and CCC @ 500 ppm showed significantly increased CGR, AGR from 60 to 90 DAS, 90 to harvest, respectively and RGR and NAR significantly increased upto 60 to 90 DAS and declined thereafter.
- Specific leaf weight (SLW) increased significantly upto 90 DAS and declined thereafter due to the seed hardening techniques, use of agro-chemicals. However, higher SLW was found with seed hardening with CCC (1000 ppm).

- Leaf area duration (LAD) increased from 60 to 90 DAS to harvest due to seed hardening techniques.
- Significant differences were observed among the seed hardening treatments and use of CCC (1000 ppm and 500 ppm), $CaCl_2$ (2%), KNO_3 (1% and 2%), KCl (1% and 2%) and water soaking etc. with respect to rate of photosynthesis, stomatal conductance, water use efficiency and relative water content. The seed treatment with CCC (1000 ppm) recorded significantly higher photosynthesis, stomatal conductance and relative water content. Water use efficiency recorded highest in CCC (1000 ppm) seed treatment and significantly lowest values were recorded in control.
- There was a significant increase in total chlorophyll content due to various treatments at all the stages. However, the maximum chlorophyll content was observed in seed hardening with CCC (500 and 1000 ppm).
- There was significant increase in proline content due to various treatments at all the stages. However, the maximum proline content was observed in seed hardening with CCC (500 and 1000 ppm) at all the stages.
- Thousand grain weight, grain yield showed significantly higher values due to seed hardening techniques as with CCC (1000 and 500 ppm) as compared to other treatments and lower values were recorded in control. Based on the above results, it is concluded that the seed hardening with CCC (1000 ppm) and cycocel (500 ppm) is more effective and economical in increasing the yield in rainfed wheat (var. Bijga yellow).

Misra and Dwivedi (1980) found that wheat seeds treated with potassium and distilled water produced distinctly more plant height, enhanced tiller production, more number of green leaves, leaf area and shoot dry weight significantly as compared to the control. De *et al.* (1982) noticed that soaking of wheat seed with cycocel decreased the shoot growth and leaf area per plant progressively with increase in concentration by 10, 40 and 80 ppm, but the radical growth was increased considerably. Similarly, Dighc *et al.* (1983) indicated that cycocel (500 ppm) applied to wheat cultivars by either seed treatment or foliar spray or soil drenching decreased plant height and delayed flowering and maturation by 4 to 5 days as compared to control.

Harvest index did not change much due to the use of different seed treatments (Mandal and Basu, 1987).

Seed treatment with cycocel @ 0.1% and 5% KH_2PO_4 increased yield and nutrient uptake of late sown wheat (Bhatia and Rathore, 1986). Triazoles increased epicuticular wax and reduced the length but increased the width and thickness of leaves, decreased stomatal aperture and length of trichomes (Gao *et al.*, 1988). Seed hardening with 10^{-5}M GA_3, kinetin or IAA or 10^{-4} cycocel generally increased number of leaves per plant, root, stem and total dry matter, RGR, NAR and LWR in wheat as compared to control (Gurudev Singh *et al.*, 1991). Amaregouda *et al.* (1994) revealed that seed treatment with $CaCl_2$ (2%) in wheat increased free proline content and RWC at both 60 and 80 DAS. This treatment also increased chlorophyll b-content, accumulated maximum K-content and total sugar content in leaves. Seed hardening with 1.0% muriate of potash was found to be effective in increasing the grain yield (32 %) of rainfed wheat over control in the hills zone of Assam (Paul *et al.*, 1998).

Osmo Hardening

Alternate soaking and drying in solutions of various chemicals *viz.*, sodium chloride, sodium sulphate, potassium nitrate, calcium chloride, ammonium sulphate and KCI, accelerated germination and growth rate of wheat seedling. Bhatia and Rathore (1986) observed beneficial effect in terms of seedling height and dry matter accumulation due to treating wheat seeds with KH_2PO_4, $CaCl_2$, NaCl and distilled water as compared to control (no treatment). Amaregouda *et al.* (1994) reported that the seed treatment with two per cent calcium chloride caused yield increase by 19 per cent in wheat as compared to control (untreated).

Migahid and Sadek (1994) reported higher proline content in wheat due to seed hardening with $CaCl_2$ or $AlCl_3$ under salt stress condition than unsoaked seed (control). Narendra *et al.* (2000) reported that pre-sowing treatment to wheat seeds with calcium chloride resulted in significant improvement in chlorophyll "a" (2.73 mg/g fresh weight) and chlorophyll "b" (0.76 g mg/g fresh weight) as compared to that of untreated seeds (1.72 and 0.7 g mg.g^{-1} fresh weight of chlorophyll "a" and "b", respectively). Wheat seeds soaked in 2.5 per cent calcium chloride produced significantly higher number of effective tillers per plant or per meter row length, ear head length, number of spikelet's per ear head, number of grains per ear head and eventually the yield (4014 $kg.ha^{-1}$) over that of unsoaked seeds (3177 $kg.ha^{-1}$) (Ugale and Mungse, 2001).

4. Sorghum

Seed hardening with water resulted in increased plant height, dry matter accumulation, leaf area index and root-shoot ratio when compared with seed hardening with CCC and calcium chloride. Pre-sowing seed soaking in 1% KH_2PO_4 solution for 12 hours favourably influenced plant growth, uptake of nutrients, 1000 grain weight and grain yield (Gopalakrishnan, 1965). Seed hardening with 1% resistin resulted in enhanced root growth and increased drought resistance (AICRPDA, 1970). However, soaking twice in water reduced grain yield due to poor germination and uneven field stand (Corleto *et al.*, 1977). Seeds soaked in 1% solution of KNO_3 and KH_2PO_4 contained more sugar and KH_2PO_4 exerting the greatest effect. The increase in sugar content would stimulate drought resistance through increased osmotic pressure and water uptake. Singh *et al.* (1984) stated that soaking of sorghum seeds in water for 12 hrs overnight hastened the flowering period by 96.83 days as against 100 days for control. These four days of enhancement in flowering can be utilized for solving synchronization problem in case of seed production. Higher seed germination (76%) and vigour index (969.1) of sorghum were recorded under seed hardening with calotropis leaf extract 2% (Devarani and Rangasamy, 1998) (Table 3).

Table 3: Seed hardening on germination and vigour index of sorghum.

Seed hardening	Germination (%)	Vigour index
Control	63	587.3
Water	65	697.2
Goat urine @ 15%	66	708.4
Cow dung extract @ 15%	76	895.0
Garlic extract @ 2%	70	775.1
Calotropis leaf extract @ 2%	76	969.1
Cow urine @ 5%	58	548.2
Morinda leaf extract @ 2%	70	765.0
KH_2PO_4	70	761.2
CD 5 %	2.8	37.3

(Devarani and Rangasamy, 1998)

Presowing soaking with calcium chloride at 0.4% and cycocel at 0.2% improved germination, vigour index and root shoot ratio (Rangaswamy *et al.*, 1993). The beneficial effects of water soaking and chemical treatments in inducing growth and yield might be attributed to leaching out of toxic metabolites from the seed and to antifungal and anticatabolic effects of the treatment (Basu, 1994). The benefits of water soaking were observed when sorghum seeds pre-soaked in water for 24 hours (Corleto *et al., 1977*). Seed hardening with 2 % KH_2PO_4 followed by pelleting with DAP recorded higher seed yield and other growth parameters (Geetha and Selvakumari, 2008).

Osmo Hardening

Henckel (1964) reported that calcium chloride lead to redistribution of nutrient reserves resulting in the greater internodal length in crop. Calcium is the most important element in plant during stress, drought condition, has ability to change stomatal movement (Schwartz, 1985, De Silva *et al.*, 1986, Umashankar, 1985) and reduces the water losses and maintain high water balance (Malathi Chari, *et al.*, 1984). Patil (1987) opined that seed hardening with two per cent calcium chloride in the ratio of 1:1 (seed:solution) for four hours increased drought resistance in sorghum and also increased the grain yield by 10 per cent over that of control. Patil (1987) reported that seed treatment with two per cent calcium chloride increased proline content by about six times than that of control (without seed soaking) in sorghum under rainfed condition. Rangaswamy *et al.* (1993) reported that seed hardening with 0.4 per cent calcium chloride and 0.2 per cent CCC increased the germination percentage, vigour index and root shoot ratio in sorghum. Seed hardening with $CaCl_2$ (0.4 %) and cycocel (0.2%) increased root length, root spread, grain and stover yield of rainfed sorghum (Nirmala *et al.*, 1994). The pre-sowing seed hardening treatments with 100 ppm of $CaCl_2$, KNO_3, ascorbic acid, pyridoxine hydrochloride, IAA and GA increased chlorophyll content in *Pennisetum americanum* and *Sorghum bicolor* (Kadiri and Hussaini, 1999). Seeds were soaked in cow's urine to induce drought tolerance of seeds. Kulkarni and Chittapur (2003) stated that pre-sowing seed treatment with water, mineral solutions *viz.*, $CaCl_2$, $ZnSO_4$, cobalt sulphate, K_2SO_4, $CuSO_4$, sodium molybdate, boric acid and $MnSO_4$ or growth regulators *viz.*, ascorbic acid, kinetin, benzyl adenine, gibberellic acid and cycocel alone and in combination was found to speed up germination process and increased the rate of germination, seedling vigour, resistance to water and salinity stress and finally the crop yield in sorghum.

Bio-Hardening

Seeds hardened with aqueous solution of botanicals performed significantly better than control (Jegathambal, 1996).

5. Pearl millet

Seed hardening techniques were adopted to improve the drought tolerant characters in pearl millet (Ramachandran and Rao, 1977). Kamala Thirumalaiswamy and Sakharam Rao (1977) reported that seed treatment with distilled water, CCC @ 5 ppm and kinetin @ 5 ppm slightly increased net assimilation rate (NAR), relative growth rate (RGR) and leaf area in pearl millet and irrespective of moisture level, the size of the leaf was greatly influenced by cycocel followed by kinetin. Soaking seeds in either 2% KCl or 3 % NaCl at 1: 1 ratio for 6 hours followed by 5 hours shade drying improved germination and final yield. Kadiri and Hussaini (1999) reported that seed hardening treatment with agrochemicals viz., $CaCl_2$, KNO_3, ascorbic acid, pyridoxine hydrochloride, IAA and GA increased the chlorophyll content in bajra and sorghum as compared to untreated seeds. The seeds invigourated with 2% $KH_2 PO_4$ improved seed germination, seedling dry weight, field stand and seed yield in bajra (Bharathi *et al.*, 2004). Seed hardening with 0.1 ppm brassinolide for 6 h excelled others in improving germination, dry matter, vigour index, á-amylase activity, chlorophyll content in laboratory condition. In field study, seeds hardened with 0.1 ppm brassinoloide gave higher percentage of field emergence, no. of productive tillers, weight of the ear head and also grain yield compared to other treatments (Menaka and Vanangamudi, 2008) (Table 4).

Table 4: Effect of seed hardening on yield attributing characters in bajra.

Treatments	Field emergence (%)	Number of productive of tillers	Weight of ear head (g)	Grain yield Kg/ha
Control	78	4.0	42.5	2140
Seed hardening with 2 % KCl	92	6.2	45.3	2518
Seed hardening with 0.1 ppm brassinolide	77	5.3	44.5	2413
Seed hardening with 200 ppm s alicylic acid	75	5.1	44.3	2317
Mean	80	5.4	44.2	234
CD (p=0.05)	0.931	0.890	0.703	86.204

6. Finger millet

Rajashekhar *et al.* (1970) reported that pre-sowing seed hardening in ragi increased the root growth by 27.0 per cent than unhardened seeds. Pre-sowing seed hardening increased early germination, vigorous seedlings production and yield (Krishna Sastry *et al.,* 1969). Seeds soaked in 100 ppm Na_2HP0_4 caused remarkable improvement in plant height, tillering, LAI, LAD, CGR, DMP and yield (Maitra *et al.,*1998).

7. Bengalgram

Four varieties of bengalgram seeds were hardened by pre soaking in the vermi wash, coconut milk, cow urine and water solutions separately for 1 hour in 1:2 proportions followed by drying for 7 days under ambient conditions to original moisture content. Significant differences were also observed on root and shoot length and vigour index due to seed treatment with organic substances. In all the genotypes, significantly higher shoot and root length, seedling dry weight and vigour index were recorded in seeds treated with vermiwash followed by seeds treated with cow urine (Deshpande *et al.,* 2008).

Osmo Hardening

Soaking seeds in 0.5% $CaCl_2$ at 1:1 ratio until viable expression of embryonic growth and then shade drying to original moisture content was effective in improving germination and plant stand. Seeds soaked in water significantly increased plant height, number of tillers, shoot weight and grain yield (Dawson, 1965). Seed hardening with 2.5% calcium chloride recorded increased tillering, plant height, root length and DMP resulting in enhanced yield (Karivaratharaju and Ramakrishnan, 1985b). Dry matter production was increased due to the effect of pre-sowing hardening with 1.5% sodium chloride (Karivaradaraju and Ramakrishnan, 1985b). Maitra *et al.* (1998) noticed that seed treatment with distilled water and agrochemicals (0.25 per cent $CaCl_2$, 100 ppm KH_2PO_4 and 100 ppm Na_2HPO_4) increased the plant height, LAI, LAD and dry matter accumulation at different growth stages in finger millet. Kalarani *et al.* (2001) studied the effect of seed soaking in late sown finger millet under rainfed condition. Seed soaked with KCl + $CaCl_2$ solution recorded an yield increase of 19.6 per cent in CO 13, 17.5 per cent in PR 202 and 13.9 per cent in Indaf 9 over the control. The yield increase was due to maintenance of higher LAI (Table 5).

Table 5: Seed hardening on seedling characters of finger millet.

Treatments	Germination (%)	Root length (cm)	Shoot length (cm)	DMP (mg/ seedling)	Vigour index
Control	82.0	7.4	5.6	2.3	1074
Water	87.0	7.3	5.8	2.0	1149
KCl 1%	86.0	8.2	6.2	2.6	1240
$CaCl_2$ 1%	82.0	7.6	6.1	2.3	1125
KCl 0.5%+ $CaCl_2$ 0.5 %	82.0	7.7	6.0	2.3	1137
KCl 1% + pungam leaf powder pelleting	93.0	9.5	8.3	4.3	1664
$CaCl_2$ 1%+ pungam leaf powder pelleting	90.0	9.1	7.7	3.6	1521
CD (5%)	-	0.28	0.21	1.25	53.8

(Palanisamy and Punithavathi, 1998)

Seeds hardened with KCl 1% followed by pelleting with pungam leaf power (60 g/kg of seed) recorded higher germination (93.0%), root length (9.5 cm), shoot length (8.3 cm), dry matter production (4.3 mg/seedling), vigour index (1664) and field emergence (80.0%) in finger millet (Palanisamy and Punithavathi, 1998). Hardened and pellcted seeds of finger millet could be stored safely upto two months in polythene bag with minimum seed certification standard of 75 % germination (Vigneswari and Natesan, 2010).

In chickpea, significantly higher seed yield was recorded in seed hardening with $CaCl_2$ 2% (26..32 q per ha), followed by Cycocel @ 1000 ppm (23..54 q per ha) and succinic acid @ 20 ppm (23..44 q per ha) as compare to control (19..04 q per ha).. The increased seed yield in seed hardening with $CaCl_2$ 2% may be attributed to increase in the morpho physiological traits like plant height,, number of leaf lets,, leaf dry matter and total dry matter accumulation,, leaf area per plant and Harvest index.. Increase in the physiological traits like LAI, LAD, CGR, NAR and RGR and yield parameters like 100 seeds weight,, seed yield per plant and Harvest index was reported by Manjunath and Dhanoji (2011).

8. Green gram

Specific leaf weight (SLW), a stable character was initially low and improved subsequently and reached the maximum value at 42 days in green gram genotypes (Kalubarme and Pandey, 1979) upon seed hardening. Crop productivity is mainly determined by crop growth rate (CGR) which depends on leaf area index (LAI) and then rate of photosynthesis. The CGR was found to be low during early vegetative stage but increased with the advancement of growth in green gram (Kalubarme and Pandey, 1979). Green gram seeds given the invigoration treatment (hydration dehydration treatment) increased the DMP and seed yield as compared to the control (Dharmalingam and Basu, 1989). Sumabai *et al.* (1987) indicated that foliar spray of ascorbic acid, NAA and GA produced increased assimilate production, favoured longer reproductive growth and increase in LAI, LAD and NAR in green gram.

Priming

Seed priming improved the antioxidant enzymes activity which decreased the adverse effects of Reactive Oxygen Species (ROS) (Del Ryo *et al.*, 2002). All the seed priming significantly improved the germination rate and vigour of the mungbean seedlings (Umair *et al.*, 2010). Priming may enhance the seedling vigour of mungbean. Nutrient priming using phosphorous and osmopriming with mannitol were the most appropriate priming treatments for mungbean (*Vigna radiate L.*) (Adnan Umair *et al.*, 2012). Combination of seed hardening with $MnSO_4$ (100 ppm) and Prosopis leaf extract along with pelleting using $MnSO_4$, $FeSO_4$, $(NH_4)_2MoO_4$ gave higher pod and seed yield which was 21.4% increase over control (Srimathi *et al.*, 2007).

9. Black gram

Black gram seeds when treated with talc based formulation of *T. viride* @4g kg^{-1} + *P. fluorescens* @ 5g kg^{-1} recorded maximum germination (93.6%), shoot length (28.12 cm), root length (18.35 cm), dry matter production (9.77 mg) and vigour index (3501) compared to control. Black gram seeds hardened with 100 ppm $ZnSO_4$ for 3 h recorded higher germination, seedling vigour, total chlorophyll content, dehydrogenase and a-amylase activity under laboratory condition (Kavitha, 2005).

Seed hardening with KCl improved the seed germination and other quality parameters compared to control. Moist sand conditioning with

KH_2PO_4 or $NaH_2 PO_4$ followed by drying was effective against ageing in black gram (Sujatha *et al.*, 2005). Seed germination and seedling vigour of black gram cv. VBN 3 seed could be improved by soaking the seeds in 0.5 per cent KNO_3 for 4h duration (Surulirajan, 2007). Seaweed treated seeds when subjected to ageing process, exhibited higher activity of free radical scavenging enzyme catalase, lower amounts of leachate, amino acids, sugar and lipid peroxidation and also enhancement in germination and dry matter production (Ramamoorthy and Sujatha, 2007).

10. Cowpea

Vijaya (1996) revealed that cowpea and blackgram seeds fortified with $ZnSO_4$ + $MnSO_4$ + Na_2MO_4 registered the maximum rate of germination, field emergence, root and shoot length, dry matter production and vigour index. The better performance of cowpea soaked seeds in 0.5 and 0.75 % in sargassum could be observed on speed of germination, seedling measurements and dry matter production and the deterioration was minimum in treated seeds (Ramamoorthy *et al.*, 2006). Krishnakumary *et al.* (2008) proved the importance of duration of soaking and volume of water in cowpea seed invigoration. Seed soaking with chicory leaf extract followed by drying was found to be effective in preventing deteriorative senescence and could be recommended as a pre-storage seed treatment in cowpea.

11. Red gram

Treating the seeds in molybdenum (100 mg kg^{-1} of seed in one litre of water) increased the hundred seed weight but the differences in the number as well as weight of pods and seed yield were not expressive due to treatment (Karivaratharaju and Ramakrishnan, 1985a). Pre-sowing soaking with calcium chloride at 0.4% and cycoel at 0.2% improved germination, vigour index and root shoot ratio (Rangaswamy *et al.*, 1993). Jayarami Reddy *et al.* (1996) reported that foliar spry of NAA (10 ppm) + KNO3 (0.5%) and NAA (20ppm) + KNO3 (0.5%) recorded the maximum dry matter production in pigeon pea as compared to control.

12. Sunflower

Filatov and Frolova (1975) found that seed hardening induced heat tolerance in sunflower and the hardened plants produced higher seed yield ranging from 150 to 300 kg per ha, which was more than unhardened plants. Pre-soaking the seeds in Zn SO_4 for 12 h gave the highest seed and oil yield per hectare (Muthuvel *et al.*, 1983) (Table 6).

Table 6: Effect of seed hardening with micro nutrients in Sunflower.

Treatments	Kharif 1982		Summer 1982	
	Seed yield (kg/ha)	Oil yield (kg/ha)	Seed yield (kg/ha)	Oil yield (kg/ha)
Water soaking	1054	446	1252	552
0.5 % Zn So4	1079	446	1280	572
1.0 % ZnSo4	1181	471	1411	633
2.0 % ZnSo4	1195	488	1531	673
0.5 % Mn So4	1032	418	1237	559
1.0 % MnSo4	1160	482	1307	578
2.0 % MnSo4	1100	432	1314	597
0.5 % Sodium molybdate	1144	462	1311	602
CD (p=0.05)	27.4	25.0	102	56

(Muthuvel et al., 1983).

Seed hardening is an effective technique that can be practiced by farmers. Simple chemicals and botanicals, bioactive organisms can be used for effective hardening. The cost of seed hardening is very low. Effective extension strategy must be adopted to promote seed hardening. With 80% of area under irrigated condition, Tamil Nadu may not be suitable for this technique. However, large areas of rainfed agriculture is found in more agrarian states of Karnataka, Andra Pradesh, Madhya Pradesh, Maharastra and Chattisgarh.

References

Adnan Umair, Safdar Ali, Muhammad Javed Tareen, Ijaz Ali and Muhammad Naveed Tareen. 2012. Effects of Seed Priming on the antioxidant enzymes activity of Mungbean (*Vigna radiata*) Seedlings. *Pakistan J. Nutrition* 11 (2): 140-144.

AICRPDA. 1970. Annual Report. All India Coordinated Research Project for Dry land Agriculture, Agricultural Research Station, Kovilpatti, T.N., P: 63-66.

Amaregouda, A., M. B. Chetti and Manjunath, S. 1994. Physiological basis of yield variation due to application of different chemicals in wheat. *Annals Pl. Physiol.*, 8: 24-28.

Ananda, M. G. and V. C. Reddy. 2002. Effect of sowing method and seed treatment on grain yield of rainfed low land paddy varieties. *Mys. J Agric. Sci.*, 36: 231-236.

Ashraf,M. and M. R. Foolad. 2005. Pre-sowing seed treatment - a shortgun approach to improve germination, growth and crop yield under saline and none-saline conditions. *Adv. Agron.*, 88: 223 – 271.

Basra S. M. A, M. Farooq and R. Tabassum.2005. Physiological and biochemical aspects of seed vigor enhancement treatments in fine rice (*Oryza sativa* L.). *Seed Sci. Technol.* 33 (3):25-29.

Basu, R.N.1994. An appraisal of research on wet and dry physiological seed treatments and their applicability with special reference to tropical and subtropical countries. *Seed Sci. Technol.*, 22: 107-126.

Bharathi, A., Natesan, P., Vanangamudi, K., Bhaskaran, M., Raja, K., Prabakar,K. and V. Balabubramani. 2004. Presowing seed invogouration treatment for better crop stand establishment and seed yield in bajra (*Pennisetum typhoides*) cv. CO 7. *Madras Agric.* J., 91(4-6): 302-310.

Bhatia, D.S. and S.S. Rathore. 1986. Effect of seed soaking treatment with agrochemicals on germination and seedling attributes of wheat. *Madras Agric. J.*, 73: 378-380.

Bradford, K. J. 1986. Manipulation of seed water relations via osmotic priming to improve germination under stress conditions. *Hort. Sci. Alexandria*, 21: 1105-1112.

Bray, C. M. 1995. Biochemical process during the osmopriming of seeds. In: Kigel J., Galili G., Eds. Seed development and germination, New York. : 767-789.

Casanovas, E.M., C.A. Barassi, F.H. Andrade and R.J. Sueldo. 2003. *Azospirillum* inoculated maize plant responses to irrigation restraints imposed during flowering. *Cereal Res. Comm.*, 31 (3): 395-402.

Chikkappaiah, 2010. Seed hardening techniques to improve drought tolerance in rainfed wheat (*Triticum aestivum* L.), M.Sc. Thesis, University of Agricultural Sciences, Dharwad.

Corleto, A., D. Linsalato and M. As-squi. 1977. Results of some investigations on pre sowing of grain sorghum seed 2. Effect of pre sowing seed on transpiration and some yield characters of grain sorghum. *Revista di Agronomica*, 11: 178-181.

Dawson, M. J. 1965. Effect of seed-soaking on the growth and development of crop plants I. Finger millet *(Eleusine coracana* Gaertn.). *Indian J. Plant Physiol.*8: 52-56.

De silva, D.L.R., R. C. Cox, A. M. Hetherington and T. A. Mansfield.1986. The role of abscisic acid and calcium in determining the behaviour of adaxial and abaxial stomata. *New Phytology,* 104: 41-51.

De, R., S. Giri, G. Saran, R. K. Singh and G. S. Chaturvedi .1982. Modification of water balance of dryland wheat through the use of chloromequat chloride. *J. Agric. Sci.Cambridge.* 98: 593-597.

Del Ryo, L.A., F.J. Corpas, L.M. Sandalio, J.M. Palma, M.Gomez and J.B. Barroso, 2002. Reactive oxygen species, antioxidant systems and nitric oxide in peroxisornes. *J. Exp. Botany,* 372: 1255-1272.

Deshpande, V. K., M. B. Kurdikeri and N. M. Shakuntala. 2008. Effect of seed hardening with organics on seed quality of Bengal gram. *Karnataka J. Agric. Sci.,* 21 (1): 118-119.

Devarani, N. and A. Rangasamy. 1998. Nutrient and drought management practices for rainfed sorghum under vertisol. *Madras Agric. J.,* 85: 391-393.

Dharmalingam, C. and R. N. Basu. 1989. Seed development and maturation studies in mungbean. *Seed Res.,* 17(2):103-108.

Dighe, R.S., J. T. Thakare and S. B. Hall.1983. Effect of 2-chloroethyl trimethyl ammonium chloride on growth and yield of wheat. *Punjabrao Krishi Vidyapeet Res. J.,* 7: 12-15.

Farooq, M., S. M. A. Basra, R. Tabassum and N. Ahmad. 2006. Evaluation of seed vigour enhancement techniques on physiological and biochemical techniques on physiological basis in coarse rice (*Oryza sativa* L.). *Seed Sci. Technol.,* 34: 741-750.

Filatov, P. A. and R. I. Frolova. 1975. Pre-sowing hardening of sunflower seed for seed production. *Field Crop Abst.,* 28 : 478.

Gao, J., G. Hofstra and R. A. Fletcher. 1988. Anatomical changes induced by triazoles in wheat seedlings. *Can. J. Bot.,* 66:1178-1185.

Geetha, R. and A. Selvakumari. 2008. Effect of seed hardening and pelleting in sorghum under rainfed condition. *Int. J. Plant Sci.,* 3 (2):610-612.

Giri, G.S. and W. F. Schillinger. 2003. Seed priming winter wheat for germination, emergence and yield. *Crop Sci.,* 43: 2135-2141.

Gopalakrishnan, S. 1965. In: Advances in Agricultural Sciences. Madras Agricultural Students Union, Coimbatore. : 76-77.

Gurudev Singh, O. P. Saxena, G. Singh, K. K. Dhir, I. S. Dua and K. S. Chark. 1991. Seed treatments with bio-regulators in relation to wheat productivity, new trends in plant physiology. Proc. Nat. Symp. on Growth and Differentiation in Plants : 201-210.

Henckel, P.A.1964. Physiology of plants under drought. *Ann. Rev. Pl. Physiol.*, 15:386.

Jayarami Reddy, P., Subba Rao, D. V., Rama Rao, G. and B. K. Mahalakshmi, B. K. 1996. Effect of potassium nitrate and naphthalene acetic acid on growth and yield of pigeonpea. *J. Maharashtra Agric. Univ.* 151-153.

Jegathambal, R. 1996. Pre-sowing seed treatment to augment productivity of sorghum cv. CO 26 under rainfed agriculture. Ph.D. Thesis, Tamil Nadu Agricultural University, Coimbatore .

Kadiri,M. and M. A. Hussaini.1999. Effect of hardening pre treatments on vegetative growth enzyme activities and yield of *Pennisetum americanum* and *Sorghum bicolor*. *Global Journal of Pure and Applied Sciences*, 5:435-437.

Kalarani, M.K., R. Sivakumar, M. Vanangamudi and K. B. Sujatha.2001. Effect of seed soaking treatments in late sown finger millet under rainfed condition. *Madras Agric. J.*, 88 (7-9): 509-511.

Kalubarme, M.H. and R. K. Pandey.1979. Note on the growth analysis of greengram genotypes. *Indian J. Agric. Sci.*, 49: 975-977.

Kamala Thirumalaiswamy and J. Sakharam Rao.1977.Effect of pre-treatment of seeds and water stress on net assimilation rate, relative growth rate and leaf area of *Pennisetum typhoids* Stapf and Hubb. *Madras Agric. J.*, 64: 270-272.

Kamalam Joseph and Rajappan Nair.1989. Effect of seed hardening on germination and seedling vigour in paddy. *Seed Res.*, 17 (2): 188-190.

Kareem, A. A., R. C. Saxena, M. E. M. Boncodin, V. Krishnasami and D. V. Seshu. 1989. Neem as seed treatment for rice. *J. Econ. Ent.*, 82 (4): 1219-1223.

Karivaratharaju, T. V. and V. Ramakrishnan. 1985a. Effects of pre-soaking seed treatment with chemical growth regulants on seed yield and quality in red gram. *Madras Agric. J.*, 72 (5): 249-255.

Karivaratharaju, T.V. and V. Ramakrishnan. 1985b. Seed hardening studies in two varieties of ragi (*Eleusine coracana* Gaern). *Indian J. Pl. Physiol.*, 28: 243-248.

Kaur, S., A. K. Gupta and N. Kaur. 2002. Effect of osmo and hydropriming of chickpea seeds on seedling growth and carbohydrate metabolism under water deficit stress. *Plant Growth Regul.*, 37: 17-22.

Kavitha, S. 2005. Evaluation of seed enhancement techniques for better survival, establishment, growth and productivity of agricultural and horticultural crops under Neyveli Lignite Mine spoil condition. Ph.D. Thesis, Tamil Nadu Agricultural University, Coimbatore.

Krishna Sastry, K. S., S. Rama Rao and K. M. Appaiah. 1969. Effect of pre-treatments on seeds *Eleusine corocana* on tolerance to high concentrations of sodium chloride and mannitol. *The Mysore J. Agric. Sci.*, 3 : 47-54.

Krishnakumary, K., A. Venkatasubramaniam and C. Mini. 2008. Optimization of hydropriming techniques for cowpea seed invigoration. *Seed Res.*, 36(1): 33–36.

Kulkarni, G. N. and M. R. Eshanna.1988. Effect of pre-soaking of corn seed on seed quality. *Seed Sci. Res.*, 16: 37-40.

Kulkarni, S.S. and B.M. Chittapur. 2003. Seed treatment technology in crop production-a review. *Agric. Rev.*, 24(4): 308-12.

Kumar, L. P, S. R. Niranjana, H. S. Prakash and H. Shetty. 2000. Effect of *Pseudomonas flourescens* formulation against *Pyricularia grisea* in rice. *Crop Improv.*, 27(2): 159-166.

Maitra, S., D. C. Ghosh, G. Sounda, P.K. Jana and D. K. Roy. 1998. Effect of seed treatment on growth and productivity of finger millet under rain fed laterite belt of West Bengal. *Indian Agriculturist*, 42:37-43.

Malathi chari, T.G. Prasad and K. S. K. Sastry.1984. Diffusivity of proline and its significance. *Current Science*, 53: 496-498.

Mandal, A.K. and R. N. Basu. 1987. Effect of sowing water soaked and dried seeds on the yield of wheat with or without fertilizer application. *Ind. J. Agric. Sci.* 57(7): 517-519.

Manjunath, B. L. and M. M. Dhanoji. 2011. Effect of seed hardening on growth and yield of chickpea (*Cicer arietinum* L.). *AAB Bioflux*, 3 (1): 1-4.

Matarira, C.H., J.C. Makadho and M. Mukahanana-Sangarwe. 2004. Vulnerability and adaptation of maize production to climate change in Zimbabwe, Ministry of Environment and Tourism, Zimbabwe.

Matthews, S. and M. K. Hosseini. 2007. Length of the lag period of germination and metabolic repair explain vigour differences in seed lots of maize (*Zea mays*). *Seed Sci. Technol.*, 35: 200-212.

May, L.H., E. J. Milthorpe and E.L.Milthorpe.1962. Presowing drought hardening of plants and drought. In. An appraisal of the contribution. (Ed) P.A.Henkel. *Field Crops Abstr.* 15(2):93-98.

Menaka, C. and Vanangamudi. K.2008. Physiological and biochemical characteristics of hardened and pelleted seeds of bajra and their influence on productivity under rainfed condition. *Indian J. Agric. Res.*, 42 (4) : 278-282.

Migahid, M.M. and L. A. Sadek. 1994. Effect of pre-soaking with trace element on salt tolerance of three cultivars of wheat. *Alexandria Journal of Agricultural Research.* 39 (3): 501-515.

Misra, N. M. and D. P. Dwivedi.1980. Effect of pre-sowing seed treatment on growth and dry matter accumulation of high yielding wheat under rain-fed conditions. *Ind. J. Agric. Sci.,* 25: 230-234.

Munoz, A. M. 1988. Increasing the vigour of rice seeds with trace element application. *Arrazy,* 37: 20-27.

Muthuvel. P., B. Habeebullah and A. Chamy.1983. Effect of pre-soaking of sunflower seeds in micronutrients on seed and oil yield. *Madras Agric. J.,* 70(6): 375-376.

Nagar, R.P., M. Dadlani and S. P. Sharma. 1998. Effect of hydro-priming on field emergence and crop growth of maize genotypes. *Seed Res.,* 26 (1): 1-5.

Nagaraj, S. 1996. Seed hardening and coating studies in upland rice (*Oryza sativa* L.). M. Sc. (Ag) Thesis, Tamil Nadu Agricultural University, Coimbatore.

Narendra, K., R. Roy, S. Ashwini and K. Srivastava. 2000. Adverse effect of salt stress conditions on chlorophyll content in wheat (*Triticum aestivum*) leaves and its amelioration through pre-sowing treatments. *Ind. J. Agric. Sci.,* 70 (11): 777-778.

Nirmala, M., J. Radhakrishnan and U. Bangarusamy.1994. Effect of seed hardening and moisture regimes on germination of sorghum and cotton in vertisols. *Madras Agric. J.,* 81: 695-696.

Palanisamy, V. and N. Punithavathi. 1998. Effect of seed hardening and pelleting on seed quality in ragi. *Madras Agric. J.,* 85(10-12): 583-585.

Parera C. A. and D. J. Cantliffe. 1994. Presowing seed priming. *Horticultural reviews,* 6 : 109–141.

Patil, R. P. 1987. Physiological factors influencing growth and yield of *rabi* sorghum (*Sorghum bicolor* L. Moench) genotypes under rainfed conditions. M. Sc. (Ag.) Thesis, Univ. Agric. Sci., Dharwad, Karnataka, India.

Paul, S. R. and N. N. Sharma and D. Sharma. 1998. Effect of seed hardening and K levels on grain yield of rainfed wheat in the hills zone of Assam. *J. Potassium Res.,* 15(1-4): 54-58.

Rajashekhar, B.G., S. Rama Rao, K. S. Krishna Sastry and K. N. Mallanna.1970. Effect of seed hardening on yield of ragi (*Eleusine coracana*) at different levels of fertility. *Mysore J. Agric. Sci.,* 4 : 270-276.

Ramachandran, K. and J. S. Rao. 1977. Inducing drought tolerance in bajra *(Pennisetum typhoides* Stapf & Hubb) by pre-sowing seed treatments with special reference to the mineral nutrition in seedlings. *Madras Agric. J.* 64 (l):15-17.

Ramamoorthy, K. and K. Sujatha. 2007. Effect of seaweed extracts on the ageing of black gram seeds. *Seaweed Res. Utiln.*, 29 (1&2) : 119-123.

Ramamoorthy, K., K. Sujatha and K. Sivasubramaniam. 2005. Organic hydration-dehydration with sargassum (*Sargassum polycystum*) seaweed extract on seed vigour and viability of rice cv. MDU 5. *Indian Agric.*, 49 (3) : 261-266.

Ramamoorthy, K., K. Sujatha, K. Sivasubramaniam and K. Subburamu. 2006. Organic priming with sargassum (*Sargassum polycystum*) extract on vigor and viability in cowpea (*Vigna unguiculata* L. Walp.) cultivar CO 6. *Seaweed Res. Utilin.*, 28 (1): 85-88.

Ramamoorthy, K., N. Natarajan and A. Lakshmanan.2000. Seed biofortification with *Azospirillum* spp. for improvement of seedling vigour and productivity in rice. *Seed Sci. Technol.*, 28: 809-815.

Rangaswamy, A., S. Purushothaman and P. Devasesapathy. 1993. Seed hardening in relation to seedling quality characters of crops. *Madras Agric. J.*, 80(9): 535-537.

Salakinkop, S.R., S. I. Halikatti, R. B. Kandagave, S. I. Patil, and C. I. Hoogar.1996. Effect of agronomic practices on germination percentage, seedling growth and shoot vigour index of dry sown rice. *Advances of Agricultural Research in India*, 7:130-135.

Schwartz, A. 1985. Role of calcium and EGTA on stomatal movements in *Cammelina communis L. Plant Physiol.*, 79: 1003-1005.

Simon, E. W. and R. M. Raja Harun. 1972. Leakage during seed imbibitions. *J. Exp. Bot.*, 23:1076-108.

Singh, A. R., N. L. Bhale and S.T. Borikar. 1984. Induction of early flowering by simple water soaking treatment in sorghum. *Sorghum Newsletter*, 27 : 134.

Solaimalai, A. and K. Subburamu. 2004. Seed hardening for field crops - a review. *Agric. Rev.*, 25(2):129-140.

Srimathi, P., S. Kavitha and J. Renugadevi.2007. Influence of seed hardening and pelleting on seed yield and quality in green gram CO 6. *Indian J. Agric. Res.*, 41(2):122-126.

Sujatha, K. and K. Ramamoorthy. 2008. Herbal Hydration-dehydration of cowpea cv. CO 6 (*Vigna ungiculata*) for vigour and viability maintenance. *Ad. Plant Sci.*, 21 (1): 83-85.

Sujatha, K., K. Ramamoorthy and K. Sivasubramaniam. 2005. Seed invigouration for improved viability and productivity in blackgram. *Plant Archives*, 5(1) 273-276.

Suma, N. 2005. Studies on seed quality enhancement techniques in sesamum (*Sesamum indicum*) Cv. CO 1. M. Sc. (Ag.) Thesis, Tamil Nadu Agricultural University, Coimbatore.

Sumabai, D. I., A. T. Abraham and S. T. Mercy. 1987. Hormonal influence on crop performance in green gram. *Legume Res.*, 10(1): 49-52.

Surulirajan, E. 2007. Upgradation of seed quality through seed fortification in blackgarm cv. VBN 3. M. Sc. (Ag.) Thesis, Tamil Nadu Agricultural University, Coimbatore.

Susmitha, S. 2006. Studies on the influence of seed quality enhancement techniques for drought tolerance in rice (*Oryza sativa* L,.) cv PMK(R) 3 and IR 50. M.Sc (Ag.) Thesis, Tamil Nadu Agricultural University, Coimbatore.

Taylor, A. G. 1997. Seed storage, germination and quality. In: Wien, H.C. (ed.), The Physiology of Vegetable Crops, CAB International. Wallingford,. UK. p 1-36.

Ugale, R.B and H. B. Mungse. 2001. Effect of pre-sowing seed hardening on yield and yield components of wheat. *J. Maharashtra Agric. Univ.*, 26: 103-104.

Umashankar, R. 1985. Moisture stress induced loss of membrane integrity, accumulation of proline and their inter relationship. Ph.D. Thesis, University of Agricultural Sciences, Dharwad.

Vanangamudi, K. and M. Vanangamudi. 2003. Response of tamarind (*Tamarindus Indica*) to pre-sowing seed treatment with growth stimulants. *J. Forest Sci.*, 15(1): 6-11.

Vanitha, C. 2005. Herbal seed treatments for improving storability and productivity in selected field crops and standardization of seed production techniques in chicory (*Cichorum intybus* L.). Ph. D. Thesis, Agricultural College & Research Institute, Madurai.

Vertucci, C. W. and A. C. Leopold. 1984. Bound water in soybean seed and its relation to respiration and imbibitional damage. *Plant Physiology*, 75 (1) :114-117.

Vigneswari, R. and P. Natesan. 2010. Bio-chemical deterioration and loss of viability in hardened and pelleted seeds of finger millet cv. CO 13. *Res. On Crops*, 11(1): 166-174.

Vigneswari, R., P. Natesan, A. Bharathi, K. Vanagamudi and C.Menaka.2005. Seed hardening and pelleting on physiological and biochemical characteristics of finger millet (*Eleusine coracana*). *Madras Agric.* J. 92(4-6): 334-337.

Vijaya J. 1996. Standardisation of pre-sowing seed management technique for pulses. M. Sc. (Ag.) Thesis, Tamil Nadu Agricultural University, Coimbatore.

❑❑❑

Chapter 6

Effects of Abiotic Factors Against Storage Pests of Rice

[1]P.YASODHA, [2]T.PANDIARAJAN, [3]P.MASILAMANI AND [1] A.TAJUDDIN

[1] Agricultural Engineering College and Research Institute, Kumulur, Trichy, Tamil Nadu

[2] Agricultural Engineering College and Research Institute, Coimbatore, Tamil Nadu

[3] National Seed Research and Training Centre, Varanasi, Uttar Pradesh

E-mail : yasodhabiotech@gmail.com

Abstract

Increased insecticide resistance against fumigants by storage pests is a major threat in storage godown. Abiotic factors such as modified atmospheres, temperature, pressure, humidity may be an alternative strategy to toxic building fumigants to protect stored rice from storage pests infestations. A study was conducted to analyze the efficacy of abiotic factors such as modified atmospheres and temperature against rice storage pests of infective stage (adults of coleopterans/larvae of lepidopteran pests). Binary mixtures of modified atmospheres O_2 3% + CO_2 11.5%, O_2 4% + CO_2 10.5%, O_2 5% + CO_2 9.5%, O_2 6% + CO_2 8.5%, O_2 + 7% CO_2 7.5%, O_2 8% + CO_2 6.5% were tested at two different temperatures *viz.,* 25°C and 30°C and replicated for three times with 30 insects/replication. The effect of chronic toxicity due to modified atmospheres and temperature was studied through mortality data after 4 days and progeny reduction in exposed after 28 days and post exposed insects in the next 28 days. The results revealed that considering the coleopteran pests, maximum of 99.4 per cent mortality was recorded against *S. oryzae* with 3% O_2 + 11.5% CO_2 combination at 30°C and 100 per cent reduction in the progeny was noticed in treatment combinations such as (3% O_2 + 11.5% CO_2), (4% O_2 + 10.5% CO_2), (5% O_2 + 9.5% CO_2) at 30°C against *S. oryzae*. Considering the lepidopterans, maximum of 100 per cent mortality of *S. cerealella* and *C.cephalonica* was observed with 3% O_2 + 11.5% CO_2 combinations at 30°C while the overall mean for progeny reduction was more than 75 per cent in both lepidopteran pests tested.

Introduction

An estimate by the Ministry of Food and Civil Supplies, Government of India, puts the total preventable post harvest losses of food grains at 10 per cent of the total production, which would be enough to feed about 70–100 million people *ie.*, 1/3rd of Indias' poor (Basavaraja *et al.*, 2007). Rice, which occupies maximum space in godown has the major menace of red flour beetle (*Tribolium castaneum*), lesser grain borer (*Rhyzopertha dominica*), rice weevil (*Sitophilus oryzae*), rice moth (*Corcyra cephalonica*) and Angoumois grain moth (*Sitotroga cerealella*). For pest management, we expect a food product that is pesticide free or with much reduced residue levels. Airtight storage of damp grain preserved germination for a short period and prevented the growth of spoilage moulds during longer term storage (White *et al.*, 1993). Modified atmospheres have an important role in integrated pest management systems that are pesticide free (Banks *et al.*, 1991). Atmosphere (or earth's atmosphere) contains 78 % Nitrogen (N_2), 21% Oxygen (O_2) and 0.03 % carbon dioxide (CO_2). Generation of modified atmosphere is achieved by decreasing the O_2 or increasing CO_2 or N_2 concentration in the atmosphere thereby interfering with normal respiration of insects (Jayas *et al.*, 1991).

Controlled atmosphere storage in containers is mainly practiced during shipment in order to prevent ripening, microbial deterioration and insect pests. Controlled atmosphere is precisely maintaining the composition of selected gases such as CO_2, O_2 and N_2 at specified concentration under normal pressures or under partial vacuum. The effectiveness of modified/controlled atmosphere in controlling insects is dependent on various abiotic (gaseous composition, relative humidity, temperature, length of exposure and gas pressure) and biotic (insect species, life stage and the size and distribution of infestation) factors. All these factors must be optimized to create an environment which is lethal to the insect species found in the stored grain.

Researchers around the world have been conducting empirical studies to define atmospheric conditions for controlling various stored-grain insects (Banks *et al.*, 1991, Jayas *et al.*, 1991, Donahaye *et al.*,

1996). But, still the exact proportion of atmospheric gases to be used for tropical conditions in different commodities has to be standardized. Therefore, an attempt has been made to manage storage pests of rice through abiotic factors like modified atmospheres (at biologically relevant levels, minimum concentrations) and temperature against development and reproduction of major coleopteran and lepidopteran stored pests of rice.

Materials and Methods

A study was conducted to analyze the efficacy of abiotic factors such as modified atmospheres and temperature against rice storage pests during the period 2010 to 2012 at the Department of Agricultural Processing, Agricultural Engineering and Research Institute, TNAU, Tiruchirappalli, Tamil Nadu. Test insects required for the study were obtained from susceptible population maintained under laboratory condition. The infective stage (adults of coleopterans/larvae of lepidopteran pests) which cause damage to the paddy grains were taken separately for the exposure study. Binary mixtures of modified atmospheres O_2 3 % + CO_2 11.5%, O_2 4% + CO_2 10.5%, O_2 5% + CO_2 9.5%, O_2 6% + CO_2 8.5%, O_2 + 7% CO_2 7.5%, O_2 8% + CO_2 6.5% were tested (Plate 2) at two different temperatures *viz.*, 25°C and 30°C and replicated for three times with 30 insects/replication. The test was conducted using a polypropylene sheet of thickness measuring 150 gauge (Plate 1) and sealed properly using sealing machine (Plate 3). The effect of chronic toxicity due to modified atmospheres and temperature was studied through mortality data after 4 days and progeny reduction in exposed after 28 days and post exposed insects in the next 28 days. Per cent data on mortality and progeny reduction was transformed in to arc sin values and analysed using Agres-Agdata software.

Results

The results of effect of chronic toxicity through abiotic factors against storage pests of rice are presented below.

1. Tribolium castaneum

At 25°C, the treatment combination of 3 % and 11.5 % of O_2 and CO_2 mixtures respectively had maximum of 16.2 per cent mortality while there was maximum of 100 per cent reduction in progeny during exposure period. Even with minimum concentration of carbondioxide

(6.5%) and maximum concentration of oxygen (8%) at 30°C, there was 79.4 per cent reduction in progeny. The residual effect of abiotic factors on the exposed insects was studied during the exposure and post exposure periods *i.e.,* in the subsequent generation. At 25°C, there was 19.2 per cent progeny reduction while there was no progeny with maximum oxygen (8%) and minimum carbondioxide (6.5%) concentration in atmospheric levels. At 30°C, maximum of 76.4 per cent of mortality was recorded with 3% O_2 and 11.5% CO_2 levels while minimum of 17.2% mortality was recorded in 8% + 6.5% concentration of oxygen and carbondioxide mixtures respectively. Comparatively, mortality was higher at 30°C than 25°C. So, the effect of temperature is positively correlated with mortality of *T. castaneum* in rice (Table.1).

2. *Rhyzopertha dominica*

The lesser grain borer, *R. dominica* showed higher percentage of mortality both at 25°C and 30°C. However, the mortality decreases as carbondioxide levels decrease (Figure 1). Therefore, effect of carbondioxide is found to be positively correlated with mortality. Considering the temperature, higher temperature showed maximum mortality similar to *T. castaneum.* Regarding progeny reduction, maximum reduction was seen at 30°C than 25°C. However, during post exposure period, at least a minimum of 14 per cent progeny reduction was recorded even with minimum carbondioxide concentration at 25°C while there was no effect on progeny at 30°C.

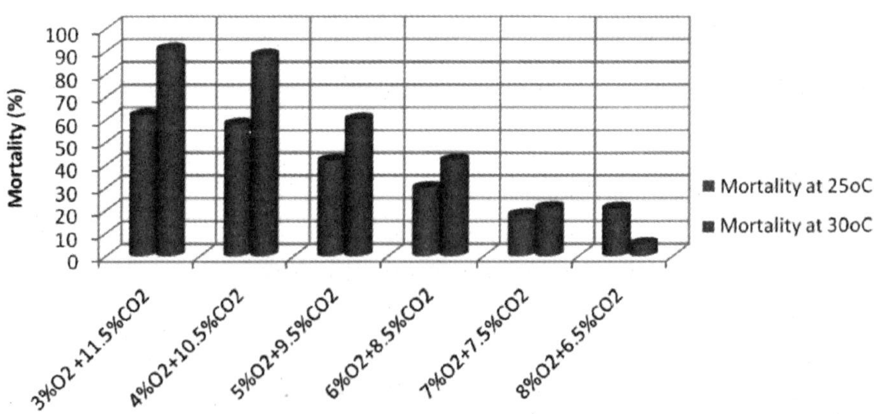

Fig. 1: Effect of minimum modified atmospheres and temperature on mortality of *Rhyzopertha dominica* during exposure period

Table 1: Effect of minimum modified atmospheres and temperature against T. castaneum.

Treatment		Mortality (%)			% progeny reduction during						
					25°C			30°C			Grand Mean
O_2 (%)	CO_2 (%)	25°C	30°C	Mean mortality	Pre exp	Post exp.	Mean	Pre exp.	Post exp.	Mean	
3	11.5	16.2 (23.73)	76.4 (60.94	46.30 (42.34)	100 (88.35)	19.2 (26.04)	59.60 (57.19)	100 (88.35)	11.2 (19.55)	55.60 (53.95)	57.60 (55.57)
4	10.5	14.4 (22.30)	43.1 (41.04)	28.75 (31.67)	82.4 (65.15)	10.0 (18.42)	46.20 (41.78)	76.2 (60.80)	10.4 (18.81)	43.3 (39.81)	44.75 (40.80)
5	9.5	13.7 (21.70)	40.3 (39.41)	27.00 (30.55)	80.3 (63.65)	9.7 (18.24)	45.00 (40.95)	71.8 (57.93)	8.0 (16.35)	39.9 (37.13)	42.45 (39.04)
6	8.5	12.0 (22.37)	29.7 (33.02)	20.85 (27.70)	78.6 (62.45)	6.1 (14.30)	42.35 (38.37)	64.1 (53.19)	6.2 (14.30)	35.15 (33.74)	42.45 (34.52)
7	7.5	13.6 (21.56)	16.1 (23.66)	14.85 (22.61)	74.2 (59.47)	6.0 (14.18)	40.10 (36.83)	61.8 (51.83)	4.8 (12.61)	33.30 (32.21)	36.7 (34.52)
8	6.5	9.2 (17.66)	11.2 (19.37)	10.20 (18.52)	79.4 (63.01)	0.0 (1.65)	39.70 (32.33)	49.7 (44.83)	2.2 (8.52)	25.95 (26.68)	32.83 (29.50)
Mean		13.18 (21.55)	36.13 (36.23)	24.68 (28.90)	82.42 (67.01)	8.50 (15.47)	70.60 (41.24)	7.13 (15.02)	7.13 (15.02)	35.72 (62.58)	35.72 (31.91)

CD (05%): T=0.88, M=1.52, TxM= 2.15
T-Temperature ; M - Mortality

CD (05%): T=0.37 ; P=0.37 ; R=0.63 ; TxP=0.52; PxR=0.90; TxR=0.90 ; TxPxR=1.30 T-Temperature; P-Progeny; R-Reduction

3. *Sitophilus oryzae*

Maximum of 99.4% mortality of rice weevil was recorded due to 3% oxygen and 11.5% carbondioxide concentration at 30°C while it was 82.4% at 25°C. However, there was no mortality with minimum levels of carbondioxide (6.5%) and maximum levels of oxygen (8%) at 30°C than 25°C which had 34.2 per cent of mortality (Table 2). Progeny reduction during pre exposure was higher and was almost on par with each other at both 25°C (98.4% to 68.1%) and 30°C (100% to 72%) than post exposure period which had only a meager effect (Figure 2).

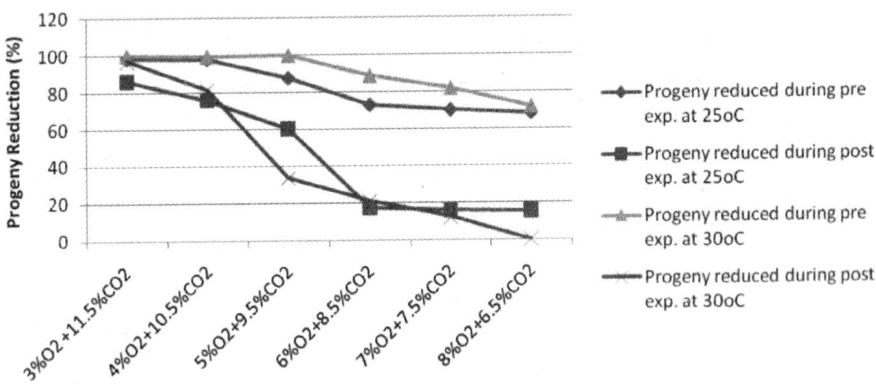

Fig. 2: Effect of modified atmospheres and temperature in Sitophilus oryzae progenies

4. *Corcyra cephalonica*

The same trend of high concentration of carbondioxide with minimum oxygen levels recorded maximum mortality of larva at both 25 and 30°C. However, 100 per cent mortality was recorded at 3% oxygen and 11.5% carbondioxide concentration at 30°C. Even up to 30 per cent of mortality was recorded even at minimum CO_2 and maximum O_2 levels at both temperatures tested (Table 4). Even the progenies of rice moth were found highly reduced both in pre and post exposure period. Therefore, it was observed that lepidopteran storage pests highly susceptible to modified atmospheres and temperature than coleopteran storage pests

5. *Sitotroga cerealella*

Higher concentration of carbondioxide (11.5%) with minimum oxygen (3%) levels recorded maximum mortality (100%) of larva at

Table 2: Effect of abiotic fators against R. dominica

Treatment		% progeny reduction during						
		25ºC			30ºC			Grand Mean
O^2(%)	CO^2(%)	Pre exp	Post exp.	Mean	Pre exp.	Post exp	Mean	
3	11.5	90 (71.62)	45.2 (42.25)	67.6 (75.07)	96.0 (78.52)	50.00 (45.00)	73.00 (43.62)	70.30 (59.35)
4	10.5	78.00 (62.04)	37.8 (37.94)	57.9 (63.51)	82.1 (64.97)	42.1 (40.46)	62.10 (39.20)	60.00 (51.35)
5	9.5	65 (53.76)	35.8 (36.75)	50.4 (55.30)	70.1 (56.85)	35.8 (36.75)	52.95 (36.75)	51.68 (46.03)
6	8.5	55 (47.91)	30.2 (33.33)	42.6 (47.14)	52.3 (46.38)	21.0 (27.27)	36.65 (30.30)	39.63 (38.72)
7	7.5	54 (47.30)	15.2 (22.95)	34.6 (46.26)	50.4 (45.23)	4.20 (11.54)	27.30 (17.24)	30.95 (31.75)
8	6.5	42 (40.40)	14.0 (21.97)	28.0 (45.03)	58.0 (49.67)	0.00 (1.65)	29.00 (11.81)	28.50 (28.42)
Mean		64.00 (53.84)	29.7 (32.53)	46.85 (43.19)	68.15 (56.94)	25.52 (27.11)	46.83 (42.02)	46.84 (42.60)

CD (05%): T=1.062 ; P=1.082 R=1.84; TxP=1.50 ; PxR=2.60 TxR=2.60; TxPxR=3.68
T-Temperature; P -Progeny; R -Reduction

Table 3: Assessment of chronic toxicity due to abiotic factors against S. oryzae

Treatment		Mortality (%)		
O2 (%)	CO 2 (%)	25ºC	30ºC	Mean mortality
3.0	11.5	82.4 (65.20)	99.4 (85.75)	90.90 (75.48)
4.0	10.5	65.6 (53.97)	86.3 (68.28)	75.95 (61.12)
5.0	9.5	52.4 (46.38)	84.5 (66.82)	68.45 (56.60)
6.0	8.5	45.3 (42.30)	76.2 (60.80)	60.75 (51.55)
7.0	7.5	35.1 (36.33)	41.9 (37.41)	38.50 (36.87)
8.0	6.5	34.2 (35.79)	0.00 (1.65)	17.10 (18.71)
Mean		52.50 (46.66)	64.72 (53.45)	58.61 (50.06)

CD (05%): T=0.59; M=1.00; TxM=1.41 T Temperature ; M Mortality

Table 4: Effect of abiotic fators against C. cephalonica

Treatment		Mortality (%)			% progeny reduction during						
					25°C			30°C			Grand Mean
O_2 (%)	CO_2 (%)	25°C	30°C	Mean mortality	Pre exp.	Post exp.	Mean	Pre exp.	Post exp.	Mean	
3	11.5	88.4 (70.09)	100.00 (88.35)	94.20 (79.22)	100.00 (88.35)	100.00 (88.35)	100.00 (88.34)	100.00 (88.35)	100.00 (88.35)	100.00 (88.35)	100.00 (88.34)
4	10.5	73.2 (58.90)	82.1 (64.98)	77.65 (61.93)	99.7 (86.89)	82.5 (65.27)	91.1 (76.08)	92.30 (73.89)	89.70 (71.29)	91.00 (72.59)	91.05 (88.35)
5	9.5	70.50 (57.11)	52.4 (46.38)	61.45 (51.74)	95.4 (77.62)	71.7 (57.27)	83.55 (67.74)	87.60 (69.38)	75.6 (60.40)	81.60 (64.89)	82.60 (74.34)
6	8.5	52.7 (46.38)	50.2 (45.12)	51.45 (45.75)	82.3 (65.12)	55.0 (47.87)	68.65 (56.50)	82.40 (65.20)	62.4 (52.18)	72.40 (58.69)	70.53 (66.32)
7	7.5	42.9 (40.92)	32.5 (34.78)	37.7 (37.84)	78.6 (62.44)	41.2 (39.93)	59.90 (51.19)	86.40 (68.36)	29.3 (32.77)	57.85 (50.57)	58.88 (57.59)
8	6.5	35.0 (35.06)	30.9 (33.77)	32.95 (34.42)	75.0 (60.00)	30.3 (33.40)	52.65 (46.70)	82.90 (65.58)	28.00 (31.95)	55.45 (48.76)	54.05 (50.88)
Mean		60.45 (51.40)	58.02 (52.22)	59.23 (51.81)	88.5 (73.41)	63.45 (55.45)	76.00 (64.43)	88.60 (71.79)	64.17 (56.16)	76.87 (63.98)	76.44 (64.20)

CD (05%): T=0.35, M=0.60, TxM= 0.85
T-Temperature ; M - Mortality

CD (05%): T=0.22;P=0.22;R=0.39; TxP= 0.31;PxR=0.51;TxR= 0.55; TxPxR=1.90 T-Temperature; P-Progeny; R-Reduction

Table 5: Assessment of chronic toxicity with minimum modified atmospheres against S. cerealella

Treatment		Mortality (%)			% progeny reduction during						
					25°C			30°C			Grand Mean
O_2 (%)	CO_2 (%)	25°C	30°C	Mean mortality	Pre exp.	Post exp.	Mean	Pre exp.	Post exp.	Mean	
3	11.5	88.40 (68.79)	100.00 (88.35)	94.20 (71.19)	100.00 (88.35)	100.00 (88.35)	100.00 (88.35)	100.00 (88.35)	100.00 (88.35)	100.00 (88.35)	100.00 (88.35)
4	10.5	73.20 (58.25)	82.10 (21.04)	77.65 (39.65)	99.70 (74.02)	82.50 (77.39)	91.10 (69.50)	92.3 (64.97)	89.7 (58.06)	91.00 (67.72)	91.05 (68.61)
5	9.5	70.50 (53.43)	52.40 (50.97)	61.45 (52.20)	95.40 (71.77)	71.70 (65.32)	83.55 (67.69)	87.6 (63.60)	75.6 (53.13)	81.60 (59.18)	82.58 (63.43)
6	8.5	52.70 (42.53)	50.20 (39.52)	51.45 (41.03)	82.30 (71.29)	55.00 (51.12)	68.65 (67.15)	82.4 (63.01)	62.4 (46.15)	72.40 (48.63)	70.53 (57.89)
7	7.5	42.90 (37.11)	32.50 (33.89)	37.70 (35.50)	78.60 (63.51)	41.20 (35.23)	59.90 (62.84)	86.4 (62.17)	29.3 (35.67)	57.85 (35.45)	58.88 (49.14)
8	6.5	35.00 (36.21)	30.90 (25.62)	32.90 (30.91)	75.00 (58.18)	30.30 (32.39)	52.65 (60.18)	82.9 (62.17)	28.00 (25.06)	55.45 (28.72)	54.05 (44.45)
Mean		60.45 (64.43)	58.02 (63.98)	59.23 (64.20)	88.50 (71.19)	63.45 (57.70)	75.98 (64.44)	88.60 (67.38)	64.17 (49.99)	76.38 (58.68)	76.19 (61.56)

CD (05%): T=1.27; M=2.20, TxM= 3.11
T-Temperature ; M - Mortality

CD (05%): T=0.54, P=0.54; R=0.95, TxP=0.77; PxR=1.34; TxR=1.34;
TxPxR=1.90 T-Temperature; P-Progeny; R-Reduction

30°C than at 25°C (88.4%) respectively. Up to a minimum of atleast 30 per cent of mortality was recorded even at minimum CO_2 (6.5%) at both temperatures tested. Even the progenies were found highly reduced both in pre and post exposure period. More than 80 per cent of progeny reduction was seen with minimum carbon dioxide concentration during the exposure period at both 25 and 30°C temperatures. Considering the progeny reduction during the post exposure period, it was more than 60 per cent even with minimum concentration of CO_2 (6.5%)(Table 5). Comparatively, *S. cerealella* was highly susceptible to abiotic factors studied than *C. cephalonica.*

Discussion

Krishnamurthy *et al.* (1986) developed simple methods for on-site generation of CO_2 based on the reaction of naturally available limestone with dilute hydrochloric acid for application in airtight storages of less than 1 tonne capacity and demonstrated storage of bagged Basmati rice (30 tonnes stacks), under CO_2 rich atmosphere for 3 months where insects were completely controlled and there was no adverse effect on the quality of Basmati rice.

In India, large scale CO_2 fumigation trials were conducted at the Food Corporation of India Buffer storage complex, Mumbai on milled raw rice stacks and in a concrete silo filled with 2200 tonnes of wheat. Complete mortality of all stages of resident infestation of *T.castaneum, Oryzaephilus surinamensis, Ephestia cautella* and psocids was achieved in rice stacks following CO_2 treatment. There was no change in colour or odour of rice and wheat.

Plate 1: Bioassay on Tribolium castaneum

Plate 2: Cylinders with gases for creating modified atmospheres

Plate 3: Sealing machine

In the present study, the per cent mortality under different binary mixtures and temperatures are presented in Tables 1, 3, 4, 5 and Figure 1. One important aspect to remember is that pure carbon dioxide is less effective on insects than carbon dioxide mixed with oxygen content of less than 1% by volume (Paul and White, 2002). Therefore, each treatment was taken atleast with a minimum oxygen in the treatments fixed.

Considering the modified atmospheres, the highest carbondioxide concentration (11.5%) was found to promote maximum mortality. When, coleopteran pests are concerned, the mean mortality of *T. castaneum* (46.30 %) was below 50 per cent even at higher concentration of carbondioxide (11.5%) whereas, the other coleopteran pests such as *R. dominica* and *S. oryzae* mortality was 75 % and 90.90 % at higher concentration of carbondioxide (11.5%) which was more than 50 per cent.

There were no changes in moisture content within the polybags. This may be due to the diminished respiration rates of both the insects and grains when they were in modified atmosphere or in low O_2 concentrations. It is also to be noted that highest per centage of mortality was seen initially and later declined. This may be due to the gradual changes in the modified atmospheres released within polybags. The same observation was also recorded by Marcelo *et al.* (2010) who depicted that the CO_2 concentration decreased with the time and initial levels of CO_2 while concentration of O_2 concentration began to increase slowly after the fifteenth day. The consumption and production of O_2 and CO_2 by vegetable tissues are metabolically interrelated, making difficult to study the effect of one gas independent on the other. The uncontrolled nature of the atmosphere in the package and the time interval required to reach the equilibrium are additional factors that complicate the interpretation of the results (Saltveit, 2003).

When lepidopteran pests are concerned, the mean mortality of both *C. cephalonica* and *S. cerealella* was 94.2 per cent at higher concentration of carbondioxide (11.5%). The use of oxygen between 3 and 5% would mean that the commodity for treatment would have to be kept at the required oxygen levels for the duration of its storage period. This is because the production of progeny by the surviving adults returned to control level as after their removal from exposure to the modified atmospheres. The concept of storage life protection has been suggested before Conyers *et al.* (2001) and previous work with hermetic storage simulations in the laboratory has shown that similar combinations of O_2 and CO_2 are efficient (Donahaye *et al.*, 1996). Spratt (1979) showed a reduction in progeny production by *S.zeamais*. Disinfestations of food grain by CO_2 treatment requires a long period of 10 days and above. However, this drawback has been overcome in developed countries by raising the treatment temperature to 30-40°C or by applying CO_2 under high pressure of 20-25 bars (Subramanyam and Hagstrum, 2000) which was seen in coherent with the present investigation that 30°C promoted enhanced mortality and progeny reduction.

Lepidopteran storage pests are relatively more sensitive to modified atmosphere than weevils or beetles. However, Alder (2001) reported that diapausing larvae of *P. interpunctella* treated at 10°C with various gas mixtures were found to be relatively more tolerant compared with normal developmental stages or adults which was seen against the present result. The mortality observed for *Sitophilus* spp. when high CO_2 concentrations were used could be associated with either O_2 depletion or CO_2 toxicity or the interaction of both (Annis and Morton, 1997). However, the natural permeability of cellular membranes appeared to be an important protection mechanism for the survival of arthropods under low O_2 or high CO_2 atmospheres (Mitcham *et al.*, 2006). Hoback and Stanley (2001) described that arthropods were hypoxia-tolerant organisms, although their mortality increased with the lower O_2 concentrations.

Temperature, time, CO_2 concentration and their interactions were the significant factors affecting mortality. Similar behavior was observed by Annis and Morton (1997), who reported that the use of time-concentration relationship was not directly applicable to explain the effects of CO_2 on *S. oryzae* because of the interdependent effects of lowered O_2 and increased CO_2 concentrations. In addition, there was a particular time interval for each species after that an irreversible damage was done to their metabolism (Conyers and Bell, 1996). In addition to the effect of the concentrations of CO_2 and O_2 and time, there are other factors that may influence the effect of these atmospheres on the arthropods, such as temperature, grain moisture, insect species and life stage (Mitcham *et al.*, 2006).

Adult mortality as expected was generally linked with the number of progeny produced in further generations. Considering the progeny reduction, it was observed at high concentration of O_2 (8%), the emergence of new insects was not repressed. Highest number of adult emergence from the surviving adults was observed at 8% O_2 + 6.5% CO_2 for 28 days of exposure in all the insects tested. However, *S.cerealella* was found highly susceptible to almost all the treatments. Adult emergence began to decrease from the fifth day in the treatment with 3% O_2 and 11.5% CO_2 in the pre exposure study of coleopteran storage pests whereas adult emergence began to decrease on the third day itself in lepidopteran storage pests.

Lepidopteran storage pests had no progeny at higher concentration of carbondioxide (11.5%) than the mean reduction in the progeny of *T.castaneum* (57.60%) and *R.dominica* (70.30%) at higher concentration of carbondioxide (11.5%). In contrast, Conyers and Bell (2007) mentioned that an increase of CO_2 to 20%, reducing O_2 to 5% was sufficient to eliminate the emergence of *Sitophilus granarius* at 20°C, but a few individuals emerged at 25°C. In addition, *S. granarius* exposed to atmospheres containing 1 to 1.6% O_2 with a balance of N_2 at 20°C and 70% RH were killed within seven days if the atmosphere also contained 10% or more CO_2 (Krishnamurthy *et al.*, 1986). Other studies have shown that the minimum concentration of CO_2 useful for the complete disinfestation of *S. oryzae* was lower than 40% but substantially above 20% CO_2 (Annis and Morton, 1997). Arthropod mortality is generally greater in response to elevated CO_2 atmospheres when compared with low O_2 atmospheres (Mitcham *et al.*, 2006).

This study showed the possibility of using modified atmosphere packaging in order to protect the paddy grains against the insect attack. It was observed that the relationship between the initial CO_2 concentration and time was important to achieve the higher percentages of mortality for adult insects. Atmospheres containing higher than 11.5 % CO_2 and lower than 3% O_2 and higher than four days of exposure were necessary to eliminate most adult insects. However a period of 28 days was necessary to achieve the total inhibition of the progeny in lepidopterans while, 56 days for coleopterans.

Conclusion

A mixture of minimum O_2 (3%) and maximum concentration of CO_2 (11.5%) at 30°C had maximum mortality and progeny reduction among the storage pests of rice. Considering the coleopteran pests, maximum of 99.4 per cent mortality was recorded against

S. oryzae with 3 % O_2 + 11.5 % CO_2 combination at 30°C and 100 per cent reduction in the progeny was noticed in treatment combinations such as (3%O_2+11.5% CO_2), (4%O_2+10.5% CO_2), (5%O_2+9.5% CO_2) at 30°C against *S. oryzae*. Considering the lepidopterans, maximum of 100 per cent mortality of *S. cerealella* and *C.cephalonica* was observed with 3%O_2+11.5% CO_2 combinations at 30°C while the overall mean for progeny reduction was more than 75 per cent in both lepidopteran pests tested.

References

Adler, C.S. 2001. Comparative efficacy of modified atmospheres against diapausing larvae of *Plodia interpunctella* and other insects in durable products. Donahaye, E.J., Navarro, S. and Leesch J.G. [Eds.] .2001. Proc. Int. Conf. Controlled Atmosphere and Fumigation in Stored Products, Fresno, CA. 29 Oct. - 3 Nov. 2000, C.S. Executive Printing Services, Clovis, CA, U.S.A. pp. 17-24.

Annis, P.C. and R. Morton. 1997. The acute mortality effects of carbon dioxide on various life stages of *Sitophilus oryzae*. *J. Stored Products Res.,* 33: 115-124.

Banks, H.J., P.C. Annis and G.R. Rigby.1991. Controlled atmosphere storage of grains: the known and the future. In: F. Lessard, P. Ducom (Eds.), Proceedings of the Fifth International Working Conference on Stored-product Protection, 9-14 Sept.1990, Bordeaux, France. Imprimerie Medocaine, Blanquefort Cedex, pp.695-706.

Basavaraja, H., S.B. Mahajanashetti and N.C. Udagatti. 2007. Economic Analysis of Post-harvest Losses in Food Grains in India: A Case Study of Karnataka. *Agricultural Economics Research Review,* 20: 117-126.

Conyers, S.T., C.H. Bell, B. Harral and B.E. Llewellin. 2001.Practical and modeling studies on the use of modified atmospheres for insect and mite control in grain stores. Home-Grown cereals Authority Project Report No.256, Home Grown Cereals Authority, London 43 pp.

Conyers, S.T. and C.H. Bell. 1996. The effect on the mortality of adult *Cryptolestes ferrugineus* (Stephens) (Coleoptera: Cucujidae), *Sitophilus granarius* (L.) (Coleoptera: Curculionidae) and *Oryzaephilus surinamensis* (L.) (Coleoptera: Silvanidae) of interrupting low oxygen exposures with periods of elevated oxygen. *J.Stored Products Res.,* 32: 187-194.

Conyers, S.T. and C. H. Bell. .2007. A novel use of modified atmospheres: Storage insect population control. *J. Stored Products Res.,* 43: 367-374.

Donahaye, E. J., S. Navarro, M. Rinder and A.Azrieli. 1996. The combined influence of temperature and modified atmospheres on *Tribolium castaneum* (Herbst) (Coleoptera: Tenebrionidae). *Journal of Stored Products Research,* 32: 225-232.

Hoback, W.W. and D.W. Stanley. 2001. Insects in hypoxia. *J. Insect Physiol.*, 47: 533–542.

Jayas, D.S., B. Khangura and N.D.G.White.1991. Controlled atmosphere storage of grains. *Post harvest News and Information*, 2: 423-427.

Krishnamoorthy, T.S., E.C. Spratt and C.H. Bell. 1986. The toxicity of carbon dioxide to adult beetles in low oxygen atmospheres. *J. Stored Products Res.*, 22: 145-151.

Marcelo D. C., B. Bresolin, C. P. Z. Noreña, I.L. and A. Brandelli. 2010. Efficacy of modified atmosphere packaging to control *Sitophilus* spp. in organic maize grain *Braz. Arch. Biol. Technol.*, 53(6): 1469-1476.

Mitcham, E.T. Martin and S. Zhou. 2006. The mode of action of insecticidal controlled atmospheres. *Bull. Entomol. Res.*, 96: 213–222.

Paul G. F. and D.G. N.White. 2002. Alternatives to methylbromide treatments for stored-product and quarantine insects. *Annu. Rev. Entomol.*, 47:331–59.

Saltveit, M.E. 2003. Is it possible to find an optimal controlled atmosphere? *Post harvest Biol. Tech.*, 27: 3-13.

Spratt, E.C., 1979. Some effects of a mixture of oxygen, carbon dioxide and nitrogen in the ratio 1:1:8 on the oviposition and development of *Sitophilus zeamais* Mots. (Coleoptera: Curculionidae). *J. Stored Products Res.*, 15: 73-80.

Subramanyam, B.H., and D.W.Hagstrum. 2000. *Alternatives to Pesticides in Stored-Product IPM*. Boston: Kluwer Academic. 437 pp.

White, N.D.G., D.S. Jayas, W.E. Muir and R.N. Sinha. 1993. Controlled atmosphere storage of cereals and oilseeds. *J. App. Zool. Res.*, 4(1): 1-12.

❑❑❑

Chapter 7

Seed Quality and Abiotic Factors Affecting Seed Quality

K. SASIKUMAR[1] AND R. AMETHA[2]
[1] Agricultural Research Station, Pattukkottai, Tamil Nadu Agricultural University.
[2] Agricultural College and Research Institute, Madurai.

Seed Quality

Seed is a crucial and basic input to increase productivity i.e. crop yield per unit area and improve the agricultural economy of a country. The concept of seed quality is a complex one, having several aspects.

Seed quality is defined as a multiple concept comprising of several components and their relative importance in different sets of circumstances and has laid much emphasis on

a) Analytical purity
b) Species purity
c) Freedom from weeds
d) Cultivar purity
e) Germination capacity
f) Vigour
g) Size
h) Uniformity
i) Health
j) Moisture content

To achieve these objectives, it is necessary that the organizations and institutions responsible for seed production, processing, storage and distribution should assess and their internal quality control system besides seeking support from statutory functionaries such as seed certification agency, law enforcement agency, seed testing laboratory etc., which are integral part of the system.

The ultimate objectives are to produce and distribute quality seeds which refers to

a) Genetic quality
b) Physical quality
c) Physiological quality
d) Health quality

Genetic Quality

Varieties are released in most cases after the plant breeders ascertain their Distinct ness, Uniformity and Stability (DUS test) and define their inherited, morphological and economic characteristics which can be maintained on reproduction in subsequent generations. For obtaining maximum yield, seeds which are used for sowing should be of one cultivar and not of a mixture of cultivars. Seed lots having larger percentage of off types give lower yield. Genetic purity is controlled in the field through the process of seed certification. For maintaining genetic purity, generation system of seed production i.e. Breeder seed to foundation seed, foundation seed to certified seed has been evolved. Genetic purity can be tested in the field plot test.

Genetic purity is expected so that the breeder seed should have cent percent genetic purity while the foundation and the certified seeds should have minimum of 99.5% and 99.0% genetic purity respectively.

In self pollinated and vegetatively propagated crop varieties, maintenance of genetic purity is not as complex as cross pollinated and often cross pollinated varieties. The packages prescribing isolation distance in relation to pollination behaviors, removal of contaminants from possible sources at appropriate stages determine a high standard of genetic purity in an individual seed lot and recurrent use of seeds from an identified crop variety in a multiplication programmes.

Physical Quality

The physical quality of seeds which include its appearances and luster is largely depending on post harvest technology and handling of a seed lot. Sometimes, new seed is qualitatively affected by insects and fungi, rain soaking etc., and by the use of mechanical devices. After the operation of cleaning and grading, each seed lot should be subjected to seed analysis for finding out its different components viz.,

1. Pure seed
2. Inert matter content
3. Other crop seeds
4. Weed seeds
5. Other distinguishable varieties of seeds

Seed analysis is done in a seed testing laboratory to ascertain whether a seed lot confirms to the prescribed seed standards. By primary cleaning and uniform sizing, the physical quality of a seed lot can be improved. These operations are not only important from the point of view of certification programme, but also useful for those farmers who utilize saved seed from their last harvest.

Physiological Quality

The physiological quality of seed relates to its vigour and viability and is affected mainly by its moisture content when it is exposed to fluctuating temperature and relative humidity conditions of the environment. Seeds remain highly vigorous and viable on plants when they reach the physiological maturity stage and decline gradually as the crop attains towards field maturity stage for harvest. To prolong the life of the seed, it is essential to bring down moisture content of seeds to the extent possible under dry and cool environmental conditions. The safe moisture level under ambient storage conditions is 10-12% in seed lots packed in non vapour proof containers; and if vapour proof containers are used, it is essential to bring down the moisture level to 6-8%. A temperature between 15° to 18° C and relative humidity between 30 to 40% keep the seed lot vigorous and viable for a longer period.

Health Quality

Seeds are known to be the carrier of deleterious pathogens causing diseases (externally or internally seed borne) and often they go as contaminants in a seed lot. Besides, the soil nematode and stored pest, fungi are equally harmful and their presence make a seed lot unfit for use.

Factors Affecting Seed Quality

Ecological influences	Production technology	Post harvest handling
Adaptability	Seed sources	Threshing technique
Light and Temperature	Isolation and protection	Drying and grading
Sun, rain and wind	Agronomical package	Seed treatment
Soils	Irrigation & nutrition	Baggage & Packaging
Insect activity	Roguing & weed control	Labeling & sealing
	Control of pest and disease	Storage & movement
	Method of harvesting	

Ecological Influences

Adaptability : This relates to better survival capacity of a variety under changes or in a given environmental conditions. Among the cereals, wheat is better adopted in winter season, whereas maize, millets and rice and certain other crop varieties perform well in monsoon and winter or summer season in one or the other climatic zone and facilitate multiplication of seed in quick succession.

Abiotic Components on Seed Quality

"Care with seeds, joy with harvest" the popular words enlighten the importance of quality seed production in agriculture. Seeds play a vital role in Agriculture and acts as a carrier of genetic potentialities of improved varieties. An improved seed is the good investment for successful agriculture that enhance good germination to produce a vigorous seedling, ensure high production and productivity. Quality seeds along with other inputs will tap the superior genetic potentiality of high yielding varieties and bring benefits to the farmers. Hence, production of high quality seeds in a planned manner is very essential for increased agricultural production, thereby enhancing the economic status of the farmers. As the quality seed is very much essential and crucial input in agriculture, utmost care should be given in the process of seed production.

Abiotic Factors on Seed Quality

Abiotic factors are geological, geographical, hydrological and climatological parameters. In general, abiotic factors include:

- Water, which is at the same time an essential element to life and a milieu.
- Air, which provides oxygen, nitrogen, and carbon dioxide to living species and allows the dissemination of pollen and spores.
- Soil, at the same time, source of nutriment and physical support.
- Soil pH, salinity, nitrogen and phosphorus content, ability to retain water.
- Temperature, which should not exceed certain extremes, even if tolerance to heat is significant for some species.
- Light, which provides energy to the ecosystem through photosynthesis.
- Global climate change
- Green house gases
- Natural disasters can also be considered abiotic factors.

Ecological Factors

Even though seed quality is influenced by numerous factors namely ecological factors, production factors and post harvest technology factors during seed production, the ecological factors (weather) are of paramount importance as mankind has no control over them. The weather parameters include temperature, light, rainfall, wind velocity, Green house gases and relative humidity, which play a vital role on the production of quality seeds.

Temperature

Of all the factors effecting seed production, temperature is considered as the most important one. In rice crop, prevalence of either high or low temperature during crop growth, affect the production of reproductive parts namely seed weight, size and seed yield. Seed dry weight in rice was stable over the temperature range from 21/16°C to 30/25° C. High temperature after heading of rice crop resulted in increased chaffiness by shortening the effective tillering stage. In wheat crop, kernel weight decreased as temperature rose above 18 °C. In maize, increased kernel abortion was reported upto 36% in high temperature condition (Cantarero *et.al.*,1999).

Temperature may be considered as a measure of intensity of heat energy. The range of maximum growth for most agricultural plants is between 15 and 40°C (59 to 14°F). The temperature of a place is largely determined by its distance from the equator (latitude) and the altitude. Based on the above, the vegetations are classified as tropical, temperate, taiga, tundra and polar. Every plant community has its own minimum, optimum and maximum temperature known as their cardinal points. The following table indicates the cardinal points of some of the common crops.

Table 1: Cardinal temperature of certain crops for germination(°C.)

Crop	Minimum	Optimum	Maximum
Wheat	4.5	20	30-32
Barley	4.5	20	29-30
Oats	4.5	20	29-30
Maize	8-10	20	40-43
Sorghum	12-13	25	40
Rice	10-12	32	36-38
Tobacco	12-14	29	35

Chilling Injury

Plants growing in hot climate are exposed to low temperature (which is above the freezing point) for some time, and are found to be killed or injured severely. Chlorotic condition or bands occur on leaves of sugarcane, sorghum and maize when exposed for 60 hours at 2 to 4°C.

Temperature Stress on Reproductive Growth

Plant reproduction is highly vulnerable to environmental factors such as temperature which may have significant consequences in agricultural crops. Decreased fruit set at higher temperature was mainly due to poor pollen viability, reduced production and poor pollen tube growth, all of which lead to poor fertilization of flowers. Flower abortion also has been attributable to the decreased seeds per plant and seed yield in other crops such as *Brassica napus, B.rapa* and *B. juncea.*

Pollen development, fertilization and asynchrony of stamen and gynoecium development are sensitive to temperatures during flowering (Croser *et al.*, 2003). The lose of pollen or stigma viability under high temperatures stress might be the primary reason for the lowered number of seeds produce in the legume (Hall, 2004). Reproductive growth leading to seed yield is often depressed by the same increase in temperature that enhances vegetative growth and development. Flower initiation was reduced by temperature >32°C and seed formation was delayed at 40-30°C.

Gan *et al.* (2004) found that the seed yield of canola decreased by 15% when high temperature stress was applied before flowering, whereas the yield reduction was 58% when the stress was delayed to the period of flowering, and further to 77% when the stress was delayed to the pod development stage.

Temperature Stress on Seed Yield and Pollen Quality

Seed vigour is an important parameter which needs to be assessed to supplement germination and viability tests to gain insight into the performance of a seed lot in the field. Seed quality has great impact on the quality of planting stock. The uniformity of seed development within the crop is a major factor through which crop production practices and growing conditions will affect seed to seed variation. During the growth of field crops, maximum seed quality is generally regarded to be attained at physiological maturity (PM), i e., at the end of seed filling (Egli, 1998). The crop is to be harvested at harvest maturity (HM), when seeds have dried to a moisture content that allows harvesting without considerable damage.

Poor seed quality can be caused by the effect of pathogens before physiological maturity, pre-harvest weathering, mechanical damage during and after harvest and deterioration during storage. Hot, dry weather during seed maturation can also result in poor quality. It is unlikely that temperature has much effect on seed quality, although that possibility has not been explored. Seed quality is however, sensitive to temperature during the seed-filling period. Because high temperature can differentially affect various processes involved in seed filling and it can also affect seed composition. In addition, seeds are smaller at high temperature, and hence their milling quality can also suffer.

In seed plants, successful fertilization requires correct regulation of pollen tube growth. At germination and during growth, the pollen tube interacts with tissue from the pistil while the pollen tube extends via tip growth. Plants growth and development, particularly reproductive processes such as pollen grain, pollen tube growth and fruit set are effected by the temperature more than by any other environmental factor when water is not limiting factor. Environmental stresses prevailing during pollen development, germination and pollen tube growth affect the functioning of the pollen and eventual fruit and seed set.

Temperature has a clear effect on pollen tube kinetics, expressed as the time required for pollen germination and the rate of pollen tube growth. While temperature affects pollen tube kinetics, information on the effect of temperatures on pollen tube dynamics, expressed as the census of the micro gametophyte population that succeeded to reach the base of the style, is missing. Pollen grains once released from anthers act as independent functional units and are exposed to ambient environment. Therefore, episodes of high temperature during flowering would more severely affect pollen than the deeply seated ovules.

Freezing Injury

This is generally caused in plants growing in temperature region. When the plants are exposed to very low temperature, water is frozen into intercellular spaces. The protoplasm of the cell is dehydrated and mechanical distortion takes place resulting in killing of the cells. Frost damage occurs to potatoes, tea etc., in winter in the hilly areas like nilgiris.

Suffocation

During winter the ice or the snow form a thick cover over the ground and the crop suffers for want of oxygen. Ice in contact with roots inhibits diffusion of carbon dioxide and the respiratory products may become harmful to plants.

Heat Injury

Very high temperature often stops growth. The plant faces incipient starvation due to high respiration rates. The plant is stunted and if such a condition persists for a long period, the plant will be killed. Direct temperature effects are noticeable in young seedlings and transplanted crops. High temperature causes sterility in flowers. The general effects of excessive heat are defoliation, premature dropping of fruit and in extreme cases death of plant.

Light

The light requirement of each kind of crop should be known otherwise, when they are grown in other conditions, their seed quality will be affected. For instance, wheat varieties in general are photosensitive while improved rice varieties are photo insensitive. In maize, the shading period from 3 to 17 days after silking has significant effect on the number of abortive seeds.

Light Intensity

In photosynthesis, about one per cent of the light energy is converted into potential chemical energy. Very low light intensity reduces the rate of photosynthesis and may even result in closing of the stomata. Very high light intensities are detrimental to plants in many ways. It causes rapid loss of water resulting in the closure of stomata. The most harmful effect of high light intensity is the phenomenon of solarization in which all the cell contents are oxidized by atmospheric oxygen. Many species produce maximum dry matter under high light intensity if water is available in plenty.

Quality of Light

The principal wave lengths absorbed in photosynthesis are in the violet-blue and the orange-red regions. Among these, short rays beyond violet such as infrared are detrimental to growth. Red light seems to be most favourable for growth followed by violet-blue.

Duration of Light

The length of the day has greater influence than the intensity. The response of plant to the relative length of day and night is known as photo periodism. Plants characters like floral development, floral initiation, bulb formation, rhizome production etc., are all influenced by photoperiodism.

Direction of Light

Shoots, roots and leaves show different orientation to the direction of light. In temperate regions the southern slopes show better growth of crops than the northern slopes due to the direction of light contributing more sunlight towards the southern side.

Rainfall

In the regions of moderate rainfall and humidity as compared to regions of high rainfall, good quality seeds can be obtained. Excessive rainfall interferes with the pollination, resulting in delayed maturity and precocious germination of vegetable seeds.

Wind Velocity

Excessive wind during pollen production will result in poor seed set and ill filled seeds. Brown seed discoloration is caused by wind at flowering and occurrence of milky white seed is increased by wind 14-21 days after heading in rice. (Ebata and Ishikawa,1989). Wind affects the growth mechanically and physiologically. The sand and dust particles carried by the wind may damage plant tissues. Emerging seedlings may be completely covered or alternatively the roots of young plants may be exposed by strong winds. Winds may also cause considerable losses by inducing lodging, breaking of stalks and shedding of grains. Hot dry winds, may however adversely affect photosynthesis and hence productivity, by causing closure of the stomata even when soil moisture is adequate. Moderate winds have beneficial effect on photosynthesis by continuously replacing the carbon dioxide absorbed by the leaf surface.

Relative Humidity

Flowering, pollination and seed setting in temperate crops needed low humidity-dry weather at seed maturity and moderate to low humidity for sub temperate and topical crop. High humidity and temperature encourage production of diseased seeds. Warm dry climate is suitable for production of disease free seed.

Global Warming

Post flowering conditions affect yield mainly by influencing ovary or seed abortion, of by changing seed filling duration (SFD) or seed filling rate (SFR). In general, SFD is more plastic than SFR. In spite of the importance of individual seed size, yield variation is more often due to changes in seed number per unit of ground area than to individual seed mass at maturity. Warming hastens crop development and therefore shortens the SFD. Seed number can be reduced via the direct effects of high temperature on reproduction, particularly pollen formation and function. Because warming speeds reproductive development, the seeds that do develop are often small. Although SFR is sometimes stimulated by warming, this effect often does not fully compensate for shortened SFD.

Season

More extreme cases of unseasonal climate usually reduce flowers and subsequently the seed set. Late spring frosts in temperate regions kill flowers and young fruits. An abnormal drought also has similar effects. Even if death or premature shed of whole fruits does not occur, a proportion of seeds may abort later.

Water

Plant growth is seriously affected by abiotic stresses such as drought, salinity or temperature. Drought stress during vegetative or early reproductive growth usually reduces yield by reducing the number of seeds, seed size and seed quality. Irrigation at critical growth phases is very much essential for the normal seed development and maintenance of seed quality. Water stress during critical phases like seed germination, seedling establishment, flowering, fruiting and seed development stages is detrimental to seed set, yield and seed quality.

Soil Nutrient

The fertility of the soil in which a plant grows influences the chemical composition of the developing seed and consequently its metabolism and vigour during germination. Nitrogen and phosphorus availability can influence seed development and seedling vigour, but their effect varies among species. It is possible that inorganic nutrients stored in seeds provide valuable reserves during early germination of seeds. Minor elements are also known to affect the seed development. Deficiency in soil boron causes 'Hollow heart' defect in pea seed. In rice seeds, copper deficiency leads to loss of viability. Calcium deficiency causes darkened plumule in groundnut. Mn/Mo deficiency is leading to 'Marshy spots' in peas grown in calcareous soil and it leads to promotion of hard coated seeds and hard seededness.

Seeds of few species cannot tolerate excessive desiccation and can still retain their viability. It is very important to harvest the seed in proper time to get the best quality of seeds. Harvesting at an early stage will cause damage to seeds during threshing and cleaning. If the harvesting of the seeds are not done in the right time and optimum moisture content and allowed to remain in the field until dead ripe stage, the seeds are likely to be exposed to weather elements like rain and temperature, besides insect and fungal pathogens and suffer mechanical damage.

Seed Drying

Excessive drying rates may cause stress cracks in the seed, because of unequal drying throughout the seed. Even if the seeds are not dried too fast, they by be injured by over drying and usually this will lead to increased susceptibility to mechanical damage of the seeds there by reduction in the quality of seed.

Moisture Content

Susceptibility to mechanical damage increases, as the moisture content decreases. However, safe moisture content varies among species. Large seed legumes are particularly sensitive and excessive injury begins to occur at moisture content below 15%.

Mechanical Damage

Mechanization of harvest, handling and other operations subject seed to physical forces that can do mechanical resistance of the seed covering and the seed as a whole will be exposed to crusting impacts, abrasion, and various types of cutting and shearing actions. The failure of the seed coverings to protect the embryonic axis and nutritive tissue affects the germination capacity of the seed.

Seed Storage

The storage sanitation is very much essential that will tell upon the quality of the seed. Based on the storability of the seeds, the storage condition should be provided. Martinez *et al.* (1994) revealed that most bean varieties stored for 180 days at 75% RH maintained high germination rates (74-98%) on seed quality.

Conclusion

Now-a-days seed has taken a new dimension and to improve the economy of the country, production of high quality seeds is highly warranted. Hence, application of favorable abiotic factors in the process of seed production will not only help to get yield of high quality seed but also make our seed production a profitable venture.

References

1. Ebata, M. and Ishikawa, M. 1989. Effects of wind and rain on the fertilization, kernel development and kernel characters of rice plant. *Jap. J. Crop Sci.* 58 (4) : 555-561.
2. Martinez, E.M., Badillo, M.E.V., Maya, R.N. and Gonzalez, J.R. 1994. Seed viability of different varieties of bean stored under low and high relative humidity. *Seed Sci & Tech.* 22: 195-202.
3. Cantarero, M.G.Cirilo, A.G. and Andrade, F.H. 1999. Night temperature at silking affects kernel set in maize. *Crop Science*, 39 (3): 703-710.
4. Croser, J.S., H. J. Clarke, K. H. M. Siddque and T. N. Khan. (2003). Low temperature stress: implications for chickpea (*Cicer arietinum* L) improvements. *Crit. Rev. Plants Sci.*, 22: 185-219.
5. Gan, Y., S.V.Angadi, H.W.Cutforth, D. Potts, V.V. Angadi and C.L. Mc Donald. (2004). Canola and mustard response to short period of high temperature and water stress at different development at stages. *Can. J. Plant Sci.*, 84: 697-704.
6. Hall, A. E. (2004). Breeding for adaptation to drought and Heat in Cowpea. *Eur. J. Agronomy*, 21: 447-454.
7. Egli, D.B., D.M. Tekrony, JJ. Heitholt and J. Rupe. (2005). Air temperature during seed filling and soybean seed germination and vigor. *Crop Sci.* 45: 1329-1335.

❑❑❑

Chapter 8

Mulching for Drought Management

R. BALASUBRAMANIAN[1] AND K. BALAKRISHNAN[2]
[1] Professor and Head, Department of Agronomy
[2] Professor & Head, Department of Seed Technology
Agricultural College & Research Institute, TNAU, Madurai-625104.

Rainfall in dry farming area is received with high intensity. Due to moderately slow rate of infiltration, the runoff is heavy. The water thus running as runoff could be stored in profile itself. Approximately 60 to 75 percent of the rainfall is lost through evaporation. Mulching is covering of the soil surface with suitable materials primarily to reduce the loss of soil moisture through evaporation, minimize the erosion of surface soil by wind and water and increase the infiltration. Mulches obstruct the solar radiation reaching soil surface responsible for formation of vapour molecules.

Mulches prevent rapid evaporation from the soil surface and reduce rapid drying. Mulching also suppress weed infestation effectively. Furthermore, it stimulated microbial activity in soil through increasing soil temperature, which improves agro-physical properties of soil. Mulching is used as means of successful crop production mainly in place where irrigation facilities are scanty. The effect of these is to cut off the upward flow of underground moisture at a point below the actual surface and to prevent its rapid escape into the air during dry weather (Macmillan, 1989; Suh and Kim, 1991). It has a unique character of reducing the maximum soil temperature and increasing the minimum temperature (Singh *et al*., 1987). Photodegradable and biodegradable mulches made of plastic, paper or other materials have been tested for their utility for annual crop production with varying degrees of success (Greer and Dole, 2003). As early as the 1920's, researchers developed acceptable production practices with paper mulches in annuals. Lippert *et al.* (1964) and Waggoner *et al.* (1960) reported that moisture distribution in the upper soil layers in mulched soil is more uniform compared to unmulched soil, root development is better in the upper soil layer, which is usually rich in nutrients and useful microorganisms. Lal (1989) also reported that biological activity of soil fauna increased with mulching thereby increases soil fertility. Mulching has been reported to increase yield by creating favorable temperature and moisture regimes.

To obtain maximum storage of moisture under any rainfall condition, the soil must absorb as much water as possible when it rains and losses due to evaporation or transpiration must be kept to a minimum. Mulching is a practice for minimizing the evaporation of moisture from soil.

Crop residues also play an important role in controlling water erosion. These dissipate the energy of falling rain drops, reduce surface sealing, increase infiltration and decrease run off velocity. Mulches not only help in reducing soil erosion but also help in conserving soil moisture and improving the yields in rainfed areas. Acharya *et al.* (1998) reported that mulches resulted in higher moisture in the seed zone when wheat was sown compared with the conventional practice of soil tillage after maize harvest. The greater the surface area covered by mulch material in a particular mode, the greater is its effectiveness in conserving soil moisture (Arora *et al.*, 2008).

Mechanisms of Mulching

1. *Rate of evaporation* : Mulching slows down the rate of evaporation from a bare wet soil considerably. It covers the soil from solar energy, lowers the proportion of solar energy absorbed by the soil and reduces the rate of evaporation. In unmulched soil, the evaporated water is rapidly removed by the air movements. But in mulched soil, the water vapour slowly diffuse through mulch. This maintains the air almost stationary and the rate of diffusion limits the rate of evaporation. In saline soils, this will reduce the rate at which salts accumulate in the soil surface due to evaporation of soil-water.

2. *Soil temperature* : Mulching affects both diurnal and seasonal fluctuation in soil temperature. Its principal effect on diurnal temperature is to reduce the midday maximum temperature of a bare soil under hot and dry conditions, without much effect on minimum temperature. With reference to seasons, mulch increases the soil temperature during winter and reduces the rate at which the temperature rises during spring or early summer. The mulch reduced the soil temperature at a depth of 10 cm resulting in increased root length and weight.

3. *Permeability* : Organic mulch increases the permeability of a poorly structured surface soil and helps to produce good stable structure. The primary effect of mulch on poorly structured soils may be due to increase in the length of time of anaerobic condition in the surface soil. It reduces the rate at which it dries after rain or in the spring and also helps to keep the soil surface permeable.

4. *Weed growth* : Mulching can have an important effect on weeds, as weed growth is slow through mulch mainly because of reduced sunlight. Black plastic or properly managed heavy organic mulch can reduce the weed growth effectively. However, transparent plastic mulch increases the growth of weed.

5. *Soil organic matter* : Mulching has certain general effects on decomposition of soil organic matter. The moist condition under the mulch favours microbiological activity, thus, increases organic matter decomposition. It also favours improvement and stabilization of soil structure through this activity.

6. *Plant nutrients* : Mulching enhances the supply of plant nutrients when nutrients present in organic mulch are added to the soil on decomposition. As organic mulching materials contain low nitrogen with more mature plant residues, its decomposition locks up ammonium and nitrate. Thus lowers the level of available nitrogen. Mulching also increases phosphate uptake by crops, partly because it contains phosphates but principally because it encourages the surface rooting of the crop and keeps the surface soil moist for a longer time. The effect of mulch on the potassium supply is considerable since the organic residues meant for mulch are high in potassium. So the potassium uptake of crop is also increased.

7. *Replacing cover crop* : In dryland areas, mulching by organic material/residue, is an effective way to replace cover crops, which competes for moisture and nutrients with main crop in dry areas. The best results can be obtained by covering 70% to 75% of the cropped area (not possible in field crops).

Benefits of Mulching

Mulching has many beneficial effects viz.,, maintaining the soil temperature, conservation of soil nutrients, building up of soil organism population, improving soil organic matter content, improving soil structure, controlling weed growth, improving the germination of seed and establishment of young plants. Hence, mulching is an important operation in farming, especially under moisture stress situation which can have positive effect on the yield and quality of the crop (Sharma *et al.*, 2010).

Limitations

1. In excess rainfall years, mulching may not be effective.
2. Residue production in dry land is inadequate to result in sustainable water conservation.

Mulching Materials

The materials like soil, paper, husk, plastic, saw dust, straw, etc. can be used as mulch. Mulching materials can broadly be divided into three groups: (1) organic mulch, (2) synthetic mulch, and (3) natural (other than organic) mulch. Generally, crop/plant residues dominate in organic mulch. Different plastic materials are used as plastic mulch. As natural mulch, dust, pebble/stone and loose soil are generally used. Tillage that makes a layer of pulverized soil on the surface acts as natural mulch. Use of mulch depends on various factors like climate, soil, crop, agronomical practices, availability of mulching materials and purpose of mulching.

A minimum ground cover of 30 percent with crop residue or similar material is necessary to keep soil loss within acceptable limit. Singer and Blackard (1978) in a study on the effects of oat straw, oak leaves and redwood litter as mulches found that the soil loss did not begin to decrease until 50 percent cover was provided. Afterwards, it declined rapidly with increase of cover.

a) Organic mulching

Mulching with crop/plant residue such as straw, crop stalks, husks, shells, dry leaves and other kinds of residue as well as standing stubbles of the previous crop can trap considerable rain or of previous crop irrigation water in the soil. This effectively reduces the velocity of runoff facilitating greater percolation. The organic mulching has high potentiality in dryland areas for successful farming. Besides increasing yield and fruit quality, organic mulches are helpful in improving the physical conditions of soil. Various organic (straw, hay organic manure, tree leaves) and inorganic (black polyethylene, white polyethylene) mulches are available which can be used for growing temperate fruits under water scarcity areas. Mulching with straw, saw dust, oak leaves or other organic matter increases the humus content and water holding capacity of soil. The higher moisture content is found throughout the growing season beneath hay or black plastic mulch than unmulched basins.

The use of crop residue of non economic value as surface mulch between the crop rows is a better practice than dust mulch. Crop residue of less economic value such a sorghum stubbles, dry grass, wheat straw or pigeonpea stalks can be utilized as surface mulch @ 5 t/ha. Before application to the field, the material should be cut to about 20 to 25 cm length. This type of mulching is specifically recommended for moisture conservation to rabi sorghum. However, the mulches were not found effective in increasing the yields of safflower, gram and rabi cotton. It should be applied immediately, after the emergence of seedlings or up to 15 days from emergence. It should be spread between two rows of crop. It prevents the loss of soil moisture through thermal insulation and prolongs the moisture retention period. Thus, the moisture retained is better during the flowering and grain formation stages of crop. The results of the experiment conducted on medium deep black soil at Dry Farming Research Station, Solapur indicated that application of organic mulch of crop residues @ 5.0 t/ha increased the grain and stover yield of sorghum substantially over control (Table 1). Similarly, coir pith mulching improved the seed cotton yield under different drip irrigation regimes (Table 2).

Table 1: Effect of different mulches on grain and stover yield of sorghum (mean of six years).

S.N	Treatment	Yield of sorghum (q/ha)	
		Grain	Stover
1	Control (without mulch)	7.28	34.3
2	Sorghum stubbles	10.35	47.2
3	Wheat straw	11.20	51.9
4	Pigeonpea stalks	11.82	56.8
5	Dry grass	12.65	62.4

(Dhopte, 2011)

Table 2: Effect of drip irrigation regimes and coirpith mulching on seed cotton yield (kg / ha).

Treatments	No mulching	Coirpith mulching
Drip irrigation at 100% PE once in 3 days	2325	2460
Drip irrigation at 80% PE once in 3 days	2048	2327
Drip irrigation at 60% PE once in 3 days	1382	1726

(Ramesh, 2003)

In the long run, organic mulch increases the permeability and helps to produce good structure of soil and stabilizes it by encouraging the activity of soil microbes, earthworms and other soil biota. The bio-soil mulch can considerably increase available soil moisture (Table 3) to the crops grown with it.

Table 3: Effect of irrigation methods and rice straw mulching on mean available soil moisture (mm) before irrigation.

Treatments	No mulch	Rice straw mulch	Mean
Drip irrigation	74.66	77.17	75.91
Skip furrow	29.86	53.96	46.91
All furrow	48.27	56.46	49.86
Alternate furrow	35.30	48.94	42.12
Borderstrip	33.95	47.83	40.89
Check basin	32.36	46.45	39.40
Mean	43.23	55.13	

(Sampathkumar, 2003)

The type of mulching also helps the plant growth by controlling the extreme soil temperature to a good extent. The average daily maximum and minimum temperature from June to September in the top 2.5 cm of soil under maize plant was 31.2°C and 17.9°C when clean weeded, and 23.6°C and 19.7°C under a mulch of 20 tones / ha of straw, respectively. The results revealed that mulching regulated the soil temperature in favour of plant growth. Experiment on the effect of mulch showed that coirpith mulch lowered the soil temperature in different drip irrigation regimes (Table 4).

Table 4: Effect of drip irrigation regimes and coirpith mulching on soil temperature (°C).

Treatments	No mulching	Coirpith mulching
Drip irrigation at 100% PE once in 3 days	32.67	30.29
Drip irrigation at 80% PE once in 3 days	38.17	30.88

(Ramesh, 2003)

Mulching reduced the first stage of drying of soil which helped to improve soil moisture status, reduced soil temperature besides checking seedling mortality and crop emergence. In West Bengal, pointed-gourd and off-season coriander (for leaf) were found to germinate better under straw mulch. It also protected the emerging seedling from impact of rain drop. This would also protect the young green part from get in touch with wet soil, thus reduce mortality significantly.

The mulch maintains a higher vapour pressure near soil surface by restricting air movement. It also shields the surface soil from detrimental effect of solar energy.

To achieve 70 to 75% cover of the soil surface with straw, an application rate of 0.5 kg/m^2 is sufficient. The less coverage reduces its effectiveness, whilst a greater covering suppresses the plant growth particularly in initial stages. Denser mulches are sometime used under young tree crops to check the weed growth. Once these plants have attained maturity, they can successfully control the weeds below by cutting the sunlight.

Mulching with organic residue can reduce soil loss effectively. On a silt loam soil with 7° slope, annual soil loss is about 2.46 kg/m^2 from uncultivated bare land, but the loss will be only 0.11 kg/m^2 in lands covered with straw mulch applied at 0.5 kg/m^2.

Benefits of Organic Mulching

1. Retains rain or irrigation water in the surface layer of soil.
2. Allows the water more time to percolate into the root-zone.
3. Maintains favourable soil temperature for plant growth
4. Reduces evaporation loss of soil moisture
5. Increases the microbial population and their activity in the soil.
6. Provides organic matter to the soil after decomposition
7. Checks weed growth
8. Improves efficiency of irrigation, particularly the drip method.
9. Reduces soil erosion.

Limitations : Organic mulching has limitations particularly in humid and sub-humid areas, as stated below:

1. It may attract pests and diseases especially termites.
2. It is not suitable in wet areas (induce gleying to crops and anaerobic condition to soil).
3. It is difficult to practice in steep lands.
4. It may create weed problem if the plants used for mulch regenerate through seeds or sets.
5. It poses extra problems in tillage, other cultural operations and sometimes in germination process. But, when combined with conservation tillage; many of these problems can be overcome.

b) Synthetic mulching

There are two types of synthetic mulching materials viz., polythene sheet and bitumen in water emulsions. Polythene sheet is spread on the soil surface between rows of vegetable crops, pineapples, melons and other high valued crops. Different kind of polythene sheet in respect of colour (white, black), thickness are used for mulching purposes. The selection of material is dependent on requirement of climate and crop.

Bitumen emulsion is sprayed over the rows after the emergence of seeding. Sometime polythene sheet, punctured at recommended spacing for a particular crop, is spread on the prepared field. The seeds or seedlings are planted through the openings. However, in dry season the openings may be placed in furrow and in wet season it may be placed on the ridge.

Being expensive, polythene mulching is used in high value crops including fruit trees. Apart from being effective in soil and water conservation, this method helps early harvest of summer vegetables thus catching high market price. In some cases, increase in yield has also been reported.

Besides, polythene, the bitumen emulsion mulching raises the soil temperature which has a positive effect in cooler climate. This may also increase the nitrate content of the soil. The heating effect of transparent polythene is greater than bitumen film. High temperature under polythene kills the germinated weeds and soil born pathogen and egg, larvae of insect pests, especially during summer.

In case of tree crops, polythene sheet may either be spread over the soil surface or it may be used round the lower portion of the trunk covering the root zone area. It looks like a conical shaped cover where sufficient opening is provided between trunk and polythene for air passage to avoid the problem of stem girdler. Polythene mulching is very effective when used with drip irrigation system.

Plastic Mulch Films

Plastic mulch is generally 1.1 to 1.5 mm thick, 4 to 6 feet wide, available in rolls of 1,000 to 4,000 feet long. It is available in a multitude of colors ranging from clear (transparent) to opaque (black or brown). Recently, colored mulches have been investigated for their influences on insect control and plant yields.

Plastic mulch functions to warm the soil, conserve moisture, and prevent nutrient leaching. Clear plastic has the highest soil warming capability (8° to 140° F over bare soil), but weed growth underneath can be extreme. An herbicide is necessary to keep weeds under control with clear mulch. Black mulch will prevent weed growth by prohibiting light transmittance to the soil and will warm the soil 3° to 5° F higher than the bare ground.

In small farms or in small fields, black, brown, or wavelength selective mulches are often preferred to eliminate the use of herbicides. Crops that respond best to mulching are those that respond to higher soil temperature (muskmelon, watermelon, cucumber, squash, tomato, pepper and sweet corn).

The use of black plastic for mulching vegetables and flowers has become quite popular.

Advantages of Using Black Plastic Mulch

1. Reduces soil moisture loss by evaporation.
2. Prevents weeds from growing due to lack of light.
3. Warms the soil (warm season crops benefit from the warmer soil).
4. Hastens maturity of crops.
5. Keeps the fruit cleaner (especially tomatoes).
6. Prevents compaction of the soil.
7. Prevents the soil surface from drying out and cracking.
8. Reduces the amount of hoeing and cultivation
9. Reduces the amount of root injury due to cultivation
10. Facilitates the development of roots in the top few inches of the soil where moisture, oxygen, temperature and nutrients are most favorable for early growth.
11. Reduces the amount of hand spading. In small gardens, hills can be worked up and then covered with plastic mulch without working all the soil between the hills.

Disadvantages of Using Black Plastic Mulch

1. The increased growth of plants grown on black plastic results in greater loss of water by the plant through transpiration. Therefore, on lighter, well-drained soils during dry seasons, it will be more necessary to irrigate mulched crops than crops that are not mulched.
2. On heavier, wetter soils with high water tables, black plastic mulch does not always produce favorable results because the moisture content of the soil under the plastic is maintained at too high level for adequate aeration.

In the case of black mulch, good uniform soil contact is essential as the soil is warmed by heat conduction. Commercially, the simplest way to apply mulch film is with a mechanical mulch layer. Hand application is an option, but applying more than a half-acre can be difficult and time consuming.

Biodegradable Plastic Mulch

Some of the first commercialized degradable mulches were photodegradable, and would break down when exposed to light. Many growers used these products reported uneven and incomplete breakdown, particularly after tillage buried the plastic fragments at the end of the season. However, biodegradable mulches made from plant starches that are broken down slowly by soil microbes into carbon dioxide and water is now available.

Woven Black Plastic Mulch

It may be a good alternative of plastic film mulch in dryland areas. Increase of temperature in summer, under plastic film mulch, is one of the main problems and to solve this problem, application of woven plastic mulch may be recommended. Black/green woven plastic mulch (75% agro shade-net can be used) effectively reduces the soil erosion and reasonably offers all the other advantages as the film mulch provided, except the increase of soil temperature.

Watering of Plants in Plastic Mulches

Drip irrigation system is the best option to irrigate crops covered by plastic mulch. However, plants can be watered with a hose by placing the hose under the plastic through the holes which were torn or punched in it. Common garden sprinkling devices could also be used. After a rain, plastic has to be checked to see where the water stands. Then punch a hole through the plastic with finger or stick wherever water is standing. This will allow the water to run through the plastic during future rains.

Gulshan Mahajan *et al.* (2007) revealed that water use efficiency was recorded maximum (Table 5) in plastic mulched treatment (30.7 kg/ha/cm), while it was minimum in un-mulched treatment (17.3 kg/ha/cm). The lowest water use efficiency was recorded in un-mulched treatment, as formation of reproductive structure of sink was not enough to give a better yield due to more infestation of weeds.

Table 5: Effect of different mulches on baby corn yield and water use efficiency of the crop.

	Baby corn yield (q/ha)			**Water use efficiency (kg/ha-cm)**		
	2003-04	**2004-05**	**Mean**	**2003-04**	**2004-05**	**Mean**
Effect of mulch						
Plastic mulch	11.1	14.2	12.6	24.6	38.4	30.7
Rice-straw mulch	9.8	11.1	10.6	21.8	30.0	25.8
Unmulched	6.1	8.0	7.1	13.6	21.7	17.3
LSD (0.05)	1.61	1.6	1.27	-	-	-

(Gulshan Mahajan et al., 2007)

C). Natural mulch

Dust mulching through interculturing

Timely interculturing operations with suitable implements are equally important in proper conservation of moisture in the soil and making it available to the crops for their satisfactory growth. The experiment conducted on medium black soils at Solapur for six years proved that interculturing with hoe at third, fifth and eighth week from sowing of rabi sorghum (M.35-1), conserved more moisture and increased the grain yield of sorghum to a reasonable degree (Table 6). Out of the three interculturings, the first interculturing should be given entirely by blade hoe and later two by toothed hoe. The first hoeing separates out the surface layer from underlying soil profile and thereby acts as dust mulch in the first instance and breaks the capillary movement of the soil moisture from top in the second instance. This results into more storage of receding soil moisture in the soil profile, which in turn is useful for boosting up the crop production. The second and third hoeing subsidize the micro cracks developed in the soil and thereby reduce the moisture loss through the micro-cracks.

Table 6: Grain yield of sorghum as influenced by interculturing (mean of six years).

S.N.	Treatment	Grain yield (q/ha)
1	One interculturing	6.36
2	Two interculturing	6.50
3	Three interculturing	6.99
4	Control	5.91
	C.D. at 5%	0.66

(Dhopte, 2011)

Pebble Mulch

In stony areas, a uniform layer of pebble on the soil surface in between two rows will prevent transfer of heat from the surrounding air to the soil, thus reducing evaporation losses. It also helps to check the run-off water effectively. This method may be used in dryland areas where soil is gravely. It is also favourable for trees in the initial stages of growth.

In the process of cooling during night, the moisture that evaporates from the soil condenses on the lower surface of the stones instead of being lost to air. As this type of mulch creates problem to cultivation and the fact that it reflects sunlight and holds and radiate heat, this method of mulching can only be used for perennial crops, especially trees. Relatively low-lying and tender plants/crops can suffer from the heat absorbed and radiated by the pebbles. In case of annual crops, pebbles and sand can be used as mulch.

Tillage Mulch

It is the tillage operation with implements (sweeps, blades or chisel plough) that undercut the soil surface, retaining most of the crop residues in the surface. This is basically a type of soil mulch, which combines with organic mulch in the form of stubble mulch. It breaks weed, insect and disease cycles but its main function is to break or disturb the continuous capillaries through which the evaporation loss of soil moisture from root zone takes place. Application of loose soil also does the same thing by way of capping of the capillary channels. When the soil surface gradually dries up, the underground water rises to the surface through the capillary pores and gets lost through evaporation. This moisture loss continues till the ground water as well as soil moisture gets completely exhausted. Mulch tillage disturbs this capillary movement and thus prevents the loss of underground water.

Live Vegetative (Cover Crop) Mulch

Creeping or climbing plants spreading along the ground and covering a good amount of area can be used as mulch instead of crop/plant residue. It is apparently termed as cover crop but it does not use the entire land it covers for its own need. Crop used as live vegetative mulch should not use more than 20 percent of mulched land for its own nutrient and moisture supply. This mulch can do almost all sorts of works better than that of the other organic mulch.

Advantages of Live Vegetative Mulch

1. Efficiently protects the soil from the impact of rain drop.
2. Canopy can retain a good amount of rainwater and provide more time for infiltration.
3. Controls the temperature and humidity of soil and just above the soil in favour of the soil microbes. It reduces the evaporation loss from soil surface.
4. Adds fresh organic matter to soil by way of adding dry / dead leaves etc.
5. Provides moderate to good amount of economic return.
6. Most suitable for tree crops and orchards.

Problems in Live Vegetative Mulch

1. Feeding, watering and other crop management aspects are to be done properly for better mulching effect and to avoid competition with main crop.
2. Increases the transpiration loss of the field and increases the surrounding humidity.
3. Acquires and uses a certain portion (less than 20 percent) of the mulched area to spread its root system.
4. Availability of suitable plant is dependent on environment and locality and an early planning is essential.
5. It is not feasible for small area, particularly for elongated small strip like area.

For tropical and sub-tropical areas, different cucurbitaceae crops having long soft stem and branches like pumpkin, cucumber etc. may be used as live vegetative mulch. These mulches may be planted in cluster or in line as per the requirement of area to be mulched with proper pruning and training.

Vertical Mulch

Vertical mulches provide a water pathway to the root zone, and can reduce the area of soil surface that is wet. In this system, narrow tranches in farm beds (Fig.1) are made and filled with dry stalks of millet, maize, sunflower, amaranth, etc. The air space created by these stalks conducts the water quickly down to plant root-zone and reduces its possibility of evaporation considerably. It simultaneously works as sub-surface irrigation system and, on decomposition, supply organic matter to soil.

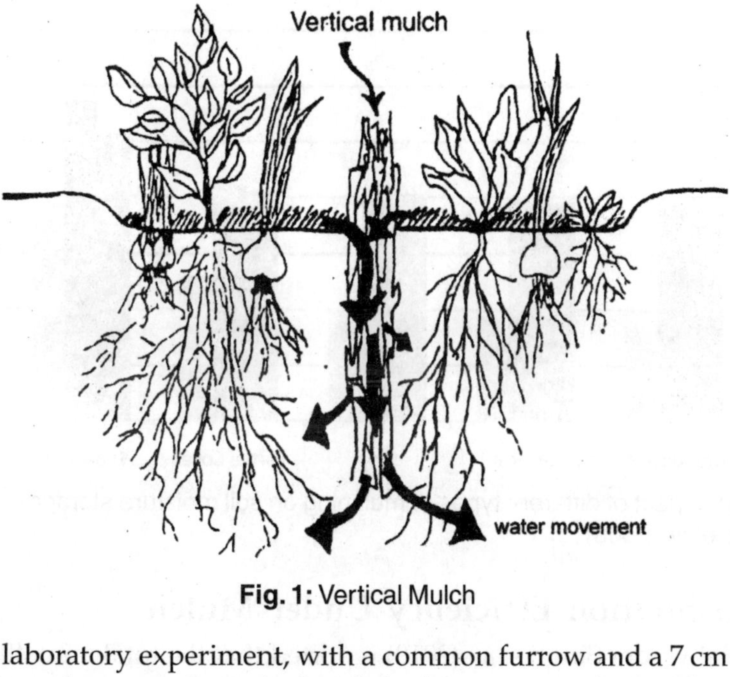

Fig. 1: Vertical Mulch

A laboratory experiment, with a common furrow and a 7 cm wide and 15 cm deep trench packed with barley straw applied with 5.5 cm of water, shows that after 17 days, 42% and 72% of the water remained in the soil in case of furrow and vertical mulch, respectively. When the applied water overflowed the vertical mulch and wetted the surroundings, 60% of it was retained after 17 days of evaporation. The subsequent experiment with sorghum crop showed that a similar mulch increased yields from 20 to 40 percent over control plots with no mulch (Ghosh and Jana, 2010).

Another experiment in gently sloping Vertisols in semi arid India shows that 15 cm wide and 30 cm deep vertical mulch with sorghum stubble, placed 2 m apart, gave 25 times more grain yield and 2 times more straw yield of an 'improved variety' of sorghum in a very dry year. The beneficial effects appeared to last 4 years, probably the result of addition of organic matter as the straw partially decomposed after the 2nd year (Ghosh and Jana, 2010).

Problem in vertical mulch : The infestation of pest and disease, especially the termites is seen. Measures against these threats should be taken at the time of implementation of this method.

The yield of dryland crops can be increased and or stabilized by adopting one of the crop life saving measures, namely mulching.

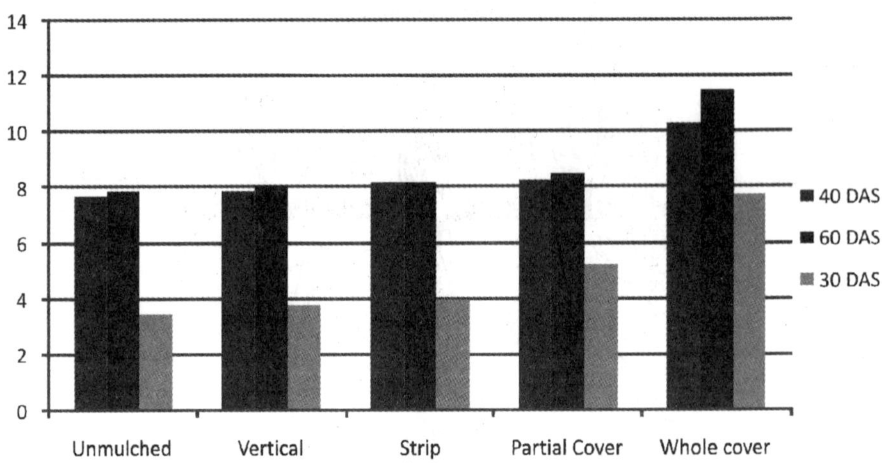

Fig. 2: Impact of different types of mulching on soil moisture storage (Arora et al., 2008)

Transpiration Efficiency Under Mulch

Balwinder Singh *et al.* (2010) reported that the total rainfall during the 2006-07 wheat season was 159 mm with well distribution and given with one post-sowing irrigation. The 2007-08 season was generally dry, with growing season rainfall of 88 mm received late in the season, and hence three post sowing irrigations were applied. Total crop ET was not affected by mulching in both years (Table 7). Transpiration was significantly higher with mulch in each year. There was a trend for higher grain yield and total biomass with mulching (Table 7). The lower spike density without mulch was associated with higher tiller mortality, especially in 2006-07. The reason for the high tiller mortality without mulching was due to soil water deficit stress for about 10 days after maximum tillering, as evidenced by tensiometer data. Each year, post-anthesis biomass accumulation and transpiration were greater in the mulched treatments.

There was a consistent trend for lower grain and total biomass transpiration efficiency with mulching, and the differences were significant in 2007-08 (Table 7). Grain transpiration efficiency was close to values observed for wheat elsewhere (Zhang *et al.* 1998). The lower transpiration efficiency with mulch may result from a low ratio of photosynthesis to transpiration at the leaf level under conditions of high water availability. Chakraborty *et al.* (2008) observed continuous low canopy temperature under mulch, and suggested that this was because the stomata remained open for longer periods, leading to higher transpiration and thus a lower photosynthesis/transpiration ratio.

Table 7: Soil water balance components and transpiration efficiency under mulch and non mulched wheat during 2006-07 and 2007-08.

	2006-07			2007-08		
	Mulch	Non-Mulch	L.SD (0.05)	Mulch	Non-Mulch	L.S.D (0.05)
Irrigation (mm)	75	75	-	225	225	-
Rain (mm)	159	159	-	88	88	-
dSWC (mm)	-107	-111	NS	-87	-91	NS
ET (mm)	341	345	NS	400	404	NS
Soil evaporation(mm)	101	135	10	121	161	17
Transpiration (mm)	240	210	26	279	242	29
Grain Yield (t/ha)	4.5	4.0	0.2	4.1	4.0	NS
Total biomass (t/ha)	10.6	9.2	0.7	10.2	10.0	NS
Grain Transpiration efficiency (kg grain/ mm/ha	18.8	19.1	NS	14.6	16.4	1.2
Total biomass transpiration efficiency (kg/mm/ha)	43.9	43.8	NS	36.6	41.4	3.1

(Balwinder Singh et al., 2010)

References

Acharye, C.L. O.C. Kapur and S.P.Dixit. 1998. Moisture conservation for rainfed wheat production with alternative mulches and conservation tillage in the hills of north–west India. *Soil and tillage Reasearch*, 46: 153-163.

Arora, S., M.S. Hadda and R. Bhat, 2008. Tillage and mulching in relation to soil moisture storage and maize yield in foot hill region. *Journal of Soil and Water Conservation*, 7: 51-56

Balwinder Singh, E. Humphreys, P.L. Eberbach and A.Katupitiya. 2010. Soil Solutions for a Changing World. 19th World Congress of Soil Science. 1 – 6, August, 2010, Brisbane, Australia, Published on DVD.

Chakraborty, D.N., S. Aggarwal, P. Gupta, V.K. Tomar, R. K. Garg, R. N. Sahoo, R. N. Sarkar, A. Chopra, K. Sarma and K. S. S. N. Kalra. 2008. Effect of mulching on soil and plant water status, and the growth and yield of wheat (*Triticum aestivum* L.) in a semi-arid environment. *Agricultural Water Management*, 95; 1323-1334.

Dhopte, A.M. 2011. Agrotechnoogy for dryland farming. Scientific Publishers (India), Jodhpur.

Ghosh, A. and P.K.Jana. 2010. Management of mulching. In: Dryland farming. Kalyani Publishers, Ludhiana. 194-203.

Greer, L. and J. M. Dole. 2003. Aluminum foil, aluminum painted, plastic and degradable mulches increase yields and decrease insect-vectored diseases of vegetables. *Hort. Technol.*, 13; 276-284.

Gulshan Mahajan, Rakesh Sharda, Ashwani Kumar and K.G. Singh. 2007. Effect of plastic mulch on economizing irrigation water and weed control in baby corn worn by different methods. *African Journal of Agricultural Research.* 2(1): 019-026.

Lal, R. 1989. Conservation tillage for sustainable agriculture: Trop, versus temperate Environ. *Adv. Agron.*, 42: 174-151.

Lippert, L.F., E.H. Takatori and F.L.Whiting. 1964. Soil moisture under bands of petroleum and polyethylene mulches. *Proc. Amer. Soc. Hort. Sci.*, 85: 541-546.

Macmillan, H.F. 1989. A handbook for tropical planting and gardening. Scientific Publishers, India, pp: 261-15.

Ramesh, K. 2003. Effect of drip irrigation regimes at various intervals and coir pith mulching on growth and yield of summer cotton cv. SVPR 2. M.Sc.(Ag) Thesis, Department of Agronomy, Agricultural College & Research Institute, TNAU, Madurai.

Sampathkumar, T. 2003. Evaluation of drip and surface irrigation methods with rice straw mulching in cotton. M.Sc. (Ag) Thesis, Department of Agronomy, Agricultural College & Research Institute, TNAU, Madurai.

Sharma, B. C., Anil Kumar., V. Sharma and A. S. Bali. 2010. Technological Innovations in dry land Farming Systems. Ibdc Publishers, Lucknow.

Singer, M.J. and J. Blackard. 1978. Effect of mulching on sediment in runoff from simulated rainfall. *J. Soil Science Soc. Amer.*, 42: 481-486.

Singh, P.N., B. P. Joshi and G. Singh. 1987. Effect of mulch on moisture conservation, irrigation requirement and yield of potato. *Ind. J. Agron.*, 32: 451-452.

Suh, J. K. and Y. S. Kim. 1991. Studies of improvement of mulching, cultivar and method in onion. *Res. Rep. Rural Dev. Adm. Hortic.*, 33: 31-36.

Waggoner, P. E., P. M. Miller and H. C. Deroot. 1960. Plastic mulching: principles and benefits. *Connecticut Agric. Exp. Station Bull.*, pp: 534 - 544.

Zhang H, T. Y. Oweis, S. Garabet and M. Pala 1998. Water-use efficiency and transpiration efficiency of wheat under rain-fed conditions and supplemental irrigation in a Mediterranean-type environment. *Plant and Soil,* 201: 295-305.

□□□

Chapter 9

Correlation and Causation Studies in Segregating Populations of Rice (*Oryza sativa* L.) Under Aerobic Condition

M. PRAKASH AND A. ANANDAN
Department of Genetics and Plant Breeding
Faculty of Agriculture, Annamalai University, Annamalai Nagar

Abstract

The study materials were evaluated in augmented design during 2010-2011 to study the two intervarietal crosses in rice namely ADT 43×MAS 964-1, BPT 5204×Kaloorandian. The genotypes were evaluated for nine biometrical traits *viz.*, days to first flowering, plant height, total number of tillers per plant, number of panicles per plant, panicle length, number of filled grains per panicle, 100 grain weight, spikelet fertility percentage, and grain yield per plant. Data were statistically analysed for correlation coefficient and path coefficient. Highly significant positive correlation was observed between filled grains per panicle and grain yield per plant for both the two crosses. Path analysis disclosed that direct effect on grain yield per plant was exerted by number of filled grains per panicle. The indirect effect on grain yield per plant was expressed by panicle length and number of panicles per plant for both the crosses.

Introduction

Globally rice is grown over an area of about 149 million ha with an annual production of 600 million tones (Bernier *et al.*, 2008). In India, rice is cultivated round the year in one or the other part of the country, in diverse ecologies spread over 44.6 M ha with a production of 132 MT of rice and average productivity of 2.96 t ha^{-1}. Asia's food security depends largely on irrigated lowland rice fields, which produce three-quarters of all the rice harvested. However, the increasing scarcity of fresh water threatens the sustainability of the irrigated rice ecosystem and increasing use of groundwater for rice cultivation has led to declines in water table by 0.1 to 1.0 m per year, resulting in water scarcity and increased cost for pumping water (Humphreys *et al.*, 2010). One way to reduce water and labor demand is to grow dry seeded rice (DSR) instead of the puddle (Yadav *et al.*, 2010) transplanted rice. A new development in water-saving technologies is the concept of aerobic rice. In aerobic rice systems, the fields remain unsaturated throughout the season. The new concept of aerobic rice may be an alternate strategy, which combines the characteristics of rice varieties adopted in upland with less water requirement and irrigated varieties with high response to inputs.

In China, the water use for aerobic rice production was 55–56% lower than the flooded rice with 1.6–1.9 times higher water productivity. Over the centuries, lowland rice has proven to be a remarkably sustainable system for rice production mostly because of its luxurious water availability. But the present day water crisis threatens the sustainability of lowland rice production and necessitates the adoption of water saving irrigation technologies. Technologies like saturated soil culture and alternate wetting and drying are receiving renewed attention by researchers. These technologies reduce water inputs only at the expense of yield. Aerobic rice is a new concept to decrease water requirements in rice production and is highly suitable for irrigated lowland rice with insufficient rainfall. Experiments on aerobic rice have shown that water requirement in aerobic rice were more than 50 per cent lower (only 470-650 mm) and water productivities were 64-88 per cent higher than the lowland rice. It indicates that aerobic rice may be a viable option where the shortage of water does not allow the growing of lowland rice. The present study was taken up to identify the role of important yield contributing traits in rice under aerobic condition in the F_2 progenies of two crosses by using correlation and causation analysis.

Materials and Methods

The study was conducted using four parents (ADT 43 (P_1), MAS 964-1 (P_2), BPT 5204 (P_3) and Kaloorandian (P_4) and their F_2 generations *viz.*, cross I (ADT 43 X MAS 964-1) and cross II (BPT 5204 x Kaloorandian). These genotypes were sown in the Plant Breeding farm in Annamalai University (11° 24′ N, 79° 43′ E, 5.79 MSL), Tamil Nadu, India. Cross I and II has 102 and 115 lines respectively. The experiment was laid out in augmented design with their parents. Field was irrigated once in a week and observations (days first flowering, plant height (cm), number of tillers plant^{-1}, number of panicles plant^{-1}, panicle length (cm), 100 seed weight (g), filled grains panicle^{-1}, grain yield (g plant^{-1}), and spikelet fertility percentage were observed at appropriate stages. Data were pooled and subjected to correlation and path analysis.

Results and Discussion

Correlation studies

Association of yield with yield component

Grain yield is a polygenically controlled character with complex inheritance. Therefore, while exercising selection pressure, breeders should have information on the nature and association of various component traits with grain yield and among themselves. Significance and negative association (r = -0.199*) of days to first flower with grain yield per plant was recorded in cross-2. Similar results were reported by Akinwale *et al.* (2011). Total number of tillers per plant with grain yield per plant showed positive and highly significant association (r = -0.199**) in cross-1. Number of panicles per plant with grain yield showed positive and significant association (r = 0.444** (cross I), r = 0.203* (cross II)) for both the two crosses. Similar results were reported by Bagheri *et al.* (2011). Number of filled grains with grain yield per plant showed positive and significant association in both the crosses. These results were in parallel with the findings of Immanuvel Selvaraj *et al.* (2011) and Akinwale *et al.* (2011). Panicle length with grain yield per plant showed positive and significant association. Immanuvel Selvaraj *et al.* (2011) and Bhadru *et al.* (2011) also obtained similar results. Plant height showed negative association with grain yield per plant in both the two crosses. This is in accordance with the findings of Mahato *et al.* (2003) and Paramesha *et al.* (2005). The association of number of filled grains per panicle showed highly positive association with grain yield per plant. Similar observations were reported by Ullah *et al.* (2011) and Mulugeta Seoyum *et al.* (2012). Hundred seed weight had positive association with grain yield per plant (r = 0.356**) in cross-2. These results are in accordance with the findings of Javed Iqbal Watto *et al.* (2010).

Table 1: Correlation analysis for the cross-1 (ADT 43 X MAS 964-1).

	Days to first flower	**Plant height**	**Total Number of tillers/ plant**	**Number of panicles/ plant**	**Panicle length**	**Number of filled grains/ panicle**	**100 seed wt**	**Spikelet fertility %**	**Single plant yield**
Days to first flower	1.000								
Plant height	0.002	1.000							
Number of tillers/plant	-0.137	0.307**	1.000						
Number of panicles/plant	0.003	0.216*	0.614**	1.000					
Panicle length	0.208*	0.201*	0.247*	0.323**	1.000				
Number of filled grains/panicle	0.194	0.157	0.338**	0.389**	0.506**	1.000			
100 seed wt	0.002	0.208*	0.014	0.112	0.135	0.214*	1.000		
Spikelet fertility %	-0.070	0.170	0.083	0.201*	0.108	0.517**	0.258**	1.000	
Single plant yield	-0.113	0.054	0.453**	0.444**	0.219*	0.465**	0.132	0.145	1.000

*Significant at 5% level ; **Significant at 1% level

Table 2: Correlation analysis for the cross-2 (BPT 5204 X Kaloorandian).

	Days to first flower	Plant height	Number of tillers/plant	Number of panicles/plant	Panicle length	Number of filled grains/panicle	100 seed weight	Spikelet fertility percentage
Plant height	0.128	1.000						
Number of tillers/plant	0.004	0.414**	1.000					
Number of panicles/plant	0.098	0.236**	0.600**	1.000				
Panicle length	0.171	0.459**	0.417**	0.197*	1.000			
Number of filled grains/panicle	-0.148	0.158	0.092	0.119	-0.002	1.000		
100 seed wt	-0.001	0.213*	-0.013	0.123	-0.119	0.402**	1.000	
Spikelet fertility %	-0.086	0.096	-0.045	0.020	-0.093	0.757**	0.428**	1.000
Single plant yield	-0.199*	0.106	0.006	0.203*	-0.041	0.764**	0.356**	0.496**

*Significant at 5% level ; **Significant at 1% level

Table 3: Path co efficient analysis for nine characters in rice in cross-1 (ADT 43 X MAS 964-1).

	Days to first flower	Plant height	Number of tillers per plant	Number of panicles/ plant	Panicle length	Number of filled grains/panicle	100 seed weight	Spikelet fertility percentage	Single plant yield
Days to first flower	-0.172	0.000	-0.028	0.001	-0.015	0.091	0.000	0.011	-0.113
Plant height	0.000	-0.101	0.063	0.044	-0.014	0.073	0.016	-0.028	0.054
Number of tillers/plant	0.024	-0.031	0.206	0.126	-0.018	0.158	0.001	-0.013	0.453**
Number of panicles/plant	-0.001	-0.022	0.127	0.205	-0.023	0.182	0.009	-0.033	0.444**
Panicle length	-0.036	-0.020	0.051	0.066	-0.072	0.236	0.011	-0.017	0.219*
Number of filled grains/panicle	-0.033	-0.016	0.070	0.080	-0.036	0.468	0.017	-0.084	0.465**
100 seed weight	0.000	-0.021	0.003	0.023	-0.010	0.100	0.079	-0.042	0.132
Spikelet fertility percentage	0.012	-0.017	0.017	0.041	-0.008	0.242	0.020	-0.162	0.145

Table 4: Path co-efficient analysis for nine characters in rice in Cross -2 (BPT 5204 X Kaloorandian).

	Days to first flower	Plant height	Number of tillers/ plant	Number of panicles/ plant	Panicle length	Number of filled grains/panicle	100 seed weight	Spikelet fertility percentage	Single plant yield
Days to first flower	-0.097	0.003	-0.001	0.008	0.000	-0.132	0.000	0.020	-0.199*
Plant height	-0.012	0.024	-0.060	0.019	-0.001	0.141	0.017	-0.023	0.106
Number of tillers/plant	0.000	0.010	-0.144	0.049	-0.001	0.082	-0.001	0.011	0.006
Number of panicles/plant	-0.010	0.006	-0.086	0.082	0.000	0.106	0.010	-0.005	0.103
Panicle length	-0.017	0.011	-0.060	0.016	-0.002	-0.002	-0.009	0.022	-0.041
Number of filled grains/panicle	0.014	0.004	-0.013	0.010	0.000	0.896	0.032	-0.178	0.764**
100 seed weight	0.000	0.005	0.002	0.010	0.000	0.360	0.079	-0.101	0.356**
Spikelet fertility percentage	0.008	0.002	0.007	0.002	0.000	0.678	0.034	-0.235	0.496**

Inter Association Among Yield Components

Association of yield and its components alone are not adequate in any selection programme. The inter relationship existing between the individual characters which ultimately influence the yield component showed the following relationship. In the present study, days to first flower recorded significant and positive correlation (r = 0.208*) with panicle length in the cross-1. Similar results were reported by Shivani and Sree Rama reddy (2000) and Raju *et al.* (2003). Plant height exhibited significant and positive correlation with total number of tillers per plant, total number of panicles per plant, panicle length and 100 seed weight in cross-1. These results were in confirmation with Chakraborty *et al.* (2001) and Latha *et al.* (2003). Positive and significant association with number of panicles per plant with panicle length was in conformation with the findings of Somnath Bhattacharya and Shyamal Ghosh (2004). In Cross-1, the total number of tillers per plant exhibited positive and significant association with number of panicles per plant, panicle length and number of filled grains per panicle. These results were in agreement with Raju *et al.* (2003) and Latha *et al.* (2003). In the cross-2, number of filled grains per panicle recorded positive and significant association with 100 seed weight and spikelet fertility indicating the importance of these traits for yield improvement in rice cultivated under aerobic condition. These results were in conformity with Mulugeta Seyoum *et al.* (2012). Number of panicles per plant showed positive and significant association (r = 197*) with panicle length in cross-2. These results were in agreement with Raju *et al.* (2003) and Latha *et al.* (2003).

Path Coefficient Analysis

The correlation coefficient is insufficient to explain the relationship for effective manipulation of characters. But path coefficient analysis is the best method to evaluate the cause and effect relationship between the yield and yield components. Path coefficient analysis provides a clear cut picture about the direct and indirect effect of component character on yield. The results of such analysis are discussed here under.

Direct Effect

The positive and direct effect of grain yield per plant with number of filled grains per panicle was noted in cross-2. These results were in conformity with the earlier findings of Javed Iqbal Watto *et al.* (2010). The positive and direct effect of grain yield per plant with number of tillers per plant was observed in the cross-1. Similar results were reported by Bhageri *et al.* (2011).The trait, number of panicles per plant showed positive and direct effect on grain yield per plant which was observed in cross-1. These results were in conformity with the earlier findings of Javed Iqbal Wato *et al.* (2010).

Indirect effect

The indirect effect of filled grains per panicle with grain yield was exhibited in cross-2. Similar results were obtained by Kannan (2002). Hundred seed weight had positive effect on number of filled grains per panicle in the cross-1. Similar results were reported by Mulegeta Seyoum *et al.* (2012). Spikelet fertility percentage showed high positive indirect effect for number of filled grains per panicle, and grain yield per plant.

Conclusion

The present study indicates that highly significant positive correlation was observed between filled grains per panicle and grain yield per plant for both the two crosses. Path analysis disclosed that direct effect on grain yield per plant was exerted by number of filled grains per panicle. The indirect effect on grain yield per plant was expressed by panicle length and number of panicles per plant for both the crosses. This suggested that these characters deserve greater weightage while formulating selection indices in rice.

References

Akinwale, M. G., G. Gregorio, F. Nwilene, B.O. Akinyele, S.A. Ogunbayo and A.C. Odiyl. 2011. Heritability and correlation coefficient analysis for yield and its components in rice (*Oryza sativa* L.). *African J. Plant Sci.*, 5(3): 207-21.

Bagheri, N., N. Babaeian-Jelodar and A. Pasha. 2011. Path coefficient analysis for yield and yield components in diverse rice (*Oryza sativa* L.) genotypes. *Bih. Biol.*, 5: 32-35.

Bernier J., G.N. Atlin, R. Serraj, A. Kumar and D. Spaner, D. 2008. Breeding upland rice for drought resistance. *J. Sci., Food and Agric.* 88: 927-939.

Bhadru, D., D. Lokanadha Reddy and M.S. Ramesha. 2011. Correlation and path coefficient analysis of yield and yield contributing traits in rice hybrids and their parental lines. *Elec. J. Plant Breed.*, 2(1): 112-116.

Chakraborty, S., P. K. Das, B. Guha, B. Barman and K.K. Sarmah. 2001. Heritability, correlation and path analysis of yield components in boro rice. *Oryza*, 38(3 &4): 99-100.

Humphreys, E., S.S. Kukal, E.W. Christen, G.S. Hira, B. Singh, S. Yadav and R.K. Sharma. 2010. Halting the groundwater decline in north-west India - Which crop technologies will be winners?. *Advances in Agronomy*, 109: 155-217.

Immanuvel Selvaraj, C. Pothiraj Nagarajan, K. Thiyagarajan, M. Bharathi and R. Rabindran. 2011. Genetic parameters of variability, correlation and path coefficient for grain yield and other yield attributes among rice blast disease resistant genotypes of rice (*Oryza sativa* L.). *African J. Biotech.*,10 (17): 3322-3334.

Javed Iqbal Watto, Abdus Salam Khan, Zulfiqar Ali, Muhammad Babar, Muhammud Naeemullah and Nazim Hussain. 2010. Study of correlation among yield related traits and path coefficient analysis in rice (*Oryza sativa* L.). *African J. Biotech.*, 9(46): 7853-7856.

Kannan, R. 2002. Studies on variability in rice (*Oryza sativa* L.). M. Sc. (Ag.) Thesis, Annamalai University, Annamalainagar.

Latha , J., R. Venuprasad, H.E. Shashidhar and Sailaja Hittalmani. 2003. Correlation and path coefficient analysis in rice cultivars adapted to rain fed lowland of southern Karnataka. *Mysore J. Agric. Sci.*, 37(2): 115-121.

Mahato, R.N., M.S. Yadava and K.S. Mohan. 2003. Genetic variation, character association and path analysis in rain fed upland rice. *Indian J. Dry land Agric. Res., & Development*, 18(2): 196 - 198.

Mulugeta Seyoum, Sentayehu Alamerew and Kassahun Bantte. 2012. Genetic variability, heritability, correlation coefficient and path analysis for yield and yield related traits in upland rice (*Oryza sativa* L.). *J. Pl. Sci.*, 7: 13-22.

Paramesha, K., M.R. Gururaja Rao and M.P. Rajanna. 2005. Evaluation of stabilized selection of different crosses of rice (*Oryza sativa* L) for yield and yield components. *Mysore J. Sci.*, 39(1): 31-36.

Raju, C.H.S., M.V.B. Rao and A. Sudarshanam. 2003. Associations in physiological growth parameters in rice hybrids. *Madras Agric. J.*, 90(10-12): 621-621.

Shivani, D. and N. Sree Rama Reddy. 2000. Correlation and path analysis in certain rice (*Oryza sativa* L.) hybrids. *Oryza*, 37(3): 183-186.

Somnath Bhattacharyya and Shyamal, K. Ghosh. 2004. Association among yield related traits of twenty four diverse landraces of rice. *Crop Res.*, 2(1): 90-93.

Ullah, M.Z., M.K. Bashar, M.S.R. Bhuiyan, M. Khalequzzaman and M.J. Hasan. 2011. Interrelationship and cause-effect analysis among morpho-physiological traits in boro rice of Bangladesh. *Int. J. Plant Breed. Genet.,* 5: 246-254.

Yadav, S., G. Gill, E. Humphreys, S. S. Kukal and U.S. Walia. 2010. Effect of water management on dry seeded and puddle transplanted rice. Part 1. Crop performance. *Field Crops Res.,* 120: 112-122.

□□□

Chapter 10

Breeding for Drought Tolerance: Looking Back the Pathway

S. LAKSHMI NARAYANAN[1], S.M.IBRAHIM[1] AND K. BALAKRISHNAN[2]
[1]Department of Plant Breeding and Genetics
[2]Department of Seed Technology
Agricultural College and Research Institute, Tamil Nadu Agricultural University, Madurai

The population of India is rapidly expanding and as a result, food supply has to be increased greatly in the coming years. Getting maximum yield from poorer soils is important task to feed the increasing mouths of rich and poor. The yield potential of crop can be completely expressed only when it is free from biotic or abiotic stresses (Table 1).

Table 1: Estimated potential yield loss due to abiotic stresses

S.No.	Abiotic stress factors	Estimated Potential yield loss (%)
1.	High temperature	40
2.	Salinity	20
3.	Drought	17
4.	Low temperature	15
5.	Other factors	8

(Athar and Ashraf, 2009)

Though a crop cannot avoid all types of stresses, it has to at least withstand the yield affecting stress factors in such a way that the metabolic processes are not much affected. Drought, high temperature, low temperature and salinity are among the major abiotic stresses that adversely affect crop growth and productivity.

Drought is the main abiotic factor as it affects 26 per cent of the arable area. In India, almost 1/6th of the geographic area (i.e. 50 million hectares) is drought prone, comprising about 68% of the cultivable area (AIR Worldwide, 2012). Unlike other natural hazards such as floods, hurricanes, tornadoes, and earthquakes, which occur over finite periods of time and result in visually obvious damage, drought develops slowly and quietly, lacking highly visible and structural impacts. Developing drought conditions often go unnoticed until precipitation shortages become severe and impacts begin to occur. About 40 per cent of the total area is irrigated and others dependant on rain gods and suffer from drought because of the erratic or delayed monsoon. Global warming and climate change has made the situation still worsening and keeps productivity and sustainability of different food crops in danger. With climate change, growing water insecurity and renewed concerns about food security, the potential of drought tolerant crops are enormous.

At present, Indian food production is 235 million tonnes and should be increased to 350 million tonnes by 2020 (www.abdulkalam.com). Exploitation of natural resources by the ever-increasing population, by products of industrialisation like CO_2, methane, CFCs, oxides of nitrogen and ozone are the main causes of *global warming*. This is the seed for climate change which ultimately makes rainfall erratic and causes water scarcity. By 2100, Earth's mean surface temperature is expected to rise by 1.4 to 5.8°C, precipitation is projected to decrease in the subtropics, and extreme events, such as floods, droughts, and cyclones, are likely to become more frequent (FAO, 2002). Nutrient and water absorption are independent processes in roots, however the necessity for available water in both plant and soil for growth and nutrient transport makes them intimately related (Viets, 1972).

Drought

Drought has been defined as the inadequacy of water availability, including precipitation and soil moisture storage capacity, in quantity and distribution during the life cycle of crop to restrict expression of full genetic yield potential. Drought is actually a meteorological event which implies the absence of rainfall for a period of time, long enough to cause moisture depletion in soil and water deficit with a decrease of water potential in plant tissues. Under drought conditions, water stress develops in the plant as the demand exceeds supply of water.

This may occur either/or both due to atmospheric or soil conditions, and is reflected in a gradient of water potential developed between the soil or soil-root interface and leaf, the transpiring organs.

Among the abiotic stresses, drought significantly reduces both quality and yield of the crop yield. Droughts cause large scale damage to our agriculture and economy. Drought is particularly more harmful because the regions affected by drought already suffer from poor and unstable yields and wide spread poverty. Drought further aggravates the severity of these conditions and causes immense suffering to the population.

Drought stress poses an increasing threat to agriculture. The failure in rainfall because of weak monsoon, as well as erratic distribution of rainfall will result in prolonged drought. This makes yield reduction even upto 100 percent in certain cases. Hence, the mechanism underlying the drought resistance and the physiological, biochemical processes that prevent yield reduction in crops need to be thoroughly understood. Very strong genetic and agronomic improvements are immediately needed.

Physiological Mechanism

Drought causes dehydration of cells and tissues that leads to considerable damage to the crop plants. Most of the plant biochemical and physiological pathways occur in solutions with water being the predominant component. To mention a few, transport of nutrients and photosynthesis are severely affected by water stress. Whenever a plant is subjected to drought stress, the physiological changes viz., closure of stomata, reduction in transpiration rate, reduction in water potential of different tissues, reduction in photosynthetic rate take place. Moreover reactive oxygen species (ROS) accumulate during abiotic stress.

Osmoprotectants (Table 2) occur in all the organisms ranging from archaebacteria to higher plants (Mc Neil *et al.*, 1999).

Table 2: Common Osmoprotectants.

Broad groups	Biochemicals
Sugars	Sucrose, Raffinose, Trehalose
Aminoacids	Proline
Sugar alcohols	Sorbitol, Mannitol
Amines	Glycine betaine, polyamines

These compounds are highly soluble and accumulate in the plant cells during stress conditions. These intra cellular compounds are non-toxic and help stabilize the cell proteins and cell membranes against the adverse effect caused by stress (Soren *et al.*, 2010). It increases the osmotic pressure and enables plant to tolerate, adapt to different stress conditions in such a way the normal metabolic processes are not affected to a greater extent. These compounds help to stabilize the intra cellular proteins and the membranes under unfavourable environmental conditions (Yancey, 1994). The collaboration of plant breeders, physiologists and molecular geneticists is absolutely necessary to engineer the metabolic pathways for the synthesis of desired Osmoprotectants in various crops to achieve the improvement in tolerance to various drought and other stresses. Thorough knowledge of the different biochemical pathways leading to the production of all these compounds is absolutely essential for a plant breeder to effectively engineer the processes to get desirable outcome. The production of hormone ABA is also triggered under drought stress.

Drought Resistance

Drought is the condition of soil moisture deficiency. The crop plants that are said to be resistant to drought will have the ability to grow well under moisture stress condition. The mechanism by which plant withstands the moisture stress during dry weather is termed drought resistance. Henckel (1964) defined drought resistant plants are those plants which in the process of ontogenesis are able to adapt to the effect of drought and which can normally grow, develop and reproduce under drought conditions because of a number of properties acquired in the process of evolution. Drought resistance is most likely dependent on one of the following components (Sheeba, 2005).

- Moderated use of water through reduced leaf area and shorter growth duration.
- Ability of roots to exploit deep soil moisture to provide for evapotranspirational demand.
- Capacity for osmotic adjustment which allows to retain turgor and protect meristems from extreme dessication.
- Control over non-stomatal water loss from leaves.

Genetics of Drought Tolerance

Drought limits the agricultural production by preventing the crop plants from expressing their full genetic potential. Drought is altogether a complex character determined by the interaction of morphological, physiological and biochemical characters and are both monogenically and polygenically inherited. Incomplete knowledge about genetic basis of drought resistance, negative correlation of drought resistance traits with productivity and unavailability of appropriate genes to obtain transgenic plants are the main constraints for genetic improvement of drought resistance (Mitra, 2001). Knowledge of genetics of drought tolerance is vital for rapid improvement of rainfed agriculture. Mechanisms underlying drought should be thoroughly understood to help manipulate the physiological pathways towards developing better genotypes capable of withstanding drought stress. Crop plants' tolerance to drought is a genetically controlled physiological property.

Breeding for Drought Tolerance

Drought affects various stages of crops in purely rainfed areas. Even in irrigated areas, water deficit is very frequent because of falling water tables. Hence, the breeders should prioritize and design their work that leads to the development of drought tolerant varieties. There is an ample scope to sustain the productivity by evolving new crop varieties which would tolerate drought.

Yield improvement without affecting any other factor under severe drought stress condition is a tedious process and has to be done meticulously in order not to compromise on quality or other aspects of crop growth. Methods normally used for drought tolerance breeding are

1. Introduction
2. Selection
3. Heterosis breeding
4. Mutation breeding
5. Recurrent selection
6. Backcross breeding
7. Synthetics
8. Interspecific hybridisation

Conventional Approach

Selection of plants can be done, during the breeding process, for those physiological and biochemical attributes correlated with yield potential under severe moisture stress conditions. The physiological and biochemical traits that can be utilised are level of proline, glycine betaine, abscissic acid, nitrate reductase, stomatal resistance, water potential and transpiration rate. The perfect relationship between them and yield potential when worked out will help bring a suitable tolerant variety. Several scientists have reported the genetics of proline, ABA, glycine betaine, nitrate reductase in various crop plants at different growth stages. These characters have to be intelligently used by the breeders to identify the donor for drought tolerance for further improvement in yield potential. The drought tolerant maize varieties have been developed using conventional methods; selectively cross breeding strains that cope well in drier conditions and produce good yields for which Department for International Development (DFID) has been awarded the Best Technological Breakthrough at the UK Climate Awards, 2012 (www.cimmyt.org).

Grain yield is the *primary trait*, the trait which is directly affected by drought, for selection in breeding programme for drought tolerance. However, other plant characteristics that are associated with crop yield under drought stress condition are

A *secondary trait* has the following characteristics

- It should be genetically correlated well with grain yield under stress situation.
- It should not be affected by environment and should be highly heritable.
- Enough variation should be present for the trait among the lines.
- It should not be associated with poor yields in the unstressed environment.
- We must be able to measure quickly and economically.

Some of the secondary traits associated to drought tolerance are flowering/maturity date, flowering delay, spikelet fertility, panicle harvest index (PHI), leaf death score and canopy temperature (Lafitte *et al.*, 2003).

Putative traits for drought are those traits that are hypothesized to be of value, generally accepted to be of having significance in selection for drought stress environment. Some of the putative traits for drought tolerance are given below.

Putative traits for drought tolerance

- Deeper, thicker roots
- Osmotic adjustment
- Membrane stability
- Leaf rolling score
- Leaf relative water content
- Water use efficiency
- Root characteristics such as root thickness, depth of rooting, root length, root pulling force, root penetration ability, branching of root system, root to shoot ration, root density & dry weight of roots below 30cm
- Drought Response Index (DRI)

(Fischer et al., 2003, Sheeba, 2005).

Case Study

A drought tolerant variety, ANNA 4, has been developed by crossing Pantdhan10 and IET9911 and selection for drought tolerance with good yield performance. It is a semi dwarf, short duration, non-lodging with high rice recovery (62.1%) and long slender rice. It is having the yield potential of 4.5 tonnes per hectare.

For the past several decades, improvement of drought tolerance in crop plants has been the challenge of researches and industrialists. (Tables 3 and 4).

Table 3: Drought tolerant varieties in major crops.

Crops	Drought tolerant varieties
Rice	TKM 9, TKM 10, TKM 11, PMK 1, PMK 2, PMK 3, PMK 4, MDU 5.
Maize	Navjot, Surya, KH 510, Vikram, D 765.
Sorghum	CO 4, CO 19, CO 20, CO 26, K 3, K 5, K 6, K 7, K 9, K 11, CSH 6, CSH 14, APK 1.
Pearl millet	CO 1, CO 2, X 1 to X 7, CO 4 to CO 7.
Ragi	CO 1 to CO 13, PLR 1, K 1, K 2, TRY 1, Paiyur 1.
Groundnut	CO 2, CO 3, VRI 2, TMV 7.
Sesame	TMV 3, TMV 5, CO 1, VRI 1.
Sunflower	K, Mordern, CO 2, CO 3, CO 4.

Greengram	K 1, Vamban1.
Horsegram	Paiyur 2, CO 1.
Redgram	APK1.
Blackgram	K 1, Vamban 3, Vamban 4.
Cowpea	CO 6, CO 7, Vamban 1.
Cotton	MCU 6, MCU 10, Lakshmi, KCH 2.

(Krishnasamy, 2006)

Table 4: Drought tolerant varieties released in rice around the world.

Country	Name of the variety
India	Sahbhagi Dhan (IR74371-70-1-1).
Phillippines	Sahod Ulan 1 (IR74371-54-1-1).
Nepal	Tarharra 1 (IR84011-B-49-1).

(Mackill et al., 2010)

In Vitro Approach

Plant tissue culture plays a vital role in the production of crop plants as well as in the manipulation of plants for improved agronomic performance. In vitro culturing of plants helps in understanding the various physiological and genetic processes. It also has a great potential to create variability thereby helping the breeder towards the development of improved cultivars (Wani *et al.*, 2010).

The genetic nature of variability in cultured cells is mainly due to the appearance of somaclonal variants with heterosis among the regenerated plants (www.eplantscience.com). Somaclonal variation is a natural phenomenon that occurs during the process of plant tissue culture. It is simply put as all the variations of genetic, cytogenetic and epigenetic during indirect embryogenesis or organogenesis. It is an excellent source of variation for breeders and can be effectively utilised for crop improvement programs. The stable heritable variation in qualitative or quantitative trait can be intelligently used by the breeder for developing improved cultivars. It can be a source of several valuable and desirable traits and lot of varieties have been developed using somaclonal variation in crop plants. Numerous applications in plant science emerged from somaclonal variation and to list a few; super tomatoes with high solid component, removal of oxalate from a leafy vegetable *Amaranthus gangeticus* and increase in oil and citronellon content in the medicinal plant *Citronella java.*

Molecular Breeding for Crop Improvement Under Drought Stress

Plant biotechnology is branch of science that manipulates plants in a way that leads to a desirable character by altering the genetic architecture i.e adding or deleting genes, switching on or off of one or more genes. The careful manipulation of plant genomes is a pre-requisite of a molecular breeder.

Conventional breeding requires large investments in land, labour and capital to screen thousands of progenies and is only partly successful. Moreover, there is only a marginal increase in returns and it necessitates the development of efficient methods or tools to enhance drought tolerance (Serraj *et al.,* 2005).

Conventional breeding, though helpful in developing drought tolerant variety is slow, because the availability of desirable genes are very much limited. Conventional breeding approaches have bottlenecks in the methods with difficulty in transferring drought tolerance character because of species barrier and lack of suitable screening techniques for drought tolerance in the segregating population. Stress tolerance is controlled by variety of genes both additively and synergistically (Muhammad Waseem *et al.,* 2011).

Tolerance to abiotic stress, because of its polygenic nature, is extremely complex and makes the breeders to consume a lot of time to breed suitable varieties. Through conventional breeding methods, scientists were successful in few crops. However, depending solely on conventional breeding strategy will always not yield much to the expectations of the breeder and remains largely unsuccessful in crop improvement for drought stress. Hence, blending of conventional breeding methods and molecular marker based selection will help resolve the problem (Flower and Yeo, 1995, Muhammad Waseem *et al.,* 2011). This approach is further hampered by low heritability and high Genotype × Environment interaction influenced by differences arising from soil heterogeneity and other environmental factors (Mastrangelo *et al.,* 2012).

The high yielding varieties of various crops popular among the farmers are susceptible to a varied extent for drought stress. The genetics of various yield attributing characters including drought tolerance have been studied for the past few years. QTLs have been identified for yield under drought stress and also secondary characters like root characteristics and leaf rolling; different QTLs are being detected in different studies (Lafitte *et al.*, 2003, Bernier *et al.*, 2008 and Mackill *et al.*, 2010).

Modern day technological tools like gene sequencing and bioinformatics has led to the discovery of genes involved in drought tolerance in relatively short period of time. In the global crop gene pool, enormous scope of identifying the genes involved in the physiological pathway leading to drought tolerance is now possible with the advent of molecular breeding and manipulation of pathways by genetic engineering methods for dissecting drought tolerance mechanisms in crop plants is being increasingly used by scientists worldwide.

The use of transgenic technology to improve resistance and tolerance to biotic and abiotic stress was quickly seen as an attractive target for commercial applications. The genetic engineering for developing stress tolerant plants, based on the introgression of genes previously identified as responsive to abiotic stress might assume the major role towards the development of improved plant cultivars (Cancado, 2011).

Manipulation of cellular machinery and activation of certain receptors will help the plants to survive and yield well under drought conditions (Mosquna *et al.*, 2011). Mostly the agronomic traits are quantitative in nature and the gene interaction and physiological basis can help identify the candidate genes for drought tolerance (Forster *et al.*, 2000).

To increase the pace of genetic gain over drought tolerance, marker aided selection can be used. Markers like Random Amplified Polymorphic DNA (RAPD), Restriction Fragment Length Polyphomism (RFLP), Amplified Fragment Length Polymorphism (AFLP), DNA Amplication polymorphism (DAF), Sequence Characterised Amplified Regions (SCARs), Sequence Tagged Sites (STS), Expressed Sequence Tag (ESTs), Amplicon Length Polymorphism (ALPs), Simple Sequence Repeats (SSR) are commonly used.

Manipulation of genetic architecture can be done in two ways viz., constitutive expression and drought stress specific expression of drought tolerance genes. In case of constitutive expression, the major drawback of engineering crops with drought tolerance is the reduction in plant productivity as they are negatively correlated. Hence stress inducible promoters (e.g. ABA inducible promoters) are being used for getting expression of drought tolerance genes only at times of drought (Tawanda Zidenga, 2006).

Conclusion

The success of conventional breeding for drought tolerance is quite obvious by the recently released varieties that perform well under drought stress condition with an yield advantage of 1 tonne per ha. In addition, these varieties also perform well under favourable drought stress conditions. Careful manipulation of plants through conventional, physiological and biotechnological approaches for improved agronomic performance is the need of the hour. The progress in characterizing the genetic determinants of drought resistance has been terribly slow, because it is a complex phenomenon comprising a number of physio-biochemical processes at both cellular and organismic levels at different stages of plant development (Yue *et al.*, 2006). Modern genomics and genetic approaches combined with advances in precise phenotyping and breeding methodologies will help unravel the genes and metabolic pathways that confer drought tolerance in crops (Reyazul Rauf Mir *et al.*, 2012). Continuous efforts have to be made in order to identify new genes/QTLs for stress conditions especially under drought and produce varieties/hybrids with improved performance under drought stress to make the dry land farming community self-sufficient.

References

Bernier J, Atlin, G.N., Serraj, R, Kumar, A. and Spaner, D. 2008. Breeding upland rice for drought resistance. *J. Sci. Food Agric.* 88:927–939.

Cancado Geraldo Magela de Almeida. 2011. The Importance of Genetic Diversity to Manage Abiotic Stress. Abiotic Stress in Plants - Mechanisms and Adaptations, Prof. Arun Shanker (Ed.), ISBN: 978-953-307-394-1.

FAO. 2002. World Agriculture: towards 2012-2030, Summary report. Food and Agriculture Organization, Rome, Italy.

Fischer, K.S., Lafitte, R., Fukai, S., Atlin, G. and Hardy, B. (eds). 2003. Breeding rice for drought-prone environments. International Rice Research Institute, Los Baños, Philippines.

Flowers, T,J. and Yeo, A.R. 1995. Breeding for salinity resistance in crop plants: where next?. *Aust. J. Plant Physiol.* 22: 875–884.

Forster, B.P., Ellis, R.P., Thomas, W.T.B., Waugh, R., Ivandic, V., Tuberosa, R., Talame, V., This, D., Teulat-Merah, B., El-Enein, R.A., Bahri, H., Ben Salem, M. 2000: Research developments in Genetics of drought tolerance in Barley. Proceedings of 8th Int. Barley Genetic Symposium, pp. 253-259.

Geraldo Magela de Almeida Cançado. 2011. The Importance of Genetic Diversity to Manage Abiotic Stress. In: Abiotic Stress in Plants – Mechanisms and Adaptations. (eds.) Arun Kumar Shanker and B. Venkateswarlu. pp 351-366. published by Intech, Croatia.

Athar, H.R. and M. Ashraf. 2009. Strategies for Crop Improvement Against Salinity and Drought Stress: An Overview. In: Salinity and Water Stress: Improving Crop Efficiency. (eds.) M. Ashraf, M. Ozturk and H.R. Athar pp 1-16. Springer Publications.

Henckel, P.A. 1964. Physiology of Plants Under Drought. *Annual Review of Plant Physiology*. 15: 363-386.

Krishnasamy, S. 2006. Twenty Five Years of Dryland Agricultural Research. Regional Research Station, TNAU, Aruppukkottai. P 75.

Lafitte, R., Blum, A. and Atlin, G. 2003. Using secondary traits to help identify drought tolerant genotypes. In: Breeding rice for drought-prone environments.

Mackill, D.J., A.M. Ismail, A. Kumar and G.B. Gregorio. 2010. "The role of stress-tolerant varieties for adapting to climate change." Paper from the CURE Workshop on Climate Change, May 4, 2010, Siem Reap, Cambodia.

Mastrangelo, A.M., Elisabetta Mazzucotelli, Davide Guerra, Pasquale De Vita and Luigi Cattivelli. 2012. Improvement of Drought Resistance in Crops: From Conventional Breeding to Genomic Selection. In: Crop Stress and its Management: Perspectives and Strategies, Pages 225-259.

McNeil, S.D., Michael L. Nuccio and Andrew D. Hanson. 1999. *Plant Physiology*. 120: 945-949.

Mitra J. 2001. Genetics and genetic improvement of drought resistance in crop plants. *Curr. Sci*. 80:758-762.

Mosquna, A. Francis C. Petersonb, Sang-Youl Parka, Jorge Lozano-Justea, Brian F. Volkmanb, and Sean R. Cutler. 2011. Potent and selective activation of abscisic acid receptors in vivo by mutational stabilization of their agonist-bound conformation. *PNAS*, 108: (51): 20838-20843.

Muhammad Waseem, Asghar Ali, M. Tahir, M. A. Nadeem , M. Ayub, Asif Tanveer, R. Ahmad and M. Hussain. 2011. Mechanism of Drought Tolerance in Plant and its Management Through Different Methods. 2011. *Continental J. Agricultural Science*, 5 (1): 10-25.

Reyazul Rauf Mir, Mainassara Zaman-Allah, Nese Sreenivasulu, Richard Trethowan and Rajeev K. Varshney. 2012. Integrated genomics, physiology and breeding approaches for improving drought tolerance in crops. *Theor. Appl. Genet.* 125(4):625-45.

Serraj, R., T.C. Hash, H.K. Buhariwalla, F.R. Bidinger, R.T. Folkertsma, Subhash Chandra, P.M. Gaur, Junichi Kashiwagi, S.N. Nigam, A. Rupakula and J.H. Crouch. 2005. Marker-assisted breeding for crop drought tolerance at ICRISAT: Achievements and prospects. In: Proceedings of the International Congress "In the Wake of the Double Helix: From the Green Revolution to the Gene Revolution" Tuberosa R., Phillips R.L., Gale M. (eds.), 27-31 May 2003, Bologna, Italy, 217-238.

Sheeba. 2005. Genetic Studies on Drought Tolerance and Stability of Temperature Sensitive Genic Male Sterility (TGMS) Based Rice Hybrids. Ph.D. Thesis. TNAU, Coimbatore.

Soren, R.K. Kishwar Ali, Vandana Tyagi and Aruna Tyagi. 2010. Recent Advances in Molecular Breeding of Drought Tolerance in Rice (*Oryza sativa* L.). *Indian Journal of Biotechnology*, 9: 233-251.

Tawanda Zidenga. 2006. Progress in molecular approaches to drought tolerance in crop plants. ISB News Report. March 2006 Issue.

Viets, F.G. Jr. 1972. Water deficits and nutrient availability. In: T.T.Kozlowski (ed). Water deficits and plant growth, pp.217-239, New York. Academic Press.

Wani, S.H., Sofi, P.A., Gosal and Singh, N.B. 2010. In vitro screening of rice (*Oryza sativa* L) callus for drought tolerance. *Communications in Biometry and Crop Science*.5 (2): 108–115.

Yancey, P. H. 1994. Compatible and counteracting solutes. In K. Strange (ed.), Cellular and molecular physiology of cell volume regulation, pp. 81–109. CRC Press, Boca Raton.

Yue, B., Xue, W., Xiong, L., Yu, X., Luo, L., Cui, K., Jin, D., and Zhang, Q. 2006. Genetic basis of drought resistance at reproductive stage in rice: Separation of drought tolerance from drought avoidance. *Genetics*, 172: 1213-1228.

❑❑❑

Chapter 11

Adaptive Mechanism of Drought Tolerance in Cassava (*Manihot esculenta Crantz.*)

K. NAGESWARI, M.K. KALARANI, P. KALAISELVI, K. MANIVANNAN AND V. PALANISAMY
Tapioca and Castor Research Station, Tamil Nadu Agricultural University,
Yethapur, P.G. Palayam-636 119, Salem, Tamil Nadu, India
E-mail : nageswarihort@yahoo.co.in

Introduction

The crop productivity is subjected to number of stresses and potential yields are seldom achieved with stress. All crops grown under natural environment are subjected to one or the other stresses. Abiotic stresses include drought, heat, flood, salinity, mineral deficiency, toxicity, and chilling or freezing stress. Abiotic stresses are linked with natural phenomenon and their scale varies at temporal and spatial dimension. Increasing temperature and change in rainfall both intensity and distribution pattern are two factors which affect agriculture.

According to Turner (1986), plants that are able to endure long periods of water shortages while maintaining a high tissue water potential are called drought tolerant. Cassava may fit among these types of plants. Nevertheless, Alves (2002), working with indoor-grown plants, found no significant accumulation of solutes and osmolytes in mature cassava leaves, and, hence, there was no occurrence of osmotic adjustment. This finding further confirms that cassava stomatal control over plant water relations is the predominant defense mechanism protecting the leaf from severe dehydration, and, hence, it can be considered a stress avoidance mechanism (El-Sharkawy, 2006a).

Drought stress is the most important environmental factor influencing crop growth, development and yield process. Drought influences on array of processes including physiological, growth, development, yield and quality of crop. The objective of this chapter is to provide an overview of the influences of drought stress on the above processes independently and in combination. This chapter suggests a clear need on information on interaction effects of stress particularly drought with relation to cassava. Some of the important traits of cassava associated with drought tolerance is indicated and discussed below.

In Tamil Nadu, cassava is being cultivated both in hills and plains under irrigated and rainfed conditions. In Tamil Nadu the crop faces drought up to 6 months during the initial growth period. When planted under rainfed conditions, failure of summer, pre-monsoon showers or delay in onset of monsoon causes a postponing of cassava planting (as in hilly areas of Tamil Nadu). This results in shortening of crop growing period and eventually reducing tuber/starch yield. In both Tamil Nadu and Andhra Pradesh, contingent drought (failure of rain during normal monsoon season) attributes for reduction in tuber yield. Furthermore, prolonged drought (> 2 months) in combination with high temperature (>33-40°C) tend to reduce starch content of cassava tubers as these stress conditions may tend to convert starch to lignin.

Although cassava is considered to be tolerant to drought conditions, significant reduction in tuber yield as well as in starch content occurs and varieties vary in their response to water deficit stress/high temperature stress conditions. Mild water deficit stress (WDS) is favourable for tuber growth. Cassava requires adequate soil moisture during initial one month for its establishment. In cassava, tuber (modified storage root) initiates within 1 month after planting and tuber bulking starts at 2 or 3 months after planting under adequate soil moisture conditions. Drought during this growth period delays tuber initiation as well as tuber bulking.

Various studies conducted in Central Tuber Crops Research Institute (CTCRI) on the response of cassava to drought under field conditions revealed a reduction in photosynthetic rate, total dry matter production and tuber yield under WDS conditions. About 28–42% of tuber yield was found to be reduced under drought conditions during last 4 months of crop growth period. However cassava was found to maintain nearly 50% of its photosynthetic rate under drought conditions offering a better choice of crop in future climate warming conditions. In CTCRI, 62 genotypes and 109 land races of cassava were evaluated for drought tolerance under field conditions based on total dry matter (DM) production, DM partitioning, LAI, light interception, photosynthetic rate, stomatal conductance/resistance, transpiration rate, PEG stress, % of yield reduction, tuber yield and starch content. From this study, 5 genotypes (CE-54, CE-534, CI-260, CI-308 and CI-848) and 9 land races (129, 7, 16, TP White, Narukku-3, Ci-4, Ci-60, Ci-17, Ci-80) were identified drought tolerant. High drought tolerance and productivity is associated with leaf retention during drought (El-Sharkawy *et al.,* 1992; Osiru *et al.,* 1990). Using the *ipt* gene encoding cytokinin production under the control of a senescence regulated promoter for *Arabidopsis thaliana,* transformed cassava plants were produced which showed high leaf retention.

Physiological Mechanisms

Physiological characteristics of cassava tolerance to prolonged drought in the tropics was studied by Mabrouk A. El-Sharkawy (2007). The high photosynthetic capacity of cassava leaves in favorable environments and the maintenance of reasonable photosynthetic rates throughout prolonged water deficits are the important characteristics for high and sustainable productivity of cassava. Cassava possess a tight stomatal control over leaf gas exchange that reduces water losses when plants are subjected to soil water deficits as well as to high atmospheric evaporative demands, thus protecting leaves from severe dehydration. During prolonged water deficits, cassava reduces its canopy by shedding their older leaves and forming smaller new leaves leading to less light interception, which is also an another adaptive trait to drought tolerance. Though root yield is reduced under water stress, the crop can recover when water becomes available by rapidly forming new canopy leaves with much higher photosynthetic rates compared to unstressed crops, thus compensating for yield losses with final yields approaching those in well-watered crops. Cassava can extract slowly water from deep soils, a characteristic of paramount importance in seasonally dry and semiarid environments where deeply stored water needs to be tapped.

Cassava was the first cultivated crop to be considered as a C_3-C_4 intermediate species based on:

1. A typical leaf anatomy including the presence of conspicuous thin-walled bundle-sheath cells with large granal chloroplasts, which are less developed than those in the typical C_4 Kranz leaf anatomy (e.g., El-Sharkawy and Hesketh,1965, 1986; Laetsch, 1974).

2. Close physical association of chloroplasts with numerous mitochondria and peroxisomes in bundle-sheath and mesophyll cells.
3. Low photorespiration (as determined by CO_2 release from illuminated leaves in a rapid stream of CO_2-free air, less than 10% of net photosynthesis) and low CO_2 compensation point (? = 20-30 cm^3 m^{-3}).
4. Elevated activities of the C_4 phospho*enol* pyruvate carboxylase (PEPC) in leaf extracts (10-30% of activities in maize and sorghum).
5. High percentage (30-60%) of ^{14}C labeling in C_4 dicarboxylic acids after 5-10 hrs of exposure under illumination.
6. Immunological analysis and DNA hybridization of PEPC from cassava and wild *Manihot* species against antibodies and *ppc* probes from maize (El-Sharkawy and Cock, 1987a; Cock *et al.*, 1987; Riaño *et al.*, 1987a,b; El-Sharkawy *et al.*, 1989; El-Sharkawy and Cock, 1990; Bernal, 1991; López *et al.*, 1993; Aguilar, 1995; El-Sharkawy, 2004, 2006; El-Sharkawy and de Tafur, 2007).

These characteristics collectively underpinned high photosynthetic rate in normal air (P_N > 40 μmol CO_2 m^{-2} s^{-1}) in high irradiances (> 1800 μmol m^{-2} s^{-1} of PAR), high leaf temperature from 30 to 40 °C, and in high atmospheric humidities observed in cassava grown in favorable environments (El-Sharkawy *et al.*, 1992a, 1993). Cassava also tolerates prolonged drought that often exceeded five months aided by partial closure of stomata, deep rooting systems and small leaf canopy.

Studies conducted at International Centre for Tropical Agriculture(CIAT) demonstrated the ability of cassava to assimilate carbon at very high rates under high levels of humidity, temperature and solar radiation, which correlates with productivity across all environments whether dry or humid. When it is grown in very poor soils under prolonged drought for more than 6 months, the crop reduce both its leaf canopy and transpiration water loss, but its attached leaves remain photosynthetically active, though at greatly reduced rates. The main physiological mechanism underlying this tolerance was rapid stomatal closure under both atmospheric and edaphic water stress, protecting the leaf against dehydration while the plant depletes available soil water slowly during long dry periods. This mechanism leads to high crop water use efficiency values. Although the cassava fine root system is sparse, compared to other crops, it can penetrate below 2 m soil, thus enabling the crop to exploit deep water if available. Leaves of cassava and wild Manihot possess elevated activities of the C_4 enzyme PEP carboxylase but lack the leaf Kranz anatomy typical of C_4 species (Mabrouk and El-Sharkawy 2007).

1. Capacity of cassava leaves to fix atmospheric carbon dioxide

The physiological mechanisms that underlie cassava's tolerance of severe water stress is the capacity of cassava leaves to fix atmospheric CO_2, a basic process for dry matter accumulation, during the stress period was 80% of that in the control(Fig 1.). This indicated that the photosynthetic process in cassava is not greatly inhibited by prolonged stress.

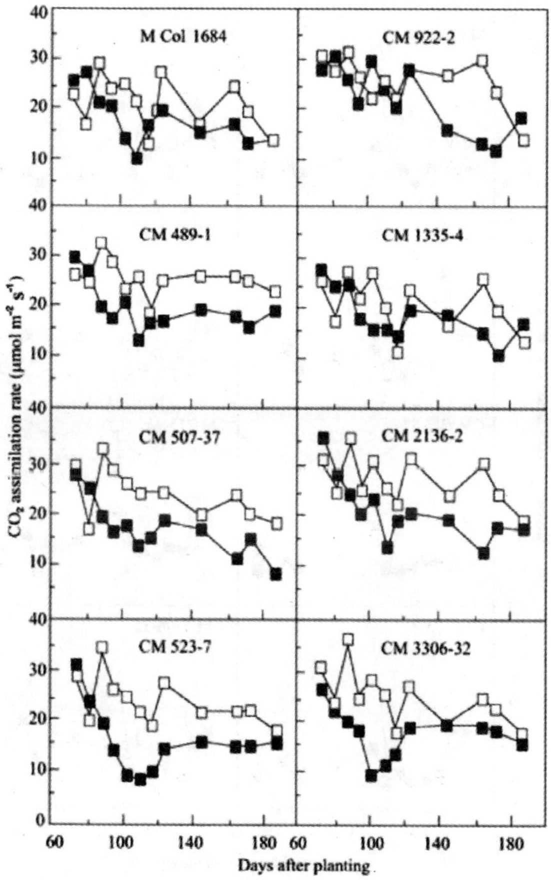

Fig. 1: Response of cassava leaf photosynthesis to prolonged water stress (120 days) imposed at 60 days after planting (Control; Stress).
Source : CIAT 1987-1989 report.

2. *Partial closure of stomata*

A second and important physiological mechanism of cassava leaves is their ability to partly close their stomata in response to water stress. In an experiment conducted at CIAT 1987-1989), there was an average reduction of 43% in leaf conductance to water vapor in stressed plots and consistent reduction over the stress period, compared to the control (Fig 2.). The partial closure of stomata enabled cassava leaves to maintain, to some extent, the midday leaf water potential at levels comparable with those of cassava leaves in the control plot. Leaf water potentials at 1400 h across all varieties were -1.13 and -1.12 MPa for the control and stressed leaves, respectively. This is another comparative advantage for cassava in dry areas, compared to other crops with poorer stomatal control.

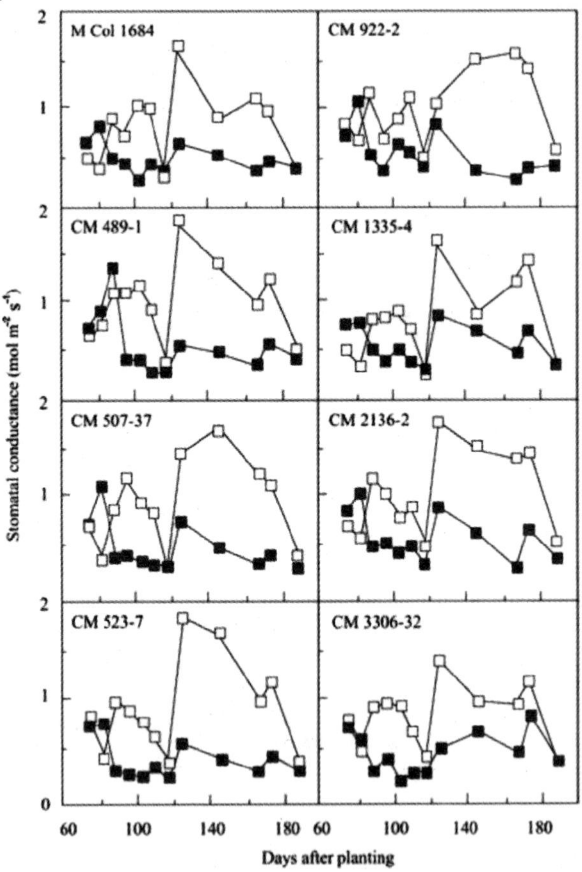

Fig. 2: Response of cassava stomata to prolonged water stress (120 days) imposed at 60 days after planting (Control; Stress).
Source : CIAT 1987-1989 report.

In addition to its beneficial effect by preventing severe leaf dehydration, and consequently preventing impairment to photosynthetic capacity of the leaf, the partial closure of stomata reduces water loss through transpiration, thereby maximizing water-use efficiency. Across all varieties, there was 39% increase in intrinsic leaf water use efficiency in stressed crops over the control.

3. *Ability to maintain predawn leaf water potential*

A third and equally important physiological mechanism that enables cassava to withstand severe water stress is its ability to maintain a predawn leaf water potential comparable with that of unstressed cassava. In an experiment, leaf water potentials at 0600 h across all varieties were -0.39 and -0.40 MPa for the control and stressed plants, respectively. This was partly achieved by reducing total leaf area as indicated by the 31% reduction in light interception in the stressed plot. Thereby reducing total canopy transpiration, and by slow withdrawal of water from the deeper layers of the soil (Connor *et al.*, 1981; Porto, 1983; CIAT, 1987-1989; El-Sharkawy and Cock, 1987b; El-Sharkawy *et al.*, 1992b; de Tafur *et al.*, 1997a; Cadavid *et al.*, 1998; El-Sharkawy, 2006). During water stress, cassava fine roots extend for more than 2 m into deeper, wetter soil from where cassava can extract between 20% and 40% of its total water uptake (CIAT, 1987-1989; El-Sharkawy *et al.*, 1992b).

4. *Ability of cassava to compensate partly for previous losses in dry matter production.*

A fourth mechanism underlying tolerance to drought is the ability of cassava to compensate partly for previous losses in dry matter production, due to water stress, by an increase in leaf canopy area (El-Sharkawy and Cock, 1987b; El-Sharkawy *et al.*, 1992b) and by higher photosynthetic rates in the newly developed leaves after recovery, as compared to the unstressed plants (Fig 3.) (El-Sharkawy, 1993, Cayón *et al.*, 1997; de Tafur *et al.*, 1997a; El-Sharkawy, 2006). These higher photosynthetic rates in new leaves of previously stressed cassava were associated with higher leaf conductances, higher nutrient contents, as well as with stronger sinks for carbohydrate in storage roots (Cayon *et al.*, 1997).

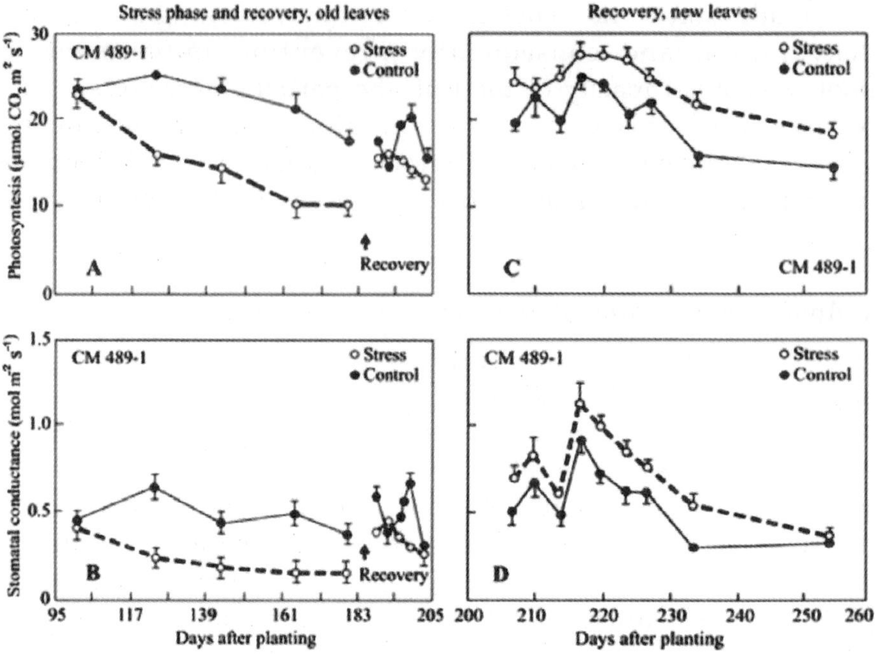

Fig. 3: Responses of field grown cassava to prolonged midseason water stress. Plants recovered from stress (indicated by arrows in A and B) under rainfall and supplemented irrigation for the rest of the growth cycle. The control plants received normal rainfall plus supplemented irrigation. A. Leaf Photosynthesis during stress and recovery for leaves deprived under stress. B. Stomatal conductance .C. and D. Photosynthesis and stomatal conductance for new leaves developed after stress was terminated.
Source : CIAT, Cassava physiology section database (1991) Annual report, El–Sharkawy(1993).

Not only can cassava tolerate long periods of soil water deficits aided with the above-mentioned inherent mechanisms, but it can also react to changes in atmospheric humidity (Connor and Palta, 1981; El-Sharkawy and Cock, 1984, 1986, 1990; El-Sharkawy *et al.*, 1984, 1985, 1989; Cock *et al.*, 1985; Berg *et al.*, 1986; El-Sharkawy, 1990, 1993, 2004, 2006; Oguntunde, 2005, Oguntunde and Alatise, 2007). Cassava leaf stomata are sensitive to air humidity, irrespective of soil water content; they close rapidly in dry air when evaporation is high under field conditions, which may be translated into high leaf water potential. This mechanism enables cassava to maximize its WUE during periods of prolonged drought. When air humidity is high (e.g., early in the morning and during rainy periods), the stomata remains open.

Thus, in a humid atmosphere and in the presence of soil water deficits, cassava leaves remain photosynthetically active and the crop can produce well.

Under prolonged water deficit condition (WDS>20 days), both vegetative growth as well as tuber bulking ceases, plants shed about 80% of leaves and become dormant. When conditions become favorable (adequate soil moisture), the plant resurrects and resumes its growth, develop canopy consuming considerable amount of starch from thetubers and the tuber continue to bulk.

Cassava being a C_3 plant has a unique, partial Krantz anatomical feature in leaf. It has 15-20 % phosphoenol pyruvate carboxylase (PEP case) activity of maize and sorghum.The PEP case activity of cassava is greater than typical C_3 plants and equivalent to that of C_3-C_4 intermediates. The PEP case activity of cassava increased by 13% whereas the RUBPcase activity decreased by 42% under water deficit stress. Thus, cassava possess a greater advantage to respond to elevated CO_2 concentrations. Elevated CO_2 has been also reported to alter and partition more dry matter towards root system (in other crops) which may provide an additional advantage for cassava. Despite this, cassava is unable to take the advantage of the rising atmospheric CO_2 because of their sensitivity to water deficit and heat stress condition. In cassava sink capacity is important for achieving maximum tuber yield. Because elevated CO_2 has been found to alter source- sink relationship, diverting more dry matter to underground organs or producing feed back effects via a carbohydrate built up. Sink strength is an important factor as it has to accommodate the excess carbohydrate formed due to CO_2 enrichment. Therefore, identification/ development of new genotypes with high sink potential is necessary.

Tim L. Setter *et al.*(2007) conducted an experiment to determine which traits contribute to its maintenance of yield in water stress environments (ws) in cassava. Results showed that in controlled environments in cassava, stomata close in response to slight decreases in leaf water status and maintain leaf water potential at values near those of well-watered controls (ww).This is associated with rapid and large increases in abscisic acid (ABA) (Fig.4).

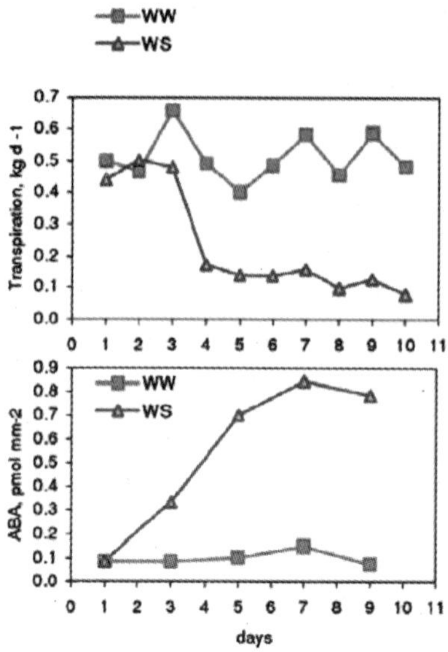

Stomata close on Day4 after soil water is depleted

ABA is rapidly produced at an early phase of turgor

Fig. 4: Stomata are responsive to small decreases in water status.
Source : Tim L. Setter et al (2007).

Also, as stress continues, a substantial fraction of leaves abscise, thereby decreasing transpirational surface area and further conserving water during stress periods. New leaf production and expansion growth is also highly sensitive to water deficit, due to inhibition of leaf cell division and cell expansion. However, growth recovers rapidly after renewed water supply, thereby permitting rapid re-establishment of leaf area (Fig 5).

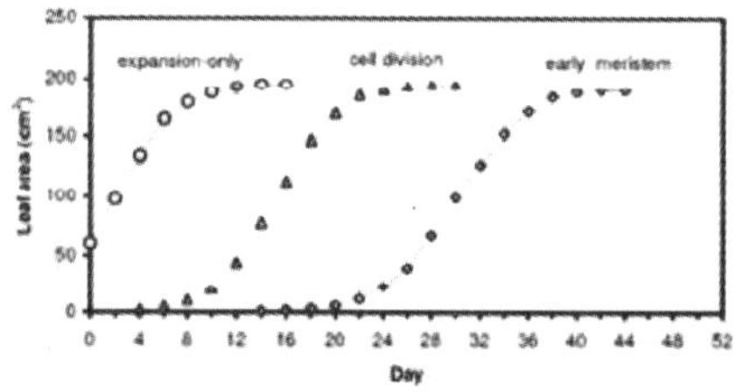

Representative leaves were examined during each of three phases in leaf development

Stressed during final phase of expansion

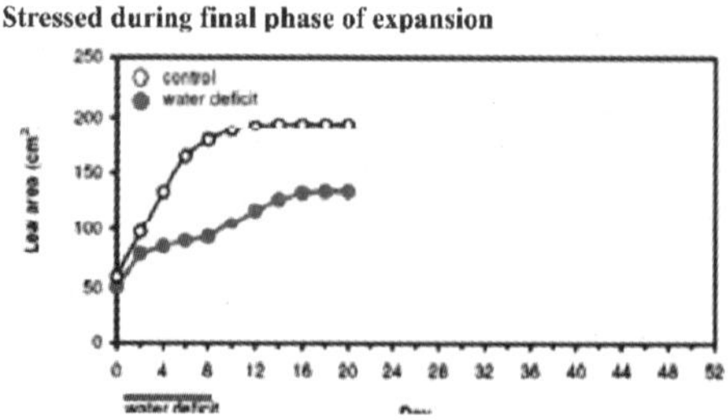

- Halt expansion growth during stress
- Recovery (partial) after rewatering

contd....

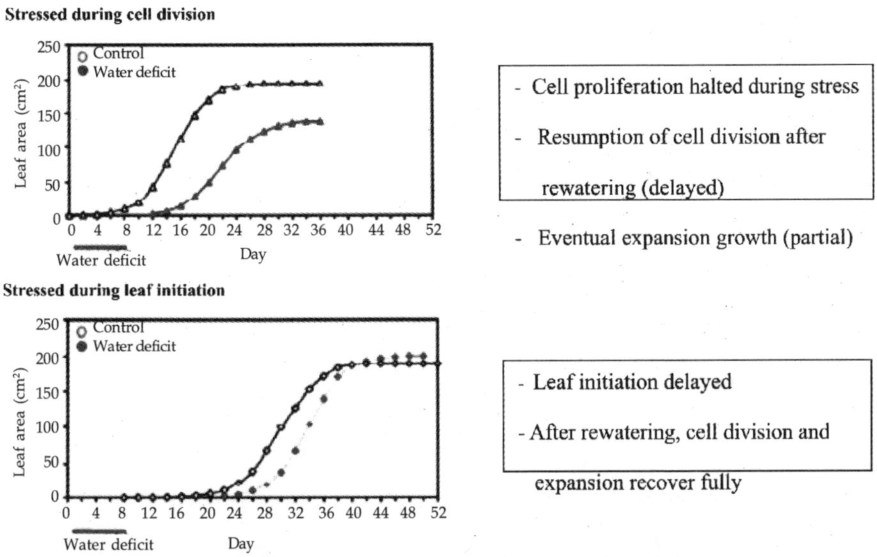

Fig. 5: Flexible leaf growth: responsive to water loss, but capable of rapid recovery. Source : Tim L. Setter et al. (2007).

Carbon use is down regulated by limiting growth. Accumulation of sugars and other osmotically active solutes is not substantially stimulated by stress (Fig 6.).

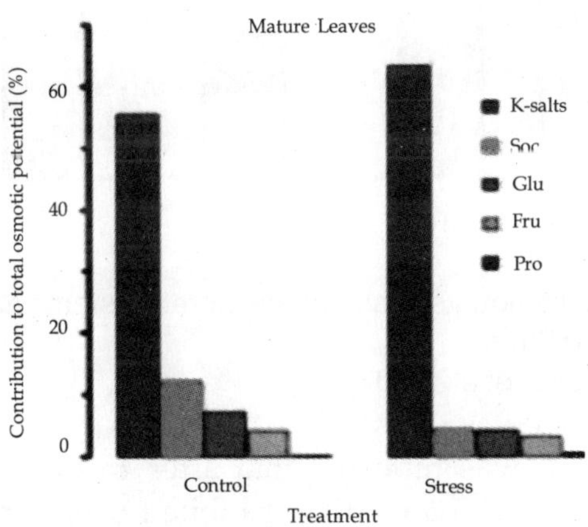

Fig. 6: Osmoticum accumulation, primarily K+, is limited. Source : Tim L. Setter et al. (2007).

Also, petiole and stem carbohydrate reserves are gradually utilized and translocated to sinks throughout the plant. The amount of starch stored in stems is considerable, representing a large share of the total non-structural carbohydrate in a plant at the initial period of storage root growth (Fig 7).

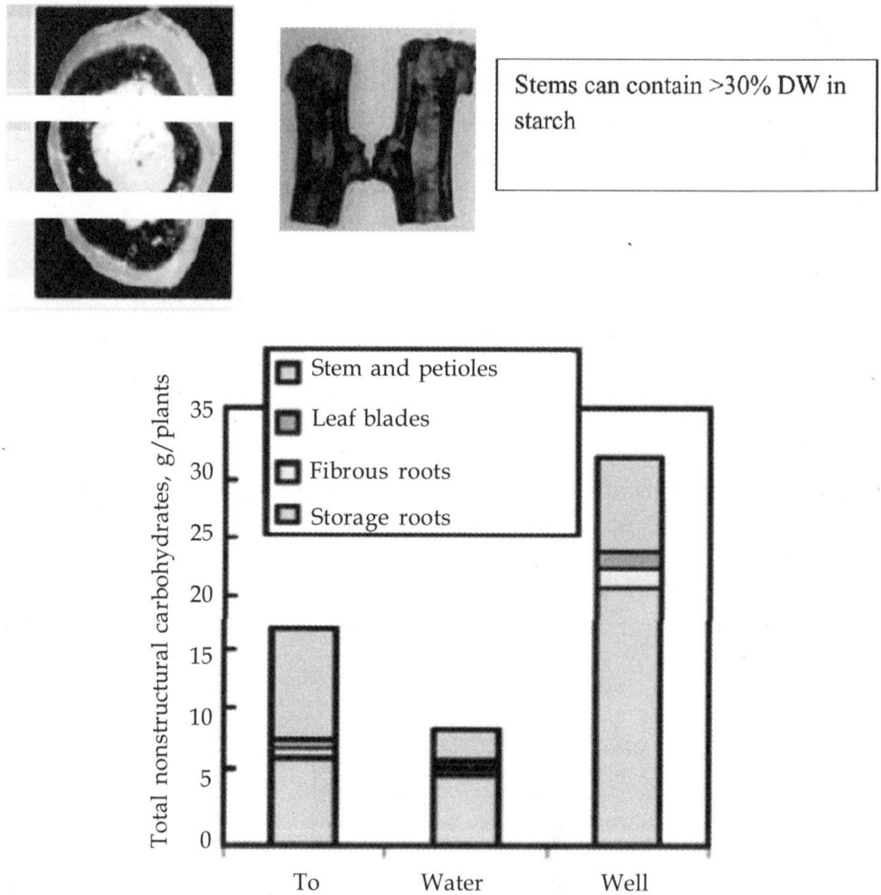

Fig. 7: Substantial amounts of starch are stored in stems and petioles which is remobilized during Stress.
Source : Tim L. Setter et al. (2007).

In a study, it was reported that after 45 days of limited water supply, *Manihot esculenta* Crantz, exhibited pronounced reduction in shoot growth, high leaf fall and decreased stomatal conductance. However, the water status of the remaining leaves was unaffected. This was combined with an amplified heliotropic response and drooping which minimizes radiant energy interception at mid-day, suggesting that leaves are sensitive to high irradiance (*I*). In well-irrigated plants, CO_2-saturated oxygen evolution and net photosynthetic rate (P_N) in air were markedly higher (5- fold) in young (expanding) leaves than in mature leaves. Water limitation did not strongly modify CO_2 -saturated oxygen evolution but it altered P_N in air for both types of leaves, although differently. The mature leaves of drought- adapted plants displayed residual rate of P_N and deteriorated photosystem 2 (PS2). In young leaves at moderate *I*, P_N was depressed by only 66% in stressed plants. Moreover, the photochemical quenching of Chl *a* fluorescence and the quantum efficiency of PS2 photochemistry in young leaves were comparable in both control and stressed plants. In contrast at high *I*, P_N was almost null and marked decreases in the two fluorescence parameters were apparent. Hence the strong heliotropic response and drooping displayed by young leaves under water limitation is an important strategy for avoiding inactivation of P_N by high *I* and therefore for cassava tolerance to drought (Calatayud *et al.*, 2000).

To study the impact of water stress on fresh tuber yield and dry matter content of cassava (*Manihot esculenta* Crantz) in Cote d'lvoire, experiments were carried out in the experimental field where the soils are moderately unsaturated ferrallitic sandy soil and characterized by high percentages of fine sands and a low thickness of humus. Fourteen (14) improved cassava varieties and one (1) local cultivar control were used as planting material and planted in June 06th, 2002 and on September 12th, 2002. June plantings provided root yields higher than those obtained with September plantings (Bakayoka *et al.* 2009). This result is in agreement with the observations according to which the highest fresh roots corresponding to planting dates were due to highest total amount of water received (Santisopasri *et al.*, 2001). According to Lal (1980), the excess of ground moisture can unfavorably affect the development of the root growth, water consumption and its useful effect. The unfavorable effects are accentuated when the root development is inhibited by the strong compactness of the ground or its low total porosity. The conditions immediately before harvesting the roots are important and the effects are demonstrated in root dry matter content and in root dry matter yield. Both root dry matter content and root dry matter yield are highest when the roots are harvested in September at the beginning of the small rainy season. It is obvious that environmental conditions during both the initial and late growth are crucial to root productivity. Percentage root dry matter content is high when the water stress does not exceed one period in the first 6 months. This study confirms the importance of rainfall during cassava growth, especially in the early development stage (Sriroth *et al.*, 1999). The distribution pattern of dry matter among the different organs of cassava plant change markedly during the growth cycle, with shoot having a dominance in the first 3-5 months while storage roots become major sink for photo assimilates during the rest of the growth cycle (El-Sharkawy, 2004).

It is apparent that water stress affects the root dry matter yield which is low when roots are harvested in December or March. Those two periods are preceded by two months considered like water stress periods. The root dry matter yield is highest with June planting when roots are harvested in September. Before September, the two months are not water stress period. That suggests that the root dry matter yield decreases under water stress conditions. This is in agreement with Schulthess *et al.* (1991) who observed that the effect of drought caused the breaking of apical dominance, leading to lateral shoot formation which use reserves from roots and stems. A high percentage root dry matter in cassava is important. It also shows that the delay in harvesting after the physiological maturity of the crop may thus be done to take advantage of this additional yield increase. Although cassava can remain in the ground and be harvested progressively with the needs, harvest in only once in 12 months, after September planting. Indeed , the reduction of the leaf area and the stomata closing slow down the growth rate of plant during the drought periods (Lenis *et al.* 2006). The leaf area index most favourable for the accumulation of the dry matter ranges between 3 and 5 (Osiru, 1990). When the foliar index exceeds these values, the tubers growth rate drops, in particular in consequence of the shade. The limiting factors to the dry matter accumulation are especially stimulated when the vegetation conditions, in particular the rainfall and the sun light, are not moderated just before the harvest period.

Cassava varieties CM 507-37, M Col 1684(EL-Sharkawy and Cock, 1987a) and M.Mex 59, M Col 22 (Connor *et al.*, 1981) were compared with control under stress for a period of 2-3 months. The treatments were imposed after 3-4 months of planting. The final root yields of the stressed crop were increased in vegetatively vigorous varieties (eg. M.Mex 59) and reduced in less vigorous varieties (M Col 22) but slightly reduced or remained unchanged in varieties with intermediate vigour (eg. the parent M Col 1684 and its hybrid CM 507-37).

These responses were closely related to leaf area characteristics (i.e., peak LAI and leaf area duration over the growing cycle), and to patterns of photo assimilate partitioning between top growth (stems and leaves) and storage roots. The vigorous types responded positively to stress because top growth was reduced and the harvest index was increased, whereas the less vigorous types responded negatively because leaf area was drastically reduced to levels below optimal LAI for root yield (EL-Sharkawy and Cock, 1987b).

Osmolyte Accumulation

Plants accumulate different types of organic and inorganic solutes in the cytosol to lower osmotic potential thereby maintaining cell turgor (Rhodes and Samaras, 1994). Under drought, the maintenance of leaf turgor may also be achieved by the way of osmotic adjustment in response to the accumulation of proline, sucrose, soluble carbohydrates, glycinebetaine, and other solutes in cytoplasm improving water uptake from drying soil. The process of accumulation of such solutes under drought stress is known as osmotic adjustment which strongly depends on the rate of plant water stress. Of these solutes, proline is the most widely studied because of its considerable importance in the stress tolerance. Proline accumulation is the first response of plants exposed to water-deficit stress in order to reduce injury to cells. The proline content increase as the drought stress progressed and reached a peak as recorded after 10 days stress, and then decreased under severe water stress as observed after 15 days of stress (Anjum *et al.*, 2011b).

Proline can act as a signaling molecule to modulate mitochondrial functions, influence cell proliferation or cell death and trigger specific gene expression, which can be essential for plant recovery from stress (Szabados and Savoure, 2009). Accumulation of proline under stress in many plant species has been correlated with stress tolerance, and its concentration has been shown to be generally higher in stress-tolerant than in stress-sensitive plants. It in?uences protein solvation and preserves the quarternary structure of complex proteins, maintains membrane integrity under dehydration stress and reduces oxidation of lipid membranes or photoinhibition (Demiral and Turkan, 2004). Furthermore, it also contributes to stabilizing sub-cellular structures, scavenging free radicals, and buffering cellular redox potential under stress conditions (Ashraf and Foolad, 2007).

Biochemical Mechanisms

Reactive oxygen species (ROS)

The generation of reactive oxygen species (ROS) is one of the earliest biochemical responses of eukaryotic cells to biotic and abiotic stresses. The production of ROS in plants, known as the oxidative burst, is an early event of plant defense response to water-stress and acts as a secondary messenger to trigger subsequent defense reaction in plants. ROS, which include oxygen ions, free radicals and peroxides, form as a natural byproduct of the normal metabolism of oxygen and have important role in cell signaling. However, during environmental stress such as drought, ROS levels increase dramatically resulting in oxidative damage to proteins, DNA and lipids (Apel and Hirt, 2004). Being highly reactive, ROS can seriously damage plants by increasing lipid peroxidation, protein degradation, DNA fragmentation and ultimately cell death.

Drought induces oxidative stress in plants by generation of reactive oxygen species (ROS) (Farooq *et al.,* 2009). The ROS such as O_2-, H_2O_2 and •OH radicals, can directly attack membrane lipids and increase lipid peroxidation (Mittler, 2002). Drought-induced overproduction of ROS increases the content of malondialdehyde (MDA).The content of MDA has been considered as an indicator of oxidative damage (Moller *et al.,* 2007). MDA is considered as a suitable marker for membrane lipid peroxidation. A decrease in membrane stability reflects the extent of lipid peroxidation caused by ROS. Furthermore, lipid peroxidation is an indicator of the prevalence of free radical reaction in tissues. Moreover, oxygen uptake loading on the tissues as both processes generate reactive oxygen species, particularly H_2O_2 that produced at very high rates by the glycollate oxidase reaction in the peroxisomes in photorespiration.

Antioxidant Enzymes

There is a defensive system in plants, that is to say, plants have an internal protective enzyme-catalyzed clean up system, which is fine and elaborate enough to avoid injuries of active oxygen, thus guaranteeing normal cellular function (Horvath *et al.,* 2007). The balance between ROS production and activities of antioxidative enzyme determines whether oxidative signaling and/or damage will occur (Moller *et al.,* 2007). To minimize the effect of oxidative stress, plants have evolved a complex enzymatic and non-enzymatic antioxidant system, such as low-molecular mass antioxidants (glutathione, ascorbate, carotenoids) and ROS-scavenging enzymes (superoxide dismutase (SOD), peroxidase (POD), catalase (CAT), ascorbate peroxidase (APX) (Apel and Hirt, 2004). Non-enzymatic antioxidants cooperate to maintain the integrity of the photosynthetic membranes under oxidative stress. The enzymatic components may directly scavenge ROS or may act by producing a non-enzymatic antioxidant. The capability of antioxidant enzymes to scavenge ROS and reduce the damaging effects may correlate with the drought resistance of plants.

Molecular Mechanisms

A study was conducted to find out the tools for under standing and manipulating drought tolerance in cassava. Expressed sequence tags (ESTs) were generated from normalized cDNA libraries prepared from dehydration-stressed and control well-watered tissues. Totally 18,166 ESTs were analyzed and identified 5383 singletons and 3194 clusters. Functional categories could be assigned to 63 % of the unigenes while another-11% were homologous to hypothetical genes with unclear functions. The remaining-26% were not significantly homologous to sequences in public databases suggesting that some may be novel and putatively specific to cassava. The dehydration-stressed library uncovered numerous ESTs with recognized roles in drought-responses including those that encode late-embryogenesis-abundant proteins thought to confer osmoprotective functions during water stress, transcription factors, heat shock proteins as well as proteins involved in signal transduction and oxidative stress. The unigene clusters were screened for short tandem repeats for further development to microsatellite markers. A total of 592 clusters contained 646 repeats, representing 3.3% of the ESTs queried. The ESTs presented here are the first dehydrated stress transcriptome of cassava and can be utilized for the development of micro arrays and gene derived molecular markers to further dissect the molecular basis of drought tolerance in cassava (Lokko *et al.*, 2007).

In a study to identify drought tolerance candidate genes in cassava, one drought tolerant and one susceptible cassava genotypes were grown in glasshouse under well-watered and water stressed conditions. Ten (10) genes that have been biologically validated and reported to be associated with drought tolerance in other plants such as Arabidopsis and rice were analysed for their expression in cassava in response to water stress using Real-Time PCR. All the ten genes were differentially expressed in response to water stress, four of which (*MeALDH, MeZFP, MeMSD* and *MeRD28)* were identified as drought tolerance associated genes because they were exclusively up-regulated in drought tolerant genotype. Studies to biologically validate these four genes and to identify other drought tolerance associated genes will advance understanding of molecular mechanisms of drought tolerance and provide useful information for gene manipulation in cassava (Turyagyenda Laban Frank *et al.,* 2010).

Anatomical Mechanisms

Anatomy of polyploid cassava and its interspecific hybrids was studied by Dalva Graciano-Ribeiro *et al.,* (2008). The stem anatomy of an interspecific hybrid (*Manihot esculenta x M. oligantha*) and its tetraploid was studied. In the cross sections, the hybrid stems appeared to be circular or cylindrical. Presence of phellogen and vascular cambium characterized secondary growth (Fig 8-A and B). In the stem, the bark epidermis was interrupted by periderm and cork (phellem) that varied in thickness. Sporadic lenticels were observed. The cortical parenchyma showed idioblasts consisting of druses, prismatic crystals and some laticifers. The collenchyma tissue was angular, consisting of up to five cell layers, within which were the parenchyma cells containing prismatic crystals, druses and starch granules, which form a starch sheath. Later, pericyclic fibers formed an interrupted cylinder with a cell wall of variable thickness (Fig. 8-B and E). Diploid stems exhibited more granules.

Rays of phloem parenchyma were continuous with those in the secondary xylem (Fig.8 - C and F). The diploid sample presented more secondary phloem cell layers (11 layers), and its laticifers occurred in groups (2-7), distributed among vessel elements and companion cells. The tetraploid sample had about 7-9 layers and its laticifers were encountered in groups of 3-5, being more abundant than in the diploid sample. Vascular region was easily identified with many cell layers undergoing differentiation (Fig. 8-A and B).

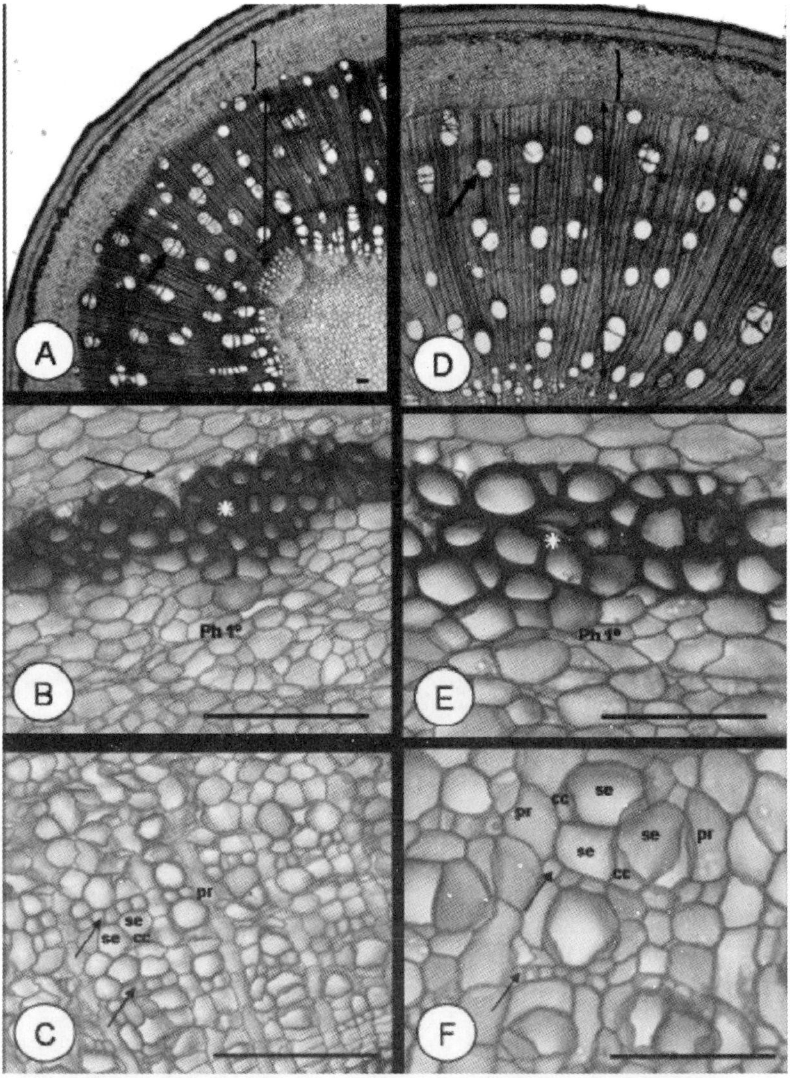

Fig. 8: Stem cross-sections of diploid hybrid (A,B,C) and tetraploid hybrid (D,E,F). A–Stem overview, primary phloem and secondary phloem in detail (}), and secondary xylem (2) and vessels with tyloses (®); B - Starch sheath in detail (®) with crystals and thickened cell wall of pericyclic fibers (*) and primary phloem; C - Secondary phloem in detail with ial parenchyma (pr), sieve elements (se) and laticifers (®); D – Stem overview, showing greater diameter and vessels without tyloses; E - Starch sheath in detail (®) without crystals and thinned cell wall pericyclic fibers (*) and primary phloem. These areas are more developed. F – Secondary phloem in detail with radial parenchyma (pr), sieve elements (etc) and laticifers (®). These cells are bigger. Bar = 0.1mm.
Source : Dalva Graciano – Ribeiro et al. (2008).

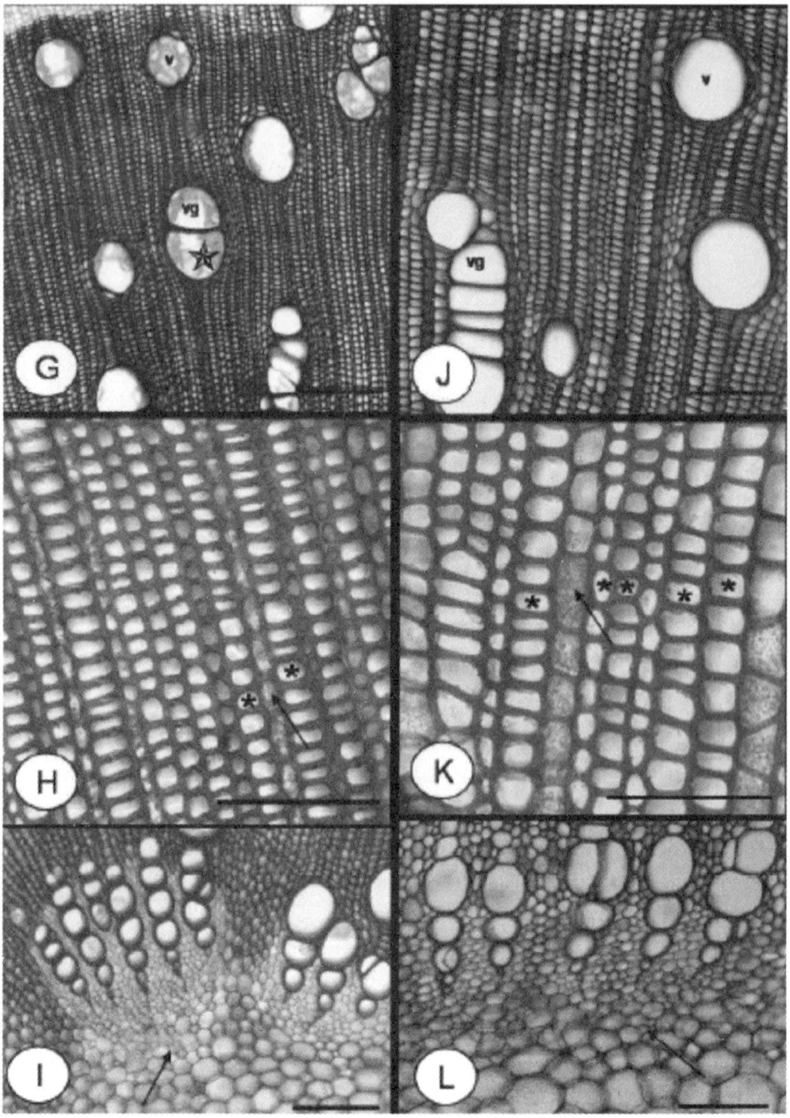

Fig. 9: Stem cross-sections of diploid hybrid (G,H,I) and tetraploid hybrid (J,K,L). G – Secondary xylem in detail, showing solitary (v) and vessel grouping (vg) with tyloses (*). H - Secondary xylem in detail, thickened cell wall of fibers (*) and radial parenchyma cells with starch (®). I – Primary xylem and primary phloem in detail (®); J - Secondary xylem in detail, showing more solitary vessel and vessel groupings (vg) without tyloses. K - Secondary xylem in detail, thinner cell wall fibers (*) and parenchyma cells with starch (®). These cells are bigger. L - Primary xylem and primary phloem in detail (®). Bar = 0.1mm.
Source: Dalva Graciano – Ribeiro et al. (2008).

The medullar area appeared to be divided into two regions, as evidenced by cell shape, content and number of layers. The peripheral area close to the primary xylem, contained isodiametric cells, whereas the central area displayed 5-6-sided polygonal cells. Content and the number of cell layers differed according to stem ploidy (Fig. 10–M and N).The peripheral medulla of the tetraploid had about six layers of isodiametric to polygonal (4-7 faces) parenchyma cells,which contained little starch. Meanwhile, up to seven parenchyma layers were noted in the diploid, and they contained both starch and druses. In the central medulla of the tetraploid, parenchyma cells apparently void of cytoplasm were polygonal, hexagonal ones prevailing, these being narrower and more elongated with rare cytoplasm content (Fig. 10-N). In the diploid hybrid, these cells were broader but smaller, containing druses, starch and rare prismatic crystals (Fig. 10-M).

The difference between the diploid and tetraploid stem was apparent in the cortex region. The tetraploid hybrid showed more prismatic crystals and druses in the cortical parenchyma. The starch sheath contained starch and rarely druses. Pericyclic fibers had slender walls. However, the main structural differences were noted in the primary and secondary phloems, primary and secondary xylems and in the medulla. The abundance of prismatic crystals in the diploid hybrid starch sheath (Fig. 8–B and E) deserves special mention. The primary and secondary phloem cells were 3-4 fold bigger in the tetraploids (Fig. 8–C and F). The number of layers and starch content were also higher in the tetraploid plant than in the diploid (Fig. 8–A and D).

The tetraploid, radial parenchyma cells of the secondary xylem were rectangular, sometimes elongated, the walls being thinner and containing abundant starch. Vessel elements were observed in solitary groupings exhibiting tyloses (Fig. 9 G and K). The primary xylem was reduced, and contained no starch (Fig. 9-L). The fibers were rectangular with thin walls and not apparently empty (Fig. 9-K). The radial parenchyma cells of the secondary xylem are normally elongated the diploid, with narrow thick cell walls and abundant starch. Vessel elements were more abundant with tyloses present (Fig. 9- G and H). The primary xylem contained starch and had no tyloses (Fig. 9- I). The fibers were about 50% smaller with densely thickened walls and they contained starch (Fig. 9-H). While growth rings were present in the diploid, they were absent in the tetraploid (Fig. 8–A and D).

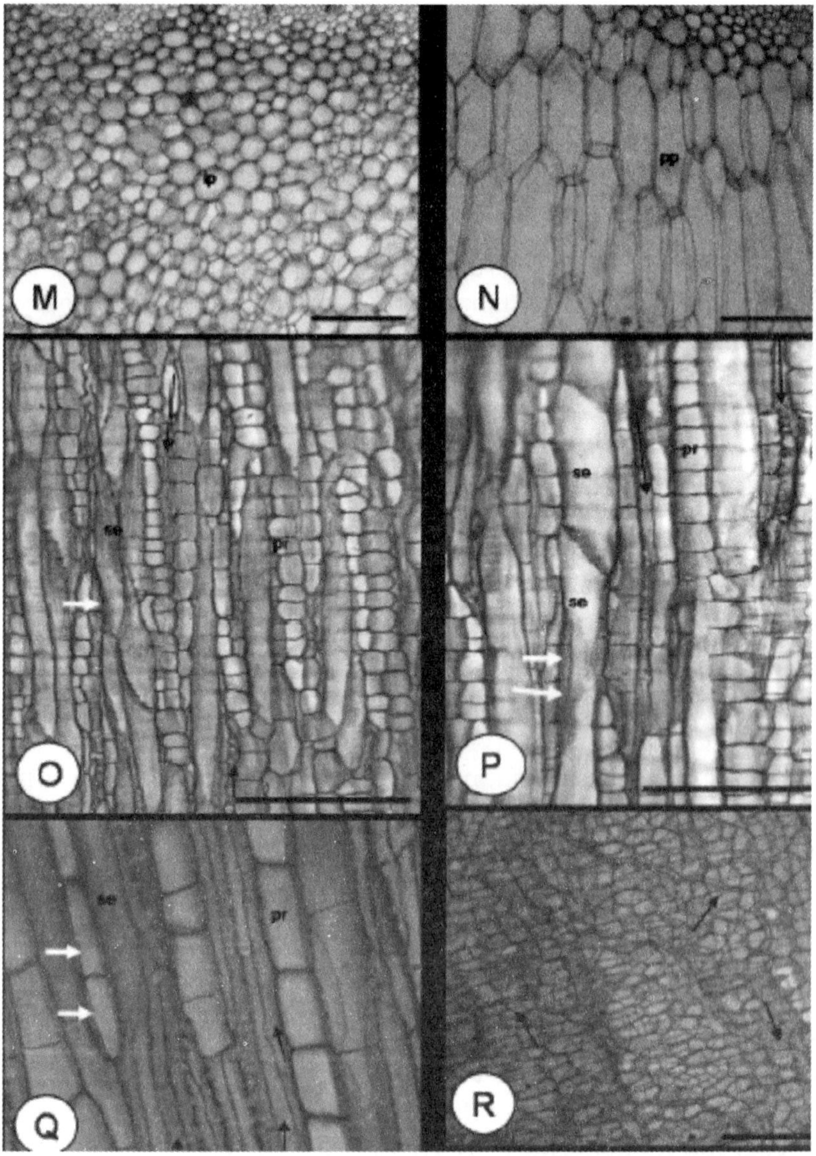

Fig. 10: Stem longitudinal sections of diploid hybrid (O,R) and tetraploid hybrid (P,Q). O – Secondary phloem in detail with radial parenchyma (pr), sieve elements (se), companion cells (white®) and laticifers (black ®). P–Secondary phloem in detail, laticifers (black ®), radial parenchyma (rp), sieve elements (se). These cells are bigger, and there is no starch inside. Q–Laticifers in detail (®) in secondary phloem tissue. R–Articulated branched laticifers in detail (®) in secondary phloem, sieve elements (se), companion cells (black®) and radial parenchyma (rp). Source: Dalva Graciano –Ribeiro et al.(2008).

The main differences, in quantitative terms, were found in the tetraploid, that had more developed secondary vascular tissues than the diploid, as much in relation to the secondary phloem as to the secondary xylem (Fig. 8–A and D). Longitudinal cross-sections of the diploid showed articulated branched laticifers in the cortex of the primary and secondary phloems as well as in the inner primary phloem (Figs. 10 and 11–O to T). Laticifers were scarce in the cortical parenchyma and abundant in the secondary phloem. In contrast, the tetraploid hybrid displayed more laticifers in all the tissues.

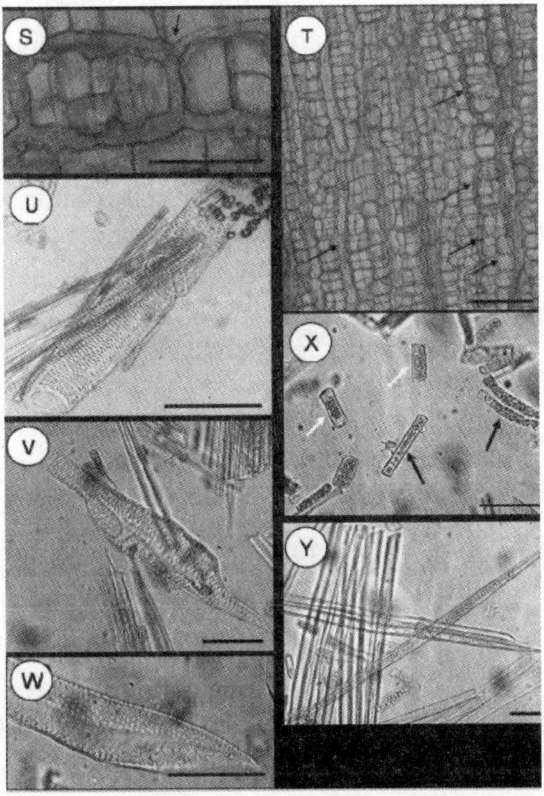

Fig. 11: Macerated tissue of diploid hybrid stem (U,X,Y) and tetraploid hybrid (V,W). U - Vessel elements of reticulate type with medium size and simple perforation plates with straight end. V - Small vessel elements of reticulate type; simple perforation plates with oblique end with appendices at both extremities. W-Bigger vessel elements of reticulate type; reticulate perforation plates with oblique end. X–Radial parenchyma cells (black ®) and axial parenchyma cells with starch content (white®). Y–Libriform fibers with starch content in detail. Bar = 0.1mm.

Source: Dalva Graciano–Ribeiro et al. (2008).

Vessel elements were observed for the type of perforation, size and extremity. In the diploid hybrid, the vessel elements were of the reticulate type and medium in size; perforation plates were generally simple with an oblique end (Fig. 11-U);the fibers were mostly libriform with abundant starch content (Fig. 11-Y); radial and axial parenchyma likewise had abundant starch content (Fig. 11-X). In the tetraploid, the vessel elements were of the reticulate type, being of all sizes, medium and big ones predominating. Starch content was low (Fig. 11–V and W). In the diploid hybrid, prismatic crystals and druses were abundant in the cortical parenchyma, secondary phloem, radial parenchyma, xylematic fibers and peripheral medulla; whereas in the tetraploid hybrid, their occurrence was rare.

Greater cell ploidy is often related to increased organ size (Stebbins, 1971; Sugiyama, 2005). Cavalier-Smith (1985) discussed the significance of the proportional increase in cell volume with increasing DNA content. Cultivars from the same species can have different lignin increments (Akin, 1989; Silva-Lima *et al.*, 2001a, 2001b; Brito and Rodella, 2002). Cultivars from certain species may exhibit differences in density and thickness of parenchyma cells, resulting in different levels of resistance to microorganisms (Philip *et al.*, 1991). Vanucci (1985) reported laticifers occurring in leaf mesophyll of *Manihot caerulescens* while *Manihot pilosa* displayed laticifers in leaf mesophyl, petiole and phloem. Typical articulated laticifers were observed in stems of Manihot sp. aff. caerulescens (Rudall, 1994). Mendonça (1983, 1992) reported articulated and non-branched laticifers in *M. caerulescens* Pohl and *M. glaziovii* Muller von Arg.

The following are the mechanisms in general among the different ploidy levels in cassava.

- In general, the tetraploid hybrid had bigger cells and larger structures.
- This fact associated with a larger number of vessel elements probably indicating more efficient conduction in the tetraploid type.
- Laticifers were of articulated and branched type, and their occurrence was in cortical parenchyma, primary and secondary phloem.
- Prismatic crystals were abundant in the diploid hybrid; they can be involved in the strength of the tissue and as a storage deposit.

- They also may be linked to water evaporation.
- The presence of reticulate vessel elements and simple perforation in both hybrids are related to a more recent evolutionary form in plants.

Nassar *et. al.*(2010) studied the anatomical structures existing in the stem of two cultivars of cassava *viz.*, UnB 110 ,UnB99 and a wild relative *Manihot glaziovii*. UnB 110 and *Manihot glaziovii* are considered to be resistant to drought, both presenting dark green to brown stems. Wheares, the stem of a genotype (UnB 99) that has low resistance to drought and it possess light green bark. In cassava, the colour of the stem. can be an indicator of drought tolerance and shed light on the mechanism of drought tolerance.

Transverse sections of the stems analyzed show the circular shape and secondary growth with vascular cambium and periderm. In the low resistant genotype (UnB 99), there is remaining epidermis, whereas in *M. glaziovii* there is a thicker phellem layer. The cortical region (collenchyma and external and internal parenchyma layers) is thicker in *M. glaziovii*. The starch sheath is one-layer of cells in all stems. Cell content frequency (starch grains, druse and prismatic crystals) is variable. The UnB 99 genotype has more starch grains and crystals (Fig 12A).

In relation to pericycle fibers, the UnB 110 genotype has more layers (1-5). The stems have different cell wall thicknesses, thin in the UnB 99 genotype and thick (with reduced lumen) in *M. glaziovii* (Fig 12 A-C). In this region there are gelatinous fibers, which are abundant in the UnB 110 genotype (Fig 12 B) and rare in *M. glaziovii* and the UnB 99 genotype (Fig 12 A-C). External primary phloem tissue is thicker in *M. glaziovii* (12-15 layers) and thinner in the UnB 110 genotype. However, UnB 110 and UnB 99 genotypes have higher cell content frequency in this tissue for starch grains, druse and prismatic-rhomboid crystals, in descending order. Secondary phloem tissue has more layers in *M. glaziovii* (20-25). Parenchyma rays are continuous with those in the secondary xylem, but phloem rays are enlarged, and more conspicuous in *M. glaziovii*. In the UnB 99 genotype cell content, starch and druses are abundant. In all stems, laticifers are interleaved with sieve elements, companion and parenchyma cells.

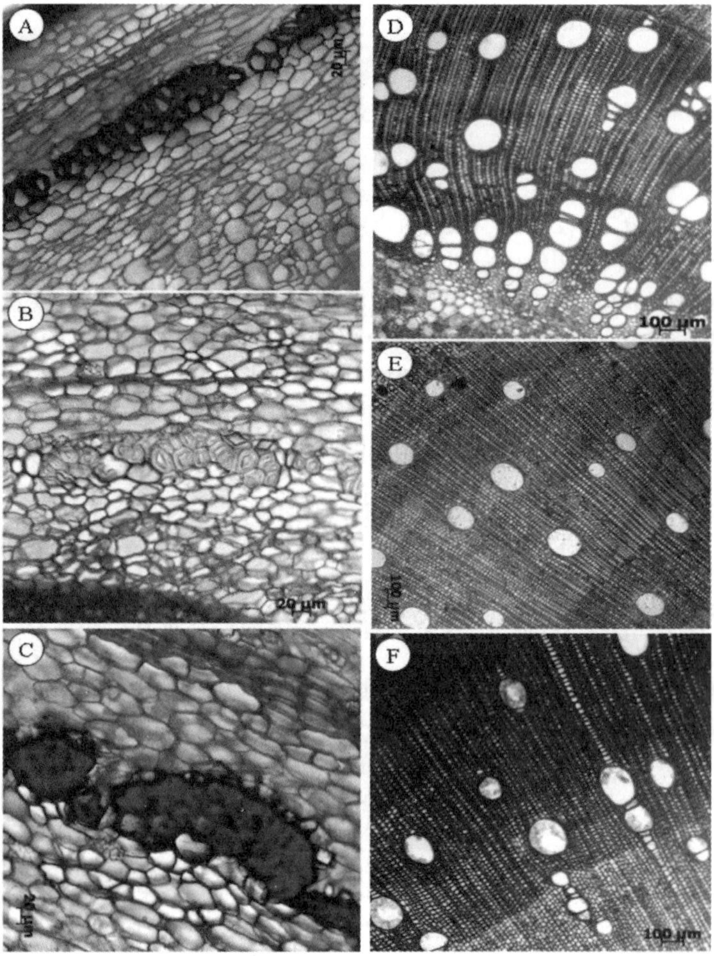

Fig. 12: Transverse sections of stem. A. and D. Genotype UnB 99. Close-up of cortical region and pericycle fibers, starch and fibers have large diameters with thin walls, gelatinous fibers are rare. D. Genotype UnB 99 Xylem growth ring with little distinct predominance of solitary vessel elements, rare geminates and multiple??? with circular shape, absence of tyloses. B. and E. Genotype UnB 110. B. Genotype UnB 110 Close-up of cortical region and pericycle, druse and prismatic/rhomboid crystals and gelatinous fibers in the pericycle. E. Genotype UnB 110 Secondary xylem with solitary vessel elements and rare multiple vessels with circular format, the growth ring is poorly differentiated and rare tyloses. C. and F. Manihot graziovii. C. Close-up of cortical region and pericycle, fibers with thick walls and rare gelatinous fibers. F. Secondary xylem with thickened walls and predominance of vessel elements in elliptic format; ring growth was very evident and there are abundant tyloses and regions with large number of layers. Source: Nassar et. al.(2010).

Growth rings are easily visible in *M. glaziovii* and less distinct in the others (Figure 12 D, E and F). *M. glaziovii* has more cell layers in xylem tissue; these cells are bigger and have thick walls (Figure 12 F and Figure 13 C and F). The secondary xylem tissue of all stems has starch grains in radial parenchyma cells and in fibers/fibrotracheids, especially in UnB 99 and UnB 110 genotypes (Figure 13 B and D). Fibers (Figure 13 A, B and C), which are thick and well lignified are abundant in xylem tissue. Together with these fibers there are gelatinous fibers in all stems, especially in *M. glaziovii*; gelatinous fibers form continuous layers in regions that correspond to early wood (Figure 13 E and F).

Solitary vessel elements (Figure 12 D, E and F) are predominant in relation to geminate and multiple ones in UnB 99 and UnB 110 genotypes; in *M. glaziovii,* there are more geminate and multiple ones, up to 12 vessels. Vessel elements are shorter and have circular shape in UnB 99 and UnB 110 genotypes (Figure 12 D-E and Figure 13 A and B); they are bigger and have elliptical shape in *M. glaziovii* (Figure 12 F). Tyloses are abundant in *M. glaziovii* (Figure 12 F) and rare in the others (Figure 12 D and E).Primary xylem is present in all stems, with more layers in the UnB 110 genotype (1-3) and *M. glaziovii* (1-9). In the UnB 110 genotype, there are tyloses in some of these cells.The parenchyma tissue that involves proto- and metaxylem has more starch grains in UnB 99 and UnB 110 genotypes (Figure 14 A and B). The non-peripheral medulla has elongated/flattened cells, with hexagonal outline in UnB 99 and UnB 110 genotypes, with cell content of starch grains (Figure 14 D and E); while in *M. glaziovii,* these cells have a polyedric outline with few druses (Figure 14F)

The peripheral medulla in *M. glaziovii* has more layers (4-10). Starch grains are found with a higher frequency in the UnB 99 genotype (Figure 14A). They are absent in *M. glaziovii*, and in the UnB 110 genotype (Figure 14B) with irregular frequency, sometimes in greater amounts, and sometimes with almost no content. However, in *M. glaziovii* this region has abundant druse and prismatic-rhomboid crystals (Figure 14C).

The non-peripheral medulla has elongated/flattened cells, with hexagonal outline in UnB 99 and UnB 110 genotypes, with cell content of starch grains (Figure 14D and E); while in *M. glaziovii* these cells have a polyedric outline with few druses (Figure 14F).

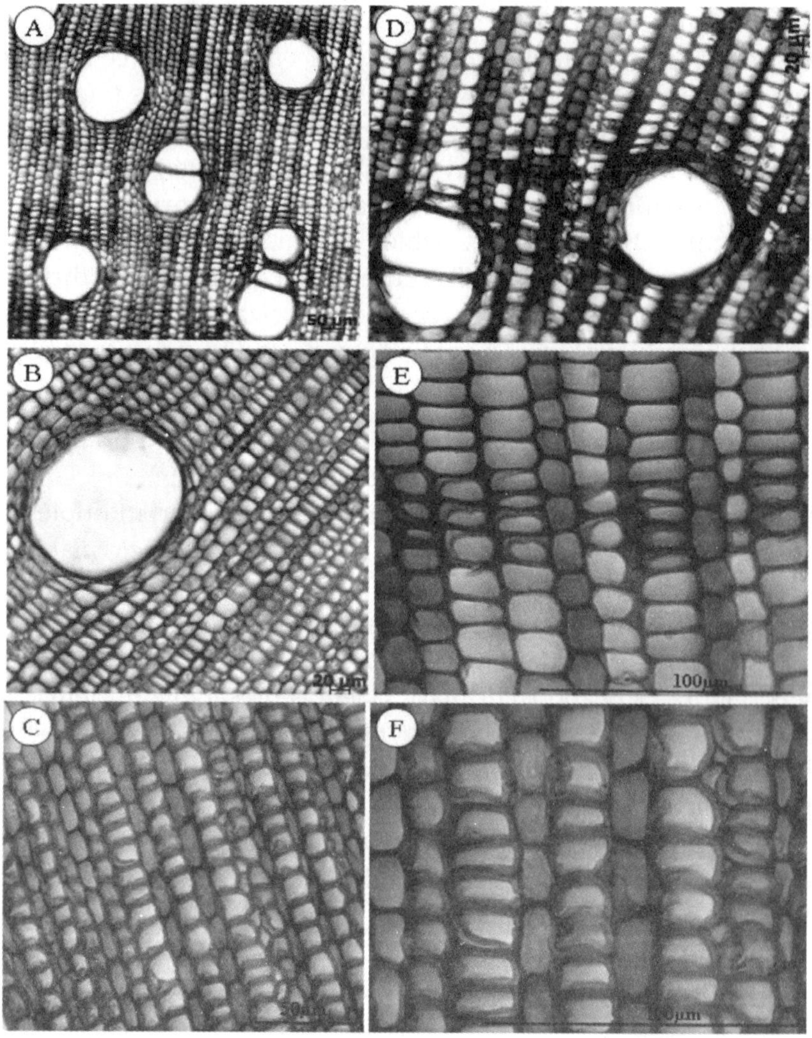

Fig.13: Transversal cross-sections of stem, close-up of secondary xylem. A. and D. Genotype UnB 99. A. Note predominance of solitary vessel elements and fibers, vessel elements are circular, fibers and rays have thin walls. D. There is abundant starch in fibers and rays, there are some gelatinous fibers. B. and E. Genotype UnB 110. Vessel element solitary circular, presence of starch in fibers and ray, cells are larger and have thicker walls. E. Presence of the gelatinous layer in gelatinous fibers internally and absence in the other fibers. C. and F. Manihot glaziovii. Observe rays and fibers of larger size with heavily thickened walls, these cells are bigger and there is occurrence of an abundance of gelatinous fibers. F. Close-up of gelatinous fibers showing the inner layer-G and the presence of starch in parenchyma cells (rays).
Source : Nassar et. al.(2010).

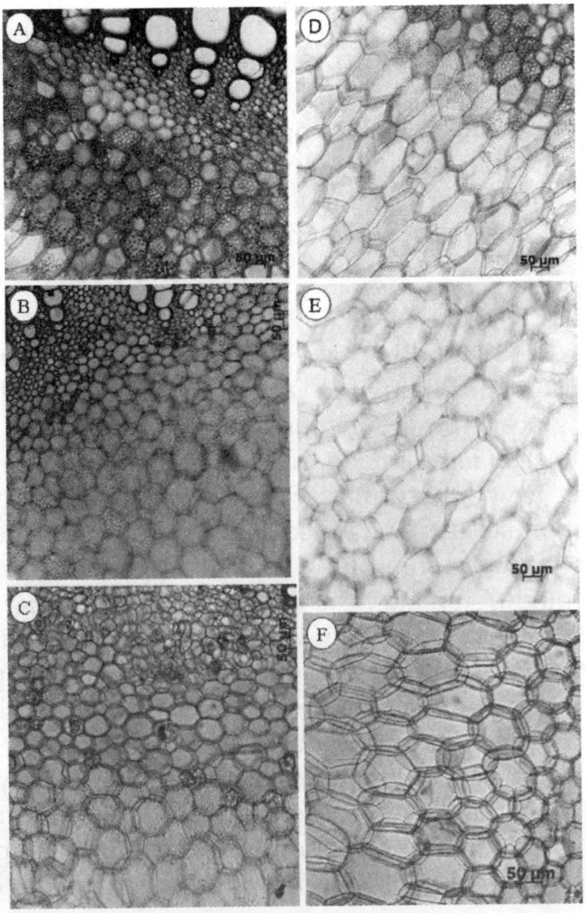

Fig. 14: Transverse sections of stem, close-up of primary xylem and medulla. A. and D. Genotype UnB 99. Abundance of starch in parenchyma cells surrounding the primary xylem and isodiametric parenchyma cells of the peripheral medulla. D. Medulla with hexagonal parenchyma cells, they are longer and narrower. B. and E. Genotype 110. B. Presence of starch in parenchyma cells surrounding the primary xylem and isodiametric parenchyma cells of the peripheral medulla. E. Medulla with hexagonal parenchyma cells, these cells are bigger. C. and F. Manihot glaziovii. C. Parenchyma cells of the primary xylem and isodiametric parenchyma cells containing prismatic and druse crystals. F. Medulla with isodiametric and polygonal parenchyma cells, there is no starch inside. The peripheral medulla in M. glaziovii has more layers (4-10). Starch grains are found with a higher frequency in the UnB 99 genotype (Figure 18A). They are absent in M. glaziovii, and in the UnB 110 genotype (Figure 18B) with irregular frequency, sometimes in greater amounts, and sometimes with almost no content. However, in M. glaziovii this region has abundant druse and prismatic-rhomboid crystals (Figure 18C). Source: Nassar et. al.(2010).

Wild species and their progeny have a dark brown color. This is noted in all clones adapted to semi-arid Northeastern Brazil, while clone UnB 99, which came from humid southern Brazil, has a white stem. This is noticeable in all indigenous clones adapted to the south of this country and to humid tropical Amazon. This leads us to believe that there is some kind of association and possibly close linkage between stem color and this specific anatomical structure.

Conclusion

Abiotic stress signaling is an important area with respect to increase in plant productivity. Drought is a worldwide problem, constraining global crop production and quality seriously, and recent global climate change has made this situation more serious. Drought stress affects growth, dry matter and harvestable yield in plants. Timing, duration, severity and speed of development undoubtedly have pivotal roles in determining how a plant responds to water deficit. Following drought, stomata close progressively with a parallel decline in net photosynthesis and water-use efficiency. In addition to other factors, changes in photosynthetic pigments are of paramount importance to drought tolerance. Protective responses at the leaf level must then be triggered quickly in response to the stress effectors to prevent the photosynthetic machinery being irreversibly damaged. Scavenging of reactive oxygen species by enzymatic and non-enzymatic systems, cell membrane stability, expression of aquaporins and stress proteins are vital mechanisms of drought tolerance.

References

Aguilar, L. P. 1995. Ultraestructura foliar y fotosíntesis de yuca en diferentes cultivares (*Manihot esculenta* Crantz). Popayán, Universidad de Cauca. BSc Thesis.

Akin, D. E. 1989. Histological and physical factors affecting digestibility of forages. *Agronomic Journal*. 81: 17-25.

Alves, AAC. 2002. Cassava botany and physiology. In: Hillocks, R.J., Thresh, J.M., Bellotti, A.C. (eds), Cassava: Biology, Production and Utilization, pp.67-89. CABI Publishing, New York.

Anjum, S.A., L.C. Wang, M. Farooq, I. Khan and L. L. Xue. 2011b. Methyl jasmonate-induced alteration in lipid peroxidation, antioxidative defense system and yield in soybean under drought. *J. Agron. Crop Sci.*, doi:10.1111/j.1439-037X.2010.00468.x.

Appel, K. and H. Hirt. 2004. Reactive oxygen species: metabolism, oxidative stress, and signal transduction. *Ann. Rev. Plant Biol.,* 55: 373-399.

Ashraf M. and M.R. Foolad. 2007. Roles of glycinebetaine and proline in improving plant abiotic stress resistance. *Environ. Exp. Bot.,* 59: 206-216.

Bakayoko1, S., A. Tschannen1 , C. Nindjin1, D. Dao1, O. Girardin and A. Assa. 2009. The impact of water stress on fresh tuber yield and dry matter content of cassava (*Manihot esculenta* Crantz) in Cote d'lvoire. *African Journal of Agricultural Research,* 4 (1): 021-027.

Berg, V.S., M.A. El-Sharkawy, A. Hernández, P. del and Cock, J.H. 1986. Leaf orientation and water relations in cassava. In: Annual Meeting of the American Society of Plant Physiologists, pp.186. Louisiana State University, Baton Rouge.

Bernal, L.M. 1991. Estudios sobre la actividad fosofoenolpiruvato carboxilasa en cultivares de yuca (*Manihot esculenta* Crantz). Bogota, Pontificia Universidad Javeriana. BSc. Thesis.

Brito, C. J. F. A. and R. A. Rodella. 2002. Caracterização morfoanatômica da folha e do caule de Brachiaria brizantha (Hoschst. Ex A. Rich.) Stapf e B. humidicola (Rendle) Schweick (Poaceae). *Revista Brasileira de Botânica* 25(2): 221- 228.

Cadavid, L.F., M.A. El-Sharkawy, A. Acosta and T. Sánchez. 1998. Long-term effects of mulch, fertilization and tillage on cassava grown on sandy soils in northern Colombia. *Field Crops Res.* 57:45-56.

Calatayud, P.A., E. Llovera, J. F. Bois and T. Lamaze. 2000. Photosynthesis in Drought adapted cassava. *Photosynthetica.* 38 (1): 97- 104.

Cavalier-Smith, T. 1985. Cell volume and the evolution of eukaryotic genome size. In Cavalier-Smith [eds.] The evolution of genome size, 105-184. Wiley and Sons, Chichester, UK.

Cayón, M.G., M.A. El-Sharkawy, and L.F. Cadavid. 1997. Leaf gas exchange of cassava as affected by quality of planting material and water stress. *Photosynthetica.* 34:409-418.

CIAT, 1987-1989. Cassava Program Annual Report for 1987-1989. Centro Internacional de Agricultura Tropical, Cali.

Cock, J. H., M. C. M. Porto and M.A. El-Sharkawy. 1985. Water use efficiency of cassava. III. Influence of air humidity and water stress on gas exchange of field grown cassava. *Crop Sci.* 25:265-272.

Cock JH, Riaño NM, El-Sharkawy MA, Lopéz F.Y, Bastidas G (1987) C_3-C_4 intermediate photosynthetic characteristics of cassava (*Manihot esculenta* Crantz). II. Initial products of $^{14}CO_2$ fixation. *Photosynth. Res.* 12:237-241.

Connor, D. J., J.H. Cock and G.E. Parra. 1981. Response of cassava to water shortage. I. Growth and yield. *Field Crops Res.* 4:181-200.

Connor D. J. and J. Palta. 1981. Response of cassava to water shortage. III. Stomatal control and water status. *Field Crops Res.* 4:297-311.

De Tafur S.M., M.A. El-Sharkawy and L.F. Cadavid. 1997a. Response of cassava (*Manihot esculenta* Crantz) to water stress and fertilization. *Photosynthetica* 34:233-239.

Demiral T and I. Turkan. 2004. Does exogenous glycine betaine affect antioxidative system of rice seedlings under NaCl treatment? *J. Plant Physiol.*, 161: 1089-1110.

El-Sharkawy, M and J. Hesketh. 1965. Photosynthesis among species in relation to characteristics of leaf anatomy and CO_2 diffusion resistances. *Crop Sci.* 5:517-521.

El-Sharkawy, M.A. 2004. Cassava biology physiology. *Plant Mol. Biol.* 56: 481-501.

El-Sharkawy, M.A., de Tafur, S.M. and I.F. Cadavid. 1992 a. Potential photosy - nthesis of cassava as affected by growth conditions. *Crop Sci.* 32: 1336 – 1342.

El-Sharkawy, M.A. 1990. Effect of humidity and wind on leaf conductance of field grown cassava. *Rev. Bras. Fisiol. Veg.* 2:17-22.

El-Sharkawy, M.A. 1993. Drought-tolerant cassava for Africa, Asia, and Latin America: breeding projects work to stabilize productivity without increasing pressures on limited natural resources. *Bio-Science*, 43:441-451.

El-Sharkawy, M.A. 2006a. International research on cassava photosynthesis, productivity, ecophysiology, and responses to environmental stresses in the tropics. *Photosynthetica*, 44:481-512.

El-Sharkawy, M. A. and J. H. Cock. 1984. Water use efficiency of cassava. I. Effects of humidity and water stress on stomatal conductance and gas exchange. *Crop Sci.* 24:497-502.

El-Sharkawy, M. A. and J.H. Cock. 1986. The humidity factor in stomatal control and its effects on crop productivity. In: Marcelle R, Clijsters H, Van Poucke M (eds), Biological Control of Photosynthesis, pp.187-198. Martinus Nijhoff Publishers, Dordrecht.

El-Sharkawy, M. A. and J. H. Cock. 1987a. C_3-C_4 intermediate photosynthetic characteristics of cassava (*Manihot esculenta* Crantz). I. Gas exchange. *Photosynth. Res.* 12:219-235.

El-Sharkawy, M. A. and J. H. Cock. 1987b. Response of cassava to water stress. *Plant Soil*, 100:345-360.

El-Sharkawy, M. A., J. H. Cock and KAA, Held. 1984. Water use efficiency of cassava. II. Differing sensitivity of stomata to air humidity in cassava and other warm-climate species. *Crop Sci.* 24:503-507.

El-Sharkawy, M.A., Cock, J.H., Hernández, A. and P. del. 1985. Stomatal response to air humidity and its relation to stomatal density in a wide range of warm climate species. *Photosynth. Res.* 7:137-149.

El-Sharkawy, M.A., J.H. Cock, J.K. Lynam, A. Hernández, del P. and L.F. Cadavid. 1990. Relationships between biomass, root-yield and single-leaf photosynthesis in field- grown cassava. *Field Crops Res.* 25:183-201.

El-Sharkawy, M.A., J.H. Cock and MCM. Porto. 1989. Características fotosintéticas da mandioca (*Manihot esculenta* Crantz). *Rev. Bras. Fisiol. Veg.* 1:143-154.

El-Sharkawy, M.A. and de Tafur, S.M. 2007. Genotypic and within canopy variation in leaf carbon isotope discrimination and its relation to short-term leaf gas exchange characteristics in cassava grown under rain-fed conditions in the tropics. *Photosynthetica*, 45:515-526.

El-Sharkawy, M. A., Hernández, A. del, P. and C. Hershey. 1992b. Yield stability of cassava during prolonged mid-season water stress. *Exp. Agric.* 28:165-174.

El-Sharkawy, M.A. and J. D. Hesketh. 1986. Citation Classic-Photosynthesis among species in relation to characteristics of leaf anatomy and CO_2 diffusion resistances. *Curr. Cont. Agric. Biol. Environ.* 27:14.

Farooq M., A. Wahid, N. Kobayashi, D. Fujita and SMA, Basra. 2009. Plant drought stress: effects, mechanisms and management. *Agron. Sustain. Dev.*, 29: 185-212.

Graciano – Ribeiro D. 2008. Anatomy of polyploid cassava and its interspecific hybrids. *Gene conserve.* 7: 620 – 635.

Horváth E, M. Pál, G. Szalai, E. Páldi and T. Janda. 2007. Exogenous 4-hydroxybenzoic acid and salicylic acid modulate the effect of short-term drought and freezing stress on wheat plants. *Biol. Plant.*, 51: 480-487.

Laetsch, W.M. 1974. The C_4 syndrome: a structural analysis. *Ann. Rev. Plant Physiol.* 25:27-52.

Lal, R. 1980. Effets de l'humidité et de la compacité des sols sur le développement et la production de deux cultivars de manioc. In Terry et al. (eds) Plantes-Racines Tropicales: Stratégies de recherches pour les année 1980, compte rendu du premier symposium triennal sur les plantes-racines du 8 au 12 septembre 1980, Ibadan, Nigéria, pp 110-116.

Lenis, J.I., F. Calle, G. Jaramillo, J.C. Perez, H. Ceballos and J.H. Cock. 2006. Leaf retention and cassava productivity. *Field Crop. Res.* 95:126-134.

Lokko.Y., J.V. Anderson, S. Rudd, A. Raji, D. Horvath, M.A. Mikel, R. Kim, L. Liu, A. Hernandez, A.G.O. Dixon and I.L. Ingelbrecht. 2007. Characterization of an 18, 166 EST dataset for cassava (*Manihot esculenta* Crantz) enriched for drought - responsive genes. *Plant Cell Rep.* 26 : 1605 - 1618.

López, Y, W. Veléz, M. El-Sharkawy and J. E. Meyer. 1993. Biochemical characterization of PEPC from cassava: a preliminary report. In: Roca WM, Thro AM (eds), Proceedings of the First International Scientific Meeting of the Cassava Biotechnology Network, Cartagena, Colombia, pp.340-343. Centro Internacional de Agricultura Tropical, Cali.

Mabrouk A. El-Sharkawy. 2007. Physiological characteristics of cassava tolerance to prolonged drought in the tropics: implication for breeding cultivars adapted to seasonally dry and semi arid environments. *Braz. J.Plant Physiol.*, 19(4): 257 - 286.

Mendonça, M. S. 1983. Estudo de plantas laticíferas. I. Aspectos anatômicos e distribuição de vasos laticíferos em Manihot caerulescens Pohl. *Acta Amazonica* 13: 501-517.

Mendonça, M. S. 1992. Estudo de plantas laticíferas. II. Aspectos anatômicos e distribuição de vasos laticíferos em Manihot glaziovii Muller Arg. *Acta Amazonica*, 22(3): 309-321.

Mittler, R. 2002. Oxidative stress, antioxidants and stress tolerance. *Trends Plant Sci.*, 7: 405-410.

Moller, I.M., P.E. Jensen and A. Hansson. 2007. Oxidative modifications to cellular components in plants. *Ann. Rev. Plant Biol.*, 58: 459-481.

Nassar, N.M.A., L.F.A. Abreu, D. A. P. Teodoro and D. Graciano Ribeiro. 2010. Drought tolerant stem anatomy characteristics in *Manihot esculenta* (Euphorbiaceae) and a wild relative. *Genetics and Molecular Research*, 9 (2): 1023-1031.

Oguntunde, P.G. 2005. Whole-plant water use and canopy conductance of cassava under limited available soil water and varying evaporative demand. *Plant Soil*, 278:371-383.

Oguntunde, P.G. and M.O. Alatise. 2007. Environmental regulation and modelling of cassava canopy conductance under drying root-zone soil water. *Meteorol. Appl.* 14:245-252.

Osiru, D.S.O. 1990. Morphologie et physiologie. In: Ikotun T et al. (eds) Le manioc en Afrique tropicale, IITA, Ibadan, Nigeria, pp 17-27.

Porto, M.C.M. 1983. Physiological mechanisms of drought tolerance in cassava (*Manihot esculenta* Crantz). Tucson, University of Arizona. PhD dissertation.

Rhodes D. Samaras Y. 1994. Genetic control of osmoregulation in plants. In cellular and molecular physiology of cell volume regulation. Strange, K. Boca Raton: CRC Press, pp. 347-361.

Riaño, M.A., J.H. Cock, Y. López and M. El-Sharkawy. 1987a. Anatomia Kranz, estructura y distribución de cloroplastos en hojas de yuca (*Manihot esculenta* Crantz). *Rev. Comalfi* 14:5-12.

Riaño, M.A., J.H. Cock, Y. López, M. El-Sharkawy and G. Bastidas. 1987b. Características fotosintéticas de la yuca (*Manihot esculenta* Crantz): productos iniciales de fijación con $^{14}CO_2$. *Rev. Comalfi*. 14:13-17.

Rudall, P. J. 1994. Laticifers in Crotonoideae (Euphorbiaceae): homology and evolution. *Annals of the Missouri Botanical Garden*, 81(2): 270-282.

Santisopasri V, K. Kurotjanawong, S. Chotineeranat, K. Piyachomkwan, K. Sriroth and C. G. Oates. 2001. Impact of water stress on yield and quality of cassava starch. *Ind. Crop. Prod.* 13:115-29.

Schulthess, F., J.U. Baumgärter, V. Delucchi and A.P. Gutierrez. 1991. The influence of the cassava mealybug, *Phenacoccus manihoti* Mat.-Ferr. (Hom., Pseudococcidae) on yield formation, *Manihot esculenta* Crantz. *J. Appl. Ent.* 111:155-165.

Silva-Lima, L. M., Y. Alquini, C. J. F. Brito and F. C. Deschamps. 2001a. Degradação ruminal dos tecidos vegetais e composição bromatológica de cultivares de Axonopus scoparius (Flüegge) Kuhlm. e Axonopus fissifolius (Raddi) Kuhlm. *Ciência Rural* 31(3):509-515.

Silva-Lima, L. M., Y. Alquini, C. J. F. Brito and F. C. Deschamps. 2001b. Área de tecidos de folhas e caules de Axonopus scoparius (Flüegge) Kuhlm. e Axonopus fissifolius (Raddi) Kuhlm. *Ciência Rural* 31(3):425-430.

Sriroth, K., V. Santisopasri, C. Petchalanuwat, K. Kurotjanawong, K. Piyachomkwan and C.G. Oates. 1999. Cassava starch granule structure-function properties: influence of time and conditions at harvest on four cultivars of cassava starch. *Carbohydr. Polym.* 38:161-170.

Stebbins, G. L. 1971. Chromosomal evolution in higher plants. Addison-Wesley, London, England.

Sugiyama, S. 2005. Polyploidy and cellular mechanisms changing leaf size: comparison of diploid and autotetraploid populations in two species of Lolium. *Annals of Botany* 96: 931-938.

Szabados, L. and A. Savoure´. 2009. Proline: a multifunctional amino acid. *Trends Plant Sci.*, 15: 89-97.

Turner, N.C. 1986. Crop water deficits: a decade of progress. *Adv. Agron.* 39: 1-51.

Turyagyenda Laban Frank, Elizabeth B. Kizito, Morag Ferguson, Yona baguma, JaggerHarvey, Morris Agaba and David Osiru. 2010. Identification of drought related and drought responsive genes in cassava. Second RUFORUM Biennial Meeting held from 20 - 24 September 2010, Entebbe, Uganda.

Vannucci, A. L. 1985. Anatomia foliar comparada de duas espécies nativas de Manihot Mill (Euphorbiaceae). In Anais do XXXVI Congresso Brasileiro de Botânica, 819-836. Curitiba, Parana, Brasil.

□□□

Chapter 12

Influence of Soil pH and Farm Yard Manure Addition on Germination and Seedling Growth Attributes of Multipurpose Tree Species

P. MASILAMANI , P. SARAVANAPANDIAN[1], K. ANNADURAI[2] AND R. SENTHIL KUMAR[3]
National Seed Research and Training Centre, Varanasi-221 106, Uttar Pradesh.
[1] Agricultural College and Research Institute, TNAU, Madurai, Tamil Nadu.
[2] Anbil Dharmalingam Agricultural College & Research Institute, TNAU, Tiruchirappalli, Tamil Nadu-620 009.
[3] Dept. of Trade & Intellectual Property, CARDS, TNAU, Coimbatore, Tamil Nadu

Introduction

Seed germination is affected by number of factors, of which one is pH of the soil (Roy, 1986; Lacey and Line, 1994 and Sharma *et al.*, 1998). Establishment of tree plantation in salt affected soil results in ameliorative process by having favourable influence on physical, chemical and biological properties of soil (Singh *et al.*, 1994; Masilamani *et al.*, 2011). Soil pH with relation to tree growth and survival was well-established but influence of soil pH on germination and seedling growth has not been studied so for in the following multipurpose tree species viz., Acacia (*Acacia nilotica*), Ailanthus (*Ailanthus excelsa)*, Vagai (*Albizia lebbeck*), Neem (*Azadirachta indica*), Kapok (*Ceiba pendantra*), Casuarina (*Casuarina equisetifolia*), Eucalyptus (*Eucalyptus tereticornis*), Acha *(Hardwickia binata*), Pungam (*Pongamia pinnata*) and teak (*Tectona grandis*). Against the stalemate, the present study is an attempt to investigate the influence of soil pH and farm yard manure addition on germination and seedling growth and chemical attributes of above multipurpose tree species.

Materials and Methods

The experiment was conducted at Agricultural Engineering College and Research Institute, Kumulur, Tiruchirappalli, Tamil Nadu. The soil samples of varying pH were collected from Agricultural College and Research Institute, Tiruchirappalli and Manikandam block of Tiruchirappalli district, Tamil Nadu. These sieved soil samples were analysed for chemical properties viz., pH, cation exchange capacity (C.mol.p^+ kg^{-1}), exchangeable sodium percentage (ESP) and electrical conductivity (EC_2) of the soil (dSm^{-1}) (Table. 1). The seeds of different tree species viz. Acacia (*Acacia nilotica*), Ailanthus (*Ailanthus excelsa*), Vagai (*Albizia lebbeck*), Neem (*Azadirachta indica*), Kapok (*Ceiba pendantra*), Casuarina (*Casuarina equisetifolia*), Eucalyptus (*Eucalyptus tereticornis*), Acha (*Hardwickia binata*), Pungam (*Pongamia pinnata*) and Teak (*Tectona grandis*) were collected from an existing plantation at Agricultural Engineering College and Research Institute, Kumulur, Tamil Nadu and seeds were extracted, cleaned and subjected to pre sowing treatments (Table. 2) before sowing in the nursery media in polythene bags (size 13cm x25cm) filled with different nursery media viz., red earth + sand + farm yard manure at 2:1:1 ratio (T_1), soil with pH 8.1 (T_2), soil with pH 9.0 (T_3), soil with pH 10.2 (T_4), soil with pH 10.5 (T_5), soil with pH 8.1 + sand + farm yard manure at 2:1:1 ratio (T_6), soil with pH 9.0 + sand + farm yard manure at 2:1:1 ratio (T_7), soil with pH 10.2 + sand + farm yard manure at 2:1:1 ratio (T_8) and soil with pH 10.5 + sand + farm yard manure at 2:1:1 ratio (T_9). T_1 was kept as control. The trial was setup in completely randomized block design (RBD) with 200 seeds sown in each treatment and replicated four times. Germination count was taken as per the procedure (Table. 3) and the seedling emergence was counted and the germination percentage was computed. Ten random seedlings were dried in a hot air oven at 85°C for 24 hr. and the dry weight was recorded. Vigour index was calculated as the product of germination percentage and seedling length (Abdul-Baki and Anderson, 1973). When the seedlings were 180 days old, the seedling attributes viz., survival (%), root length (cm), shoot length (cm), number of leaves/ seedling, collar diameter (cm), root and shoot dry weight (g) and, leaf area (cm^2), and biochemical constituents viz., chlorophyll contents -a, b and total (mg g^{-1}) and total N (%), P (%) and K (%) contents were also estimated. The data were subjected to analysis of variance as per the method of Panse and Sukhatme (1985) and tested for the significance of variance (p=0.05 per cent) and mean values were compared using Duncan's Multiple Range Test (Gomez and Gomez, 1984).

Table 1: Major chemical properties of the experimental soil.

pH	Cation exchange capacity (c mol $p^{+}kg^{-1}$)	Exchangeable sodium percentage	Electrical conductivity (dSm^{-1})
8.1	18.4	11.8	0.18
9.0	23.8	21.4	0.24
10.2	15.7	45.8	0.21
10.5	14.5	49.5	0.25

Note : Red earth along with farmyard manure and sand at 2:1:1 ratio was used for preparing the nursery mixture (Control treatment). Red earth was analyzed for its chemical properties viz., pH (6.8), CEC (9.2 cmol $p^{+}kg^{-1}$), ESP (2.8) and EC (0.16 dSm^{-1})

Table 2: Pre -sowing treatment of the experimental species.

S.No	Trees name	Pre-sowing Treatment	Reference
1	Acacia (*Acacia nilotica*)	Treated with sulfuric acid (200ml / kg of seeds) for 25 minutes.	Gupta *et al.* 1975
2	Ailanthus (*Ailanthus excelsa*)	No treatment	-
3	Vagai (*Albizia lebbeck*)	Treated with sulfuric acid (200ml / kg of seeds) for 20 minutes.	Gupta *et al.* 1975
4	Neem (*Azadirachta indca*)	No treatment	-
5	Kapok (*Ceiba pendantra*)	No treatment	-
6	Casuarina(*Casuarina equisetifolia*)	No treatment	-
7	Eucalyptus(*Eucalyptus tereticornis*)	No treatment	-
8	Acha (*Hardwickia binata*)	Soak in water for 12hours	Masilamani *et al.*, 1999
9	Pungam (*Pongamia pinnata*)	Soak in water for 24hours	-
10	Teak (*Tectona grandis*)	Drupes soaked in water for 24hrs. followed by drying for 72hrs.,this cycle repeated for six times.	ISTA,1993

Table 3: Germination count of the experimental species.

S.No	Trees name	Germination count (days)	Reference
1	Acacia (*Acacia nilotica*)	28 days after sowing	ISTA, 1985
2	Ailanthus (*Ailanthus excelsa)*	21 days after sowing	AOSA,1965
3	Vagai (*Albizia lebbeck*)	25 days after sowing	ISTA, 1985
4	Neem (*Azadirachta indca*)	21 days after sowing	ISTA, 1985
5	Kapok (*Ceiba pendantra*)	28 days after sowing	Gupta *et al.* 1975
6	Casuarina(*Casuarina equisetifolia*)	22 days after sowing	Gupta *et al.* 1975
7	Eucalyptus(*Eucalyptus ereticornis*)	14 days after sowing	ISTA, 1985
8	Acha *(Hardwickia binata*)	14 days after sowing	Gupta *et al.* 1975
9	Pungam (*Pongamia pinnata*)	28 days after sowing	Gupta *et al.*, 1975
10	Teak (*Tectona grandis*)	28 days after planting the stumps	ISTA, 1993

Results

a) Chemical properties of the experimental soil

The chemical properties of the experimental soils of different pH are given in Table 1. The pH of the experimental soils is ranged from 8.1 to 10.5 (slightly alkaline to strong alkaline). The CEC increased from 18.4 to 23.8, as the soil pH increased from 8.1 to 9.0. There was reduction in CEC to 15.7 and 14.5 for soils with pH of 10.2 and 10.5 respectively. The exchangeable sodium percentage (ESP) values were 11.8, 21.4, 45.8 and 49.5 for the soils with pH of 8.1, 9.0 ,10.2 and 10.5 respectively. The EC was 0.18 dsm^{-1} for soils with a pH of 8.1. After that, there was not much variation in EC.

b) Germination and initial seedling vigour

Acacia nilotica

The results revealed that 100 percent germination was observed with T_3 treatment. (soil pH of 9.0). Followed by T_2 and T_1 treatments (99.0 percent). The lowest germination of 95.0 percent was recorded where the seeds were sown in soil with a pH of 8.1 + sand + farm yard manure at 2:1:1 ratio. The highest root and shoot length was observed with T_3 treatment (soil pH of 9.0). There was reduction in root and shoot length as the soil pH was increased from 9.0 to 10.5. The highest vigour index of 2325 was recorded with the soil of 9.0 pH, which was on par (2064) with the normal nursery mixture of red earth + sand + farm yard manure at 2:1:1 ratio (Table.4).

Table 4: Effect of soil pH and organic matter on germination and initial seedling vigour of Acacia nilotica

Treatments	Germination (%)	Root length (cm)	Shoot length (cm)	Dry matter production (mg seedling^{-1})	Vigour Index
T_1	99.0^{ab}(86.59)	10.6^{abc}	10.3^{bcd}	8.6^{ab}	2064^{b}
T_2	99.0^{ab}(85.25)	10.9^{ab}	11.1^{abc}	10.1^{a}	2174^{ab}
T_3	100.0^{a}(88.19)	11.2^{a}	12.1^{a}	8.2^{abc}	2325^{a}
T_4	97.3^{abc}(82.31)	10.0^{bcd}	10.4^{bcd}	6.5^{cd}	1988^{bc}
T_5	91.5^{c}(73.03)	8.9^{e}	9.4^{d}	6.1^{d}	1677^{d}
T_6	97.2^{abc}(82.47)	10.0^{bcd}	10.7^{bcd}	8.5^{abc}	2014^{bc}
T_7	98.4^{ab}(85.83)	10.0^{bcd}	12.2^{a}	9.8^{a}	2186^{ab}
T_8	98.0^{ab}(83.28)	9.8^{cde}	11.3^{ab}	9.2^{ab}	2062^{b}
T_9	95.0^{bc}(76.62)	9.6^{de}	9.7^{cd}	7.2^{bcd}	1826^{cd}

Means followed by same letter in a column are not significantly different by DMRT. (Figures in parantheses indicate arcsine values)

Ailanthus Excelsa

The highest germination of 51.2 percent was recorded with the nursery mixture of red earth + sand + farm yard manure at 2:1:1 ratio (T_1), which was on par with soil with a pH of 9.0 mixed with sand and farm yard manure at 2:1:1 ratio (T_7). The lowest germination of 40.0 per cent was observed with the soil with a pH of 10.5 (T_5). The highest root length of 7.4 cm was recorded with T_7, where as the lowest root length of 5.7 cm was recorded with the soil with a pH of 10.5 (T_5). There was not much variation in shoot length due to different treatments tried except T_9 (soil with a pH of 10.5 + sand + farm yard manure at 2:1:1 ratio). Regarding vigour index, the highest vigour index of 737 was recorded with T_7 and the lowest values were recorded with alkaline soils with pH from 9.0 to 10.5 (Table. 5).

Table 5: Effect of Soil pH and organic matter on germination and initial seedling vigour of Ailanthus excelsa.

Treatments	Germination (%)	Root length (cm)	Shoot length (cm)	Dry matter production (mg seedling^{-1})	Vigour Index
T_1	51.2^{a}(45.67)	6.1^{cde}	7.0^{a}	170^{ab}	687^{b}
T_2	46.0^{bc}(43.40)	6.1^{cde}	7.0^{ab}	167^{abc}	600^{c}
T_3	45.0^{cd}(41.97)	5.7^{e}	7.4^{a}	159^{cd}	586^{c}
T_4	44.0^{d}(41.30)	6.3^{bcd}	7.3^{a}	161^{bc}	593^{c}
T_5	40.0^{e}(39.03)	6.0^{de}	7.0^{ab}	149^{d}	509^{d}
T_6	50.5^{ab}(44.57)	6.3^{bcd}	7.3^{a}	174^{a}	690^{b}
T_7	51.0^{a}(45.60)	7.4^{a}	7.0^{ab}	176^{a}	737^{a}
T_8	48.4^{ab}(44.07)	7.0^{b}	7.3^{a}	177^{a}	685^{b}
T_9	43.2^{d}(41.07)	6.5^{bc}	6.7^{b}	171^{ab}	568^{c}

(Figures in parantheses indicate arcsine values)

Azadirachta Indica

The highest germination of 78.0 percent was recorded with the soil with a pH of 8.1 (T_2), which was on par with T_1 and T_6. The lowest germination percentage of 33.0 was recorded with T_5 (soil with a pH of 10.5) treatment. There was not much variation in root length due to different treatments tried. The highest shoot length of 13.0 cm was recorded with T_2 (soil with a pH of 8.2) and the lowest shoot length of 6.3 cm was recorded with T_8 treatment, which was on par with T_3, T_5, T_6 T_7 and T_9 treatments. The highest vigour index of 1896 was recorded with T_2 treatment, which was on par with T_1 treatment (1870). The lowest vigour index of 524 was recorded in soil with a pH of 10.5 (Table. 6).

Albizia lebbeck

The highest germination of 84.4 percent was recorded with T_6 (soil with a pH of 8.1 mixed with sand and farm yard manure at 2:1:1 ratio), which was on par with T_1- red earth mixed with sand and farm yard manure at 2:1:1 ratio (83 .4 per cent). The highest root length of 10.9 cm was recorded with T_1, which was on par with T_2, T_4, T_8 and T_9 treatments. The lowest root length of 7.0 cm was recorded with T_5. Regarding the shoot length, a different trend was observed when compared to root length. The highest vigour index of 2150 was observed with T_1 treatment and the lowest value of 769 was observed with T_5 treatment. After that, there was a slight difference in vigour index values among the remaining treatments (Table.7).

Table 6: Effect of soil pH and organic matter on germination and initial seedling vigour of Neem.

Treatments	Germination (%)	Root length (cm)	Shoot length (cm)	Dry matter production (mg seedling^{-1})	Vigour Index
T_1	83.0^{a}(64.33)	13.0^{a}	9.7^{b}	15^{ab}	1870^{a}
T_2	78.0^{a}(62.01)	12.0^{a}	13.0^{a}	17^{a}	1896^{a}
T_3	63.0^{b}(52.34)	10.3^{b}	8.4^{bcd}	13^{abc}	1175^{c}
T_4	58.0^{b}(49.34)	13.0^{a}	9.4^{bc}	14^{abc}	1305^{bc}
T_5	33.0^{c}(34.66)	8.5^{c}	8.0^{bcd}	5^{d}	524^{d}
T_6	77.5^{a}(62.21)	13.0^{a}	7.3^{bcd}	16^{ab}	1573^{b}
T_7	64.0^{b}(53.01)	13.0^{a}	7.0^{cd}	14^{abc}	1252^{c}
T_8	60.0^{b}(50.79)	12.0^{a}	6.3^{d}	10^{c}	1095^{c}
T_9	60.0^{b}(50.77)	12.0^{a}	7.0^{cd}	10^{c}	1122^{c}

(Figures in parantheses indicate arcsine values)

Table7: Effect of soil pH and organic matter on germination and initial seedling vigour of Albizia lebbeck.

Treatments	Germination (%)	Root length (cm)	Shoot length (cm)	Dry matter production (mg seedling^{-1})	Vigour Index
T_1	83.4^{a}(66.1)	10.9^{a}	14.9^{ab}	9^{b}	2150^{a}
T_2	70.0^{bc}(56.52)	9.8^{ab}	16.8^{a}	12^{a}	1856^{ab}
T_3	67.4^{c}(55.21)	8.8^{bc}	16.8^{a}	9^{b}	1740^{b}
T_4	55.4^{de}(48.13)	10.3^{ab}	12.3^{b}	8^{b}	1255^{cd}
T_5	38.0^{f}(38.05)	7.0^{c}	13.6^{ab}	5^{c}	769^{e}
T_6	84.4^{a}(66.77)	8.9^{b}	11.7^{b}	7^{b}	1742^{b}
T_7	77.4^{ab}(61.97)	8.8^{b}	12.2^{b}	9^{b}	1659^{b}
T_8	60.0^{cd}(50.75)	9.5^{ab}	15.4^{ab}	9^{b}	1491^{bc}
T_9	49.0^{e}(44.23)	10.2^{ab}	11.7^{b}	8^{b}	1065^{de}

(Figures in parantheses indicate arcsine values)

Ceiba pentandra

The highest germination of 88.6 percent was recorded with control treatment (Red earth + sand + FYM at 2:1:1 ratio). There was reduction in germination percentage as the soil pH was increased from 8.1 to 10.5. The highest root length of 17.1 and shoot length of 9.5 was recorded with soil with a pH of 8.1 mixed with sand + FYM at 2:1:1 ratio (T_6). There was no significant influence on root length due to different soil pH samples except the soil with a pH of 10.2 mixed with sand and FYM at 2:1:1 ratio. The highest vigour index of 2296 was recorded with soil with a pH of 8.1 mixed with sand and FYM at 2:1:1 ratio. When the same soil alone was used, it recorded the Vigour Index of 1603, which was on par with soil pH of 9.2 (Table. 8).

Table 8: Effect of soil pH and organic matter on Germination and initial seedling vigour of Kapok.

Treatments	Germination (%)	Root Length (cm)	Shoot length (cm)	Dry matter production (mg seedling^{-1})	Vigour Index
T_1	88.6^{a}(70.9)	14.1^{ab}	7.2^{abc}	22^{a}	1894^{ab}
T_2	87.0^{ab}(69.2)	15.6^{ab}	5.6^{c}	17^{ab}	1857^{ab}
T_3	86.0^{abc}(68.1)	12.9^{ab}	5.7^{bc}	17^{ab}	1603^{b}
T_4	78.4^{abcd}(62.3)	14.5^{ab}	4.7^{c}	14^{b}	1502^{b}
T_5	72.0^{d}(58.0)	16.3^{ab}	7.2^{abc}	15^{b}	1696^{ab}
T_6	86.1^{abc}(68.5)	17.1^{a}	9.5^{a}	18^{ab}	2296^{a}
T_7	84.1^{abcd}(66.8)	16.2^{ab}	6.4^{bc}	20^{ab}	1910^{ab}
T_8	74.0^{cd}(59.4)	11.8^{b}	8.2^{ab}	16^{ab}	1476^{b}
T_9	76.0^{bcd}(60.8)	16.5^{ab}	9.2^{a}	15^{b}	1953^{ab}

(Figures in parantheses indicate arcsine values)

Eucalyptus tereticornis

There was no significant influence on germination among the treatments tried. Seeds sown in soil with pH 8.1 gave higher root length (2.6 cm), closely followed by seeds sown in soil pH of 10.5 + sand + FYM at 2:1:1 ratio (2.5 cm). Lowest root length of 1.9 cm was recorded with soil pH of 10.5. The seeds sown in the nursery mixture of red earth, sand and FYM at 2:1:1 ratio recorded higher shoot length of 3.0 cm, which was on par with T_6 and T_7. The highest vigour index value of 349 was recorded with T_1 and T_7 treatments. As the soil pH was increased from 8.1 to 10.5, there was a reduction in vigour index value (Table 9).

Table 9: Effect of soil pH and organic matter on germination and initial seedling vigour of Eucalyptus Terticornis.

Treatments	Germination (%)	Root length (cm)	Shoot length (cm)	Dry matter production (mg seedling^{-1})	Vigour Index
T1	63.5^{a}(53.91)	2.3^{b}	3.0^{a}	325^{bc}	349^{a}
T2	64.0^{a}(52.96)	2.6^{a}	2.4^{cd}	343^{ab}	322^{ab}
T3	67.4^{a}(55.19)	2.4^{b}	2.4^{cd}	323^{bc}	323^{ab}
T4	63.0^{a}(52.58)	2.3^{b}	2.4^{cd}	306^{c}	300^{b}
T5	56.2^{a}(48.54)	1.9^{c}	2.1^{d}	276^{d}	225^{c}
T6	64.0^{a}(52.99)	2.4^{b}	3.0^{ab}	351^{a}	340^{a}
T7	65.0^{a}(53.52)	2.4^{b}	3.0^{ab}	338^{ab}	349^{a}
T8	69.0^{a}(54.86)	2.3^{b}	2.6^{bc}	320^{bc}	332^{ab}
T9	66.0^{a}(54.10)	2.5^{ab}	2.5^{c}	273^{d}	324^{ab}

(Figures in parentheses indicate arcsine values)

Casuarina equisetifolia

The highest germination percentage of 46.0 was recorded with T_6 treatment where the seeds were sown in soil with pH of 8.1 + sand + FYM at 2:1:1 ratio. It was on par with T_1, T_2, T_3, T_4 & T_7 treatments. There was not much statistical significant influence on root length among the treatments tried. The same trend was noticed in the case of shoot length, as in root length. (Except treatment T9 that recorded the lowest shoot length of 0.53 cm). When the soil pH was increased from 8.1 up to 10.2, the vigour index was not changed. When the soil pH was increased beyond 10.2 (i.e.) 10.5, there was drastic reduction in vigour index value (Table.10).

Hardwickia binata

The highest germination percentage was recorded with T_1 treatment (89 percent), which was on par with T_6, T_2, T_3, T_7 and T_8 treatments. The T_5 treatment recorded the lowest germination percentage of 66.0 percent. There was no significant difference in root length, shoot length and dry matter production due to different treatments tried. The highest vigour index value of 1815 was recorded with T_8 treatment, which was on par with all other treatments tried, except T_9 and T_5 treatment, which recorded the vigour index values of 1177 and 1120 respectively (Table.11).

Table10: Effect of soil pH and organic matter on germination and initial seedling vigour of Casuarina equisetifolia.

Treatments	Germination (%)	Root length (cm)	Shoot length (cm)	Dry matter production (mg seedling^{-1})	Vigour Index
T_1	44.0^{a}(41.74)	2.37^{a}	1.37^{a}	1.47^{ab}	167^{a}
T_2	44.0^{a}(41.54)	2.40^{a}	0.93^{ab}	1.60^{a}	149^{ab}
T_3	41.5^{ab}(40.06)	2.50^{a}	1.23^{a}	1.33^{bc}	156^{ab}
T_4	40.4^{ab}(39.47)	2.07^{a}	1.30^{a}	1.20^{cd}	135^{abc}
T_5	16.5^{d}(23.75)	1.77^{a}	0.73^{ab}	1.10^{d}	43^{d}
T_6	46.0^{a}(42.52)	2.63^{a}	0.83^{ab}	1.63^{a}	158^{ab}
T_7	41.0^{ab}(39.58)	1.83^{a}	1.40^{a}	1.33^{bc}	129^{abc}
T_8	33.0^{bc}(34.79)	1.93^{a}	1.27^{a}	1.30^{bcd}	105^{bc}
T_9	30.0^{c}(32.89)	2.20^{a}	0.53^{b}	1.13^{cd}	81^{cd}

(Figures in parantheses indicate arcsine values)

Table 11: Effect of soil pH and organic matter on germination and initial seedling vigour of Hardwickia Binata

Treatments	Germination (%)	Root length (cm)	Shoot length (cm)	Dry matter production (mgseedling^{-1})	Vigour Index
T_1	89.0^{a}(70.41)	8.5^{a}	9.1^{a}	8.3^{a}	1568^{abc}
T_2	80.4^{ab}(64.04)	10.3^{a}	10.0^{a}	9.1^{a}	1626^{abc}
T_3	80.3^{ab}(64.10)	11.2^{a}	11.0^{a}	8.3^{a}	1758^{ab}
T_4	72.1^{bc}(58.28)	11.4^{a}	9.0^{a}	9.0^{a}	1449^{abc}
T_5	66.0^{c}(54.33)	7.4^{a}	11.0^{a}	8.3^{a}	1120^{c}
T_6	85.3^{a}(67.68)	7.0^{a}	11.0^{a}	8.3^{a}	1424^{abc}
T_7	81.0^{ab}(64.02)	11.2^{a}	10.2^{a}	9.0^{a}	1797^{ab}
T_8	82.0^{ab}(64.84)	11.1^{a}	11.0^{a}	8.3^{a}	1815^{a}
T_9	70.3^{bc}(56.71)	7.1^{a}	10.0^{a}	8.2^{a}	1177^{bc}

(Figures in parantheses indicate arcsine values)

Pongamia pinnata

The highest germination of 66.1 percent was recorded with T_1 treatment (red earth + sand + farmyard manure at 2:1:1 ratio), followed by T_2 treatment (57.0 per cent). The highest root length of 29.0 cm, dry matter production (166 mg seedling^{-1}) and vigour index (3306) was recorded with seeds sown in red earth + sand + farm yard manure at 2:1:1 ratio, whereas the shoot length of 26.0 cm was recorded with T_6 treatment (soil with a pH of 8.1 + sand + farmyard manure at 2:1:1 ratio), followed by T_1 treatment (21.4 cm) where the seeds were sown with normal nursery mixture of red earth + sand + farmyard manure at 2:1:1 ratio (Table 12).

Table 12: Effect of soil pH and organic matter on germination and initial seedling vigour of Pongamia pinnata

Treatments	Germination (%)	Root length (cm)	Shoot length (cm)	Dry matter production (mg seedling^{-1})	Vigour Index
T_1	66.1^a(54.42)	29.0^a	21.4^b	166^a	3306^a
T_2	57.0^b(48.73)	27.1^{ab}	19.1^{bc}	135^b	2643^b
T_3	49.0^{bc}(44.34)	22.1^{cd}	21.0^b	129^{bc}	2053^c
T_4	42.2^{cd}(40.52)	20.0^d	14.4^d	109^{cd}	1431^d
T_5	23.0^e(28.37)	15.0^e	15.0^{cd}	104^d	657^e
T_6	56.5^b(48.76)	27.0^{ab}	26.0^a	146^{ab}	2992^{ab}
T_7	47.0^c(43.27)	24.3^{bc}	18.0^{bcd}	147^{ab}	1987^c
T_8	36.4^d(37.12)	25.1^{abc}	18.0^{bcd}	140^b	1555^{cd}
T_9	35.0^d(36.19)	20.0^d	17.2^{bcd}	100^d	1260^d

(Figures in parantheses indicate arcsine values)

Tectona grandis

The highest number of sprouts per stump (2.52) was observed with T_5 treatment (soil with pH of 10.5) followed by T_7, T_8 and T_9 treatments (2.35). The lowest number of sprouts per stump (1.60) was recorded with T3 treatment (soil with pH 9.0). The highest number of leaves (11.10) per stump and dry matter production were found in the T6 treatment (4.11 mg seedling^{-1}), which was closely followed by T_8 treatment (stumps planted with 10.2 pH + sand + farm yard manure at 2:1:1 ratio). The lowest number of leaves (3.70) and dry matter production (1.95 mg seedling $^{-1}$) was recorded with T_5 treatment, where the stumps were planted in soil with pH of 10.5. A higher number of secondary roots per stump was recorded irrespective of the pH range of soils, when the soil is mixed with sand + farm yard manure at 2:1:1 ratio (it ranges from 2.64 to 4.55) compared with the stumps planted without addition of sand + farm yard manure in 2:1:1 ratio (Table 13).

Table 13: Effect of soil pH and organic matter on germination and initial seedling attributes of 28 days old Tectona grandis

Treatments	Number of sprout / stump	Number of leaves / stump	Number of secondary roots / stump	Dry matter production (mg seedling^{-1})	Survival percentage
T_1	2.33^{a}	8.17^{cd}	1.64^{c}	3.25^{b}	100^{a}(88.40)
T_2	2.30^{a}	10.03^{ab}	1.40^{c}	3.15^{b}	100^{a}(87.93)
T_3	1.60^{b}	9.93^{ab}	2.21^{bc}	2.00^{b}	99.5^{a}(86.59)
T_4	1.61^{b}	7.37^{d}	1.51^{c}	2.78^{b}	100^{a}(86.46)
T_5	2.52^{a}	3.70^{e}	2.22^{bc}	1.95^{c}	99.0^{a}(85.19)
T_6	2.21^{a}	11.10^{a}	4.42^{a}	4.11^{a}	99.5^{a}(86.78)
T_7	2.35^{a}	9.63^{ab}	4.55^{a}	3.87^{a}	99.2^{a}(83.49)
T_8	2.35^{a}	10.80^{ab}	3.48^{ab}	3.90^{a}	100^{a}(87.44)
T_9	2.35^{a}	9.33^{bc}	2.64^{bc}	3.17^{b}	100^{a}(87.48)

(Figures in parantheses indicate arcsine values)

c) Biometric and Biochemical observations

When the seedlings were 180 days old, the seedling attributes viz., survival percentage, root length (cm), shoot length (cm), number of leaves/seedling, collar diameter (cm), root and shoot dry weight (g) and leaf area (cm^2) and biochemical attributes viz., chlorophyll content- a, b and total (mg/g^{-1}), and total N (%), P (%) and K (%) were recorded.

The results on biometric observation at 180 days-old seedlings of different species revealed that significant differences were observed among different treatments tried. The highest survival percentage, root length (cm) and shoot length (cm) were recorded in control treatment (red earth + sand + farm yard manure 2:1:1 ratio) followed by the seeds sown in soil with pH 8.1 + sand + farm yard manure at 2:1:1 ratio, the lowest values were recorded in the seeds sown in soil with pH of 10.5. The same trend was observed in collar diameter (cm), root and shoot dry weight (g) of all the species studied. When the soil with different pH alone was used in this experiment, there was reduction in survival percentage, root length, shoot length, collar diameter, root and shoot dry weight of the seedlings when compared to the same soils mixed with sand and farm yard manure at 2:1:1 ratio (Table. 14-23).

Table 14: Effect of soil pH and organic matter on biochemical constituents of 180 days old Acacia nilotica seedling.

Treatments	Chlorophyll content (mg/g)			Total nitrogen content (%)	Total phosphorus content (%)	Total potassium content (%)
	a	b	Total			
T_1	2.57^{cd}	2.57^{d}	5.36^{d}	2.91^{a}	0.68^{a}	1.40^{f}
T_2	2.38^{d}	2.49^{d}	5.35^{d}	2.31^{c}	0.44^{d}	1.63^{e}
T_3	2.79^{c}	3.53^{b}	6.44^{b}	1.64^{e}	0.30^{f}	2.08^{b}
T_4	2.40^{d}	3.16^{c}	6.42^{b}	1.47^{f}	0.63^{b}	1.82^{c}
T_5	3.47^{b}	4.27^{a}	7.46^{a}	1.15^{g}	0.45^{d}	1.76^{cd}
T_6	2.55^{cd}	3.43^{bc}	7.46^{a}	2.66^{b}	0.37^{e}	2.22^{a}
T_7	2.75^{c}	3.27^{bc}	5.81^{c}	1.82^{d}	0.45^{d}	1.63^{e}
T_8	3.36^{b}	3.38^{bc}	6.53^{b}	1.65^{e}	0.53^{c}	1.67^{de}
T_9	4.44^{a}	3.19^{c}	7.55^{a}	1.49^{f}	0.37^{e}	1.65^{de}

Means followed by same letter in a column are not significantly different by DMRT.

Table 15: Effect of soil pH and organic matter on bio chemical constituents of 180 days old Ailanthus excelsa seedling

Treatments	Chlorophyll content (mg/g)			Total nitrogen content (%)	Total phosphorus content (%)	Total potassium content (%)
	a	b	Total			
T_1	4.05^{a}	3.49^{bc}	7.64^{ab}	1.41^{a}	0.26^{g}	1.57^{d}
T_2	3.68^{ab}	3.63^{ab}	7.47^{ab}	1.00^{c}	0.77^{c}	2.77^{b}
T_3	3.42^{b}	3.62^{ab}	8.00^{a}	0.88^{cde}	0.65^{ef}	2.70^{bc}
T_4	3.30^{b}	3.46^{bc}	7.35^{b}	0.81^{e}	0.87^{b}	2.57^{bc}
T_5	3.56^{ab}	3.69^{ab}	7.33^{b}	0.76^{e}	0.82^{f}	2.30^{a}
T_6	3.25^{b}	3.48^{bc}	7.47^{ab}	1.14^{b}	0.73^{cd}	2.43^{bc}
T_7	3.48^{ab}	3.84^{a}	7.24^{b}	0.96^{cd}	0.93^{a}	2.57^{a}
T_8	3.52^{ab}	3.37^{c}	7.56^{ab}	0.87^{cde}	0.65^{ef}	3.53^{a}
T_9	3.52^{c}	2.74^{d}	7.93^{c}	0.84^{de}	0.70^{de}	2.33^{c}

Means followed by same letter in a column are not significantly different by DMRT

Table 16: Effect of soil pH and organic matter on bio chemical constituent of 180 days old Neem seedling

Treatments	Chlorophyll content (mg/g)			Total nitrogen content (%)	Total phosphorus content (%)	Total potassium content (%)
	a	b	Total			
T_1	2.88^{cd}	3.77^{abc}	6.39^{c}	1.98^{a}	0.36^{b}	1.70^{cd}
T_2	2.66^{d}	3.56^{c}	6.37^{c}	1.64^{abc}	0.33^{b}	2.39^{b}
T_3	2.70^{d}	3.57^{c}	5.54^{d}	1.51^{bc}	0.37^{b}	2.13^{b}
T_4	2.66^{d}	3.73^{bc}	5.00^{c}	1.42^{bc}	0.43^{a}	2.27^{b}
T_5	2.85^{d}	4.21^{ab}	7.16^{b}	1.38^{c}	0.37^{b}	2.26^{b}
T_6	3.12^{bc}	4.27^{ab}	7.49^{ab}	1.66^{abc}	0.37^{b}	2.32^{b}
T_7	3.17^{b}	4.34^{a}	7.75^{a}	1.82^{ab}	0.36^{b}	1.80^{c}
T_8	3.49^{a}	4.01^{abc}	7.39^{ab}	1.82^{ab}	0.45^{a}	2.80^{a}
T_9	3.50^{a}	4.00^{abc}	7.07^{b}	1.62^{abc}	0.44^{a}	1.44^{d}

Means followed by same letter in a column are not significantly different by DMRT.

Table 17: Effect of soil pH and organic matter on bio chemical constituent of 180 days old *Albizia lebbeck* seedling

Treatments	Chlorophyll content (mg/g)			Total nitrogen content (%)	Total phosphor us content (%)	Total potassium content (%)
	a	b	Total			
T_1	4.28^{a}	4.21^{bcd}	8.49^{a}	1.41^{a}	0.73^{a}	1.90^{d}
T_2	4.49^{a}	4.30^{bc}	8.35^{a}	1.15^{b}	0.71^{a}	2.32^{c}
T_3	4.42^{a}	4.45^{abc}	8.67^{a}	0.94^{c}	0.71^{a}	2.32^{c}
T_4	4.08^{ab}	4.59^{ab}	8.39^{a}	0.94^{c}	0.47^{d}	2.14^{cd}
T_5	4.06^{ab}	3.85^{cde}	8.14^{a}	0.84^{c}	0.57^{c}	2.33^{c}
T_6	3.49^{bc}	3.87^{cde}	6.83^{bc}	1.25^{b}	0.57^{c}	2.19^{cd}
T_7	3.03^{c}	3.57^{de}	6.21^{c}	1.17^{b}	0.70^{a}	2.47^{c}
T_8	3.06^{c}	3.39^{e}	6.45^{bc}	0.94^{c}	0.65^{b}	2.92^{b}
T_9	3.14^{c}	5.01^{a}	7.00^{b}	0.94^{c}	0.57^{c}	4.64^{a}

Means followed by same letter in a column are not significantly different by DMRT

Table 18: Effect of soil pH and organic matter on bio chemical constituent of 180 days old Kapok seedling

Treatments	Chlorophyll content (mg/g)			Total nitrogen content (%)	Total phosphorus content (%)	Total potassium content (%)
	a	b	Total			
T_1	4.16^{a}	4.37^{a}	8.36^{a}	1.25^{a}	0.55^{c}	1.27^{d}
T_2	3.81^{ab}	3.46^{b}	7.52^{b}	1.17^{bc}	0.55^{c}	2.20^{bcd}
T_3	3.39^{cd}	3.72^{b}	7.35^{bc}	0.98^{d}	0.41^{d}	1.72^{bcd}
T_4	3.69^{bc}	3.42^{b}	7.65^{b}	0.84^{e}	0.37^{d}	1.89^{bcd}
T_5	3.53^{bcd}	3.59^{b}	7.30^{bc}	0.74^{f}	0.68^{a}	3.43^{a}
T_6	3.52^{bcd}	3.43^{b}	7.46^{b}	1.21^{ab}	0.61^{abc}	2.58^{ab}
T_7	3.42^{bcd}	4.51^{a}	7.44^{b}	1.14^{c}	0.64^{ab}	1.41^{cd}
T_8	3.54^{bcd}	4.45^{a}	7.42^{b}	0.97^{d}	0.57^{bc}	2.43^{abc}
T_9	3.28^{d}	4.23^{a}	7.02^{c}	0.79^{ef}	0.63^{ab}	2.11^{abc}

Means followed by same letter in a column are not significantly different by DMRT.

Table 19: Effect of soil pH and organic matter on bio chemical constituents of 180 days old Casuarina equisetifolia seedling

Treatments	Chlorophyll content (mg/g)			Total nitrogen content (%)	Total phosphorus content (%)	Total potassium content (%)
	a	b	Total			
T_1	2.60^{cde}	3.52^{ab}	6.29^{c}	0.95^{a}	0.43^{d}	2.29^{bc}
T_2	2.54^{de}	3.33^{b}	6.42^{bc}	0.83^{c}	0.54^{c}	2.84^{a}
T_3	2.78^{bcd}	3.43^{ab}	6.72^{a}	0.83^{bc}	0.71^{a}	2.40^{abc}
T_4	2.16^{e}	3.50^{ab}	6.46^{abc}	0.72^{d}	0.65^{ab}	2.17^{bc}
T_5	3.07^{abc}	3.46^{ab}	6.66^{ab}	0.74^{d}	0.69^{a}	2.53^{ab}
T_6	3.33^{a}	3.67^{a}	6.68^{ab}	0.94^{a}	0.47^{d}	1.94^{c}
T_7	3.29^{a}	3.34^{b}	6.42^{bc}	0.90^{ab}	0.58^{bc}	2.63^{ab}
T_8	3.35^{a}	3.35^{b}	6.26^{c}	0.88^{bc}	0.39^{d}	1.38^{d}
T_9	3.27^{ab}	3.35^{b}	6.62^{d}	0.84^{bc}	0.70^{a}	2.54^{ab}

Means followed by same letter in a column are not significantly different by DMRT.

Table 20: Effect of soil pH and organic matter on bio chemical constituents of 180 days old Eucalyptus terticornis seedling

Treatments	Chlorophyll content (mg/g)			Total nitrogen content	Total phosphor us content	Total potassium content
	a	b	Total	(%)	(%)	(%)
T_1	2.8[b]	3.2[c]	5.6[c]	0.84[a]	0.53[d]	1.27[b]
T_2	2.6[b]	3.2[c]	5.7[c]	0.73[b]	0.57[c]	1.28[b]
T_3	2.7[b]	3.7[b]	6.7[ab]	0.73[b]	0.57[cd]	2.49[a]
T_4	3.1[b]	3.7[b]	6.4[b]	0.73[b]	0.46[e]	1.28[b]
T_5	3.0[b]	3.5[b]	6.0[c]	0.64[c]	0.55[cd]	1.39[b]
T_6	3.4[a]	4.2[a]	6.6[b]	0.86[a]	0.46[e]	1.68[a]
T_7	3.3[a]	4.2[a]	6.5[b]	0.77[b]	0.67[b]	1.41[b]
T_8	3.0[ab]	4.4[a]	6.3[b]	0.85[a]	0.55[cd]	1.54[b]
T_9	3.0[ab]	4.5[a]	6.6[b]	0.73[b]	0.82[a]	1.37[b]

Means followed by same letter in a column are not significantly different by DMRT.

Table 21: Effect of soil pH and organic matter on bio chemical constituents of 180 days old Hardwickia binata seedling

Treatments	Chlorophyll content (mg/g)			Total nitrogen content	Total phosphor us content	Total potassium content
	a	b	Total	(%)	(%)	(%)
T_1	2.02[d]	2.74[c]	4.21[f]	0.95[b]	0.43[b]	0.68[f]
T_2	2.16[d]	2.73[c]	4.69[e]	0.94[b]	0.40[c]	1.18[d]
T_3	2.63[c]	3.24[b]	5.83[c]	0.82[cd]	0.51[a]	0.87[ef]
T_4	2.25[cd]	3.30[b]	5.44[d]	0.64[ef]	0.47[b]	1.26[cd]
T_5	3.35[b]	3.71[a]	7.21[b]	0.57[f]	0.45[b]	1.41[bc]
T_6	3.39[b]	3.73[a]	7.73[a]	1.08[a]	0.52[a]	1.52[b]
T_7	3.36[b]	3.73[a]	7.30[b]	0.85[c]	0.46[b]	1.08[de]
T_8	4.06[a]	3.39[ab]	7.40[ab]	0.76[d]	0.53[a]	1.41[bc]
T_9	3.39[b]	3.50[ab]	7.34[b]	0.66[e]	0.47[b]	2.57[a]

Means followed by same letter in a column are not significantly different by DMRT.

Table 22: Effect of soil pH and organic matter on bio chemical constituents of 180 days old Pongamia pinnata seedling

Treatments	Chlorophyll content (mg/g)			Total nitrogen content (%)	Total phosphor us content (%)	Total potassium content (%)
	a	b	Total			
T_1	3.75^{cd}	4.47^{abc}	7.57^{b}	1.03^{ab}	0.63^{bc}	0.90^{e}
T_2	3.25^{d}	4.32^{bc}	8.13^{ab}	1.02^{bc}	0.46^{d}	1.83^{b}
T_3	4.18^{abc}	4.46^{abc}	8.28^{a}	0.91^{de}	0.71^{b}	2.06^{a}
T_4	3.80^{bcd}	4.47^{abc}	8.40^{a}	0.77^{f}	0.57^{bcd}	1.94^{ab}
T_5	4.31^{abc}	3.80^{d}	8.41^{a}	0.65^{g}	0.51^{cd}	1.90^{b}
T_6	4.59^{a}	4.22^{c}	8.40^{a}	1.09^{a}	0.88^{a}	1.38^{c}
T_7	4.42^{ab}	3.82^{d}	8.51^{a}	0.96^{cd}	0.66^{bc}	1.34^{c}
T_8	4.42^{ab}	4.55^{ab}	8.60^{a}	0.87^{e}	0.65^{bc}	1.37^{c}
T_9	4.35^{abc}	4.73^{a}	8.42^{a}	0.80^{f}	0.66^{bc}	1.06^{d}

Means followed by same letter in a column are not significantly different by DMRT.

Table 23: Effect of soil pH and organic matter on bio chemical constituent of 180 days old Tectona grandis seedling

Treatments	Chlorophyll content (mg/g)			Total nitrogen content (%)	Total phosphor us content (%)	Total potassium content (%)
	a	b	Total			
T_1	4.20^{a}	4.58a	9.44^{a}	2.12^{a}	0.65^{b}	1.36^{c}
T_2	4.03^{ab}	4.30^{a}	9.42^{a}	1.76^{ab}	0.59^{bc}	1.36^{c}
T_3	3.59^{abcd}	4.36^{a}	8.62^{b}	1.58^{b}	0.84^{a}	1.27^{c}
T_4	3.46^{bcd}	3.62^{b}	7.81^{cd}	1.02^{c}	0.30^{e}	1.32^{c}
T_5	3.77^{abc}	3.53^{b}	6.94^{e}	1.03^{c}	0.40^{de}	1.30^{c}
T_6	3.39^{bcd}	3.37^{b}	8.33^{bc}	1.68^{ab}	0.46^{cde}	2.36^{c}
T_7	3.17^{cd}	4.39^{a}	7.29^{de}	1.47^{bc}	0.37^{de}	2.22^{a}
T_8	2.97^{d}	3.39^{b}	7.39^{de}	1.03^{c}	0.48^{cd}	1.70^{b}
T_9	3.08^{cd}	4.42^{a}	7.58^{de}	1.00^{c}	0.68^{b}	1.42^{c}

Means followed by same letter in a column are not significantly different by DMRT.

The results on biochemical constituents of 180 days old seedlings of different species revealed that there was not much variation in chlorophyll content(a, b and total) between the treatments tried. The highest total N content was observed with T_1 (red earth mixed sand and farm yard manure at 2:1:1 ratio) treatment. The lowest N content was recorded with T_5 treatment (soil with pH 10.5) for almost all the species studied. On the other hand, no consistent results were recorded in P content and the reduction of P content was observed when the soil pH was increased from 9.0 to 10.2. The highest K content was observed with the strong alkaline soil mixed with sand and farmyard manure at 2:1:1 ratio, whereas the normal nursery mixture recorded the lowest value for almost all the species studied. There was a significant difference in total K content among the remaining treatments (Fig.1-10).

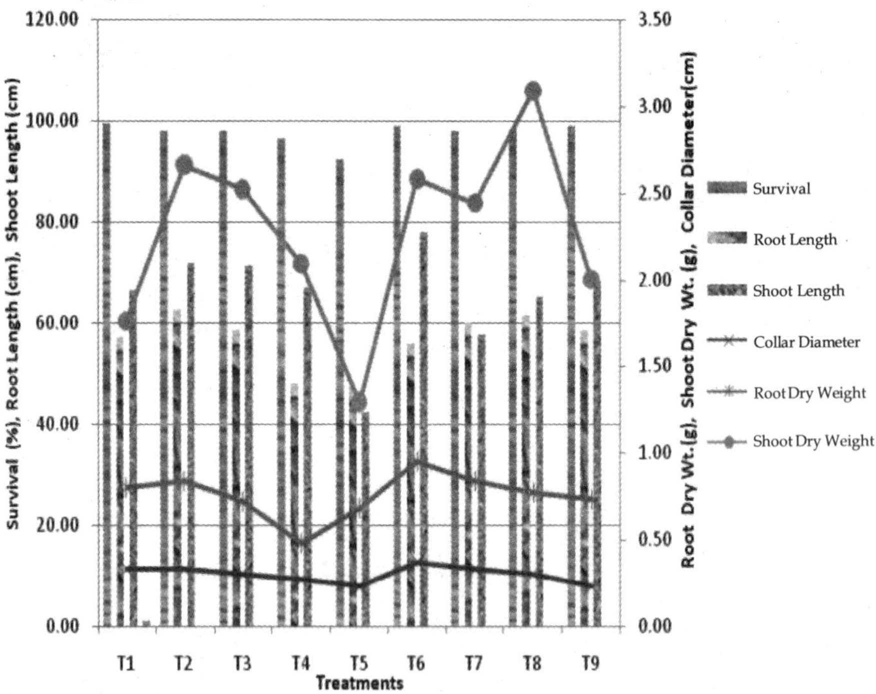

Fig. 1: Effect of soil pH and organic matter on seedling attributes of 180 days old

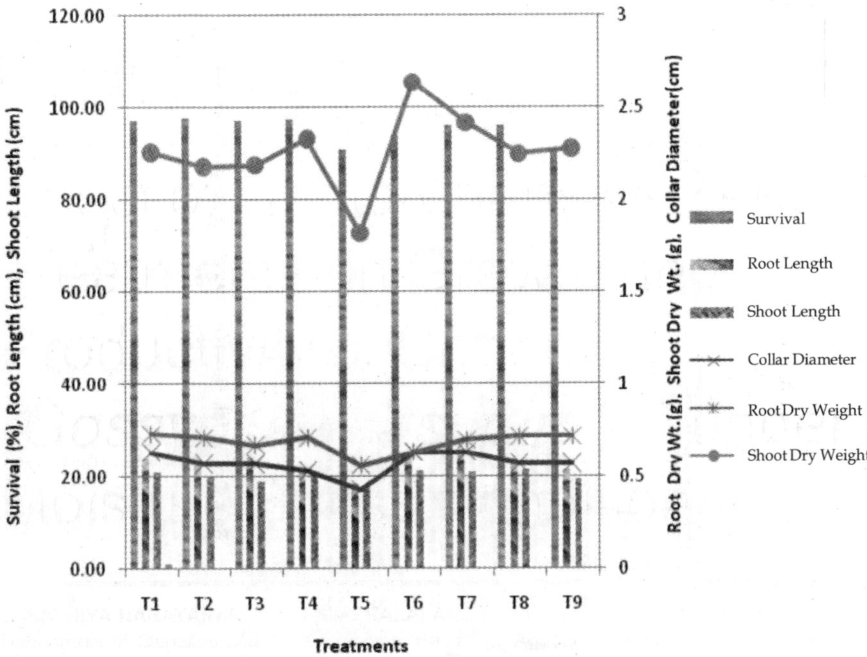

Fig. 2: Effect of soil pH and organic matter on seedling attributes of 180 days

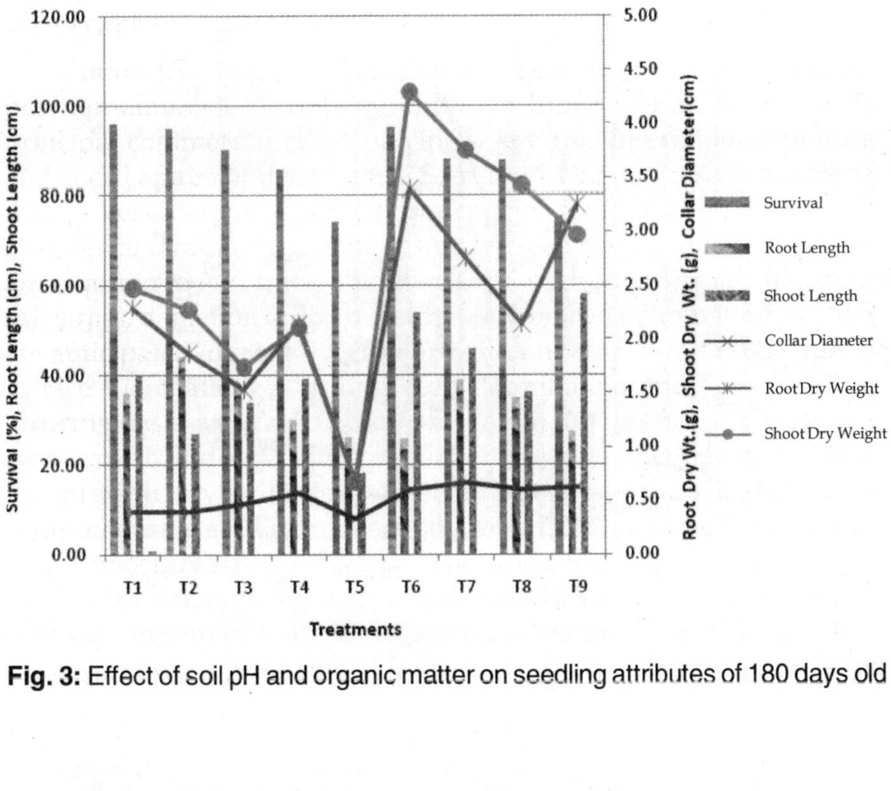

Fig. 3: Effect of soil pH and organic matter on seedling attributes of 180 days old

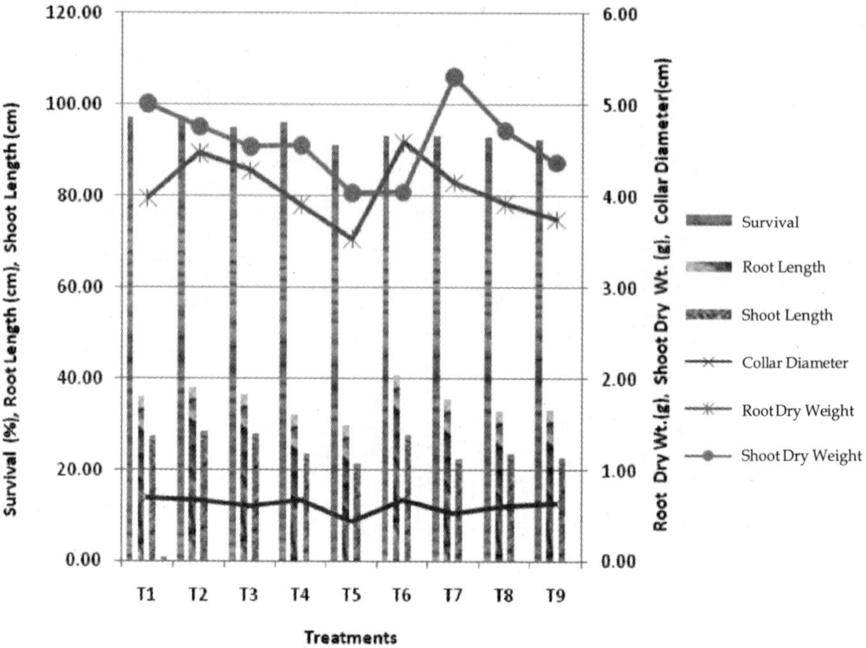

Fig. 4: Effect of soil pH and organic matter on seedling attributes of 180 days old

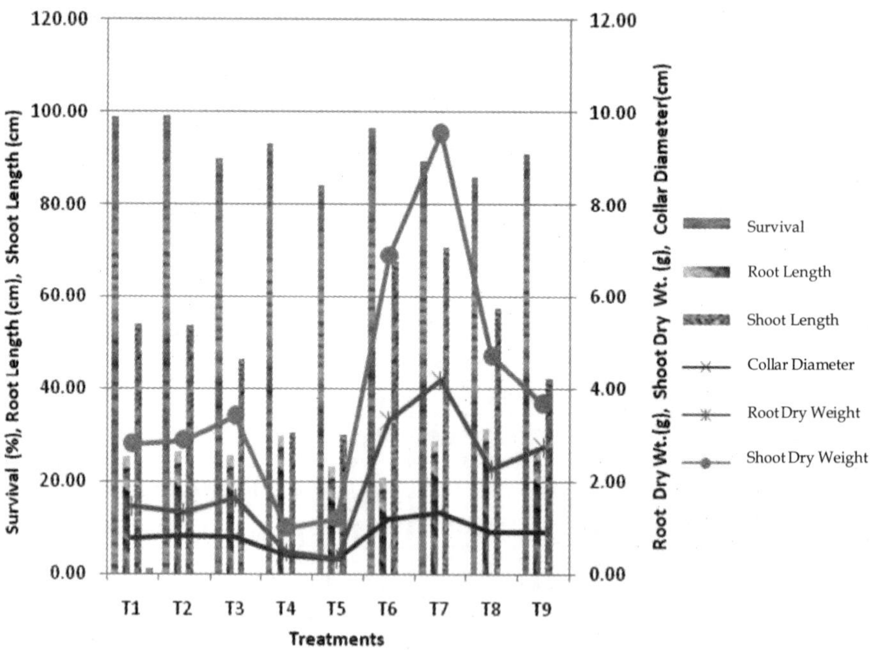

Fig. 5: Effect of soil pH and organic matter on seedling attributes of 180 days old

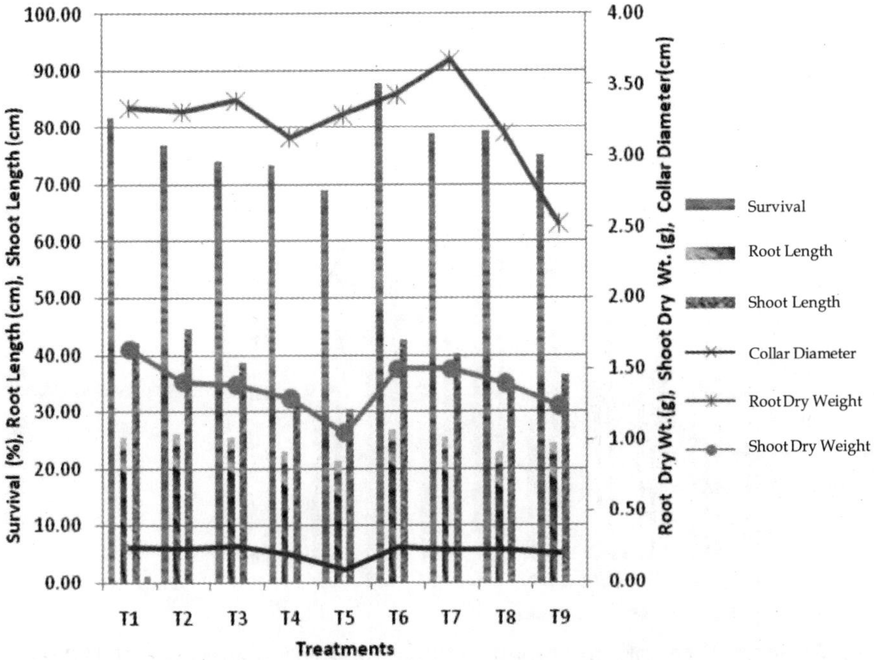

Fig. 6: Effect of soil pH and organic matter on seedling attributes of 180 days old Casuarina equisetifolia

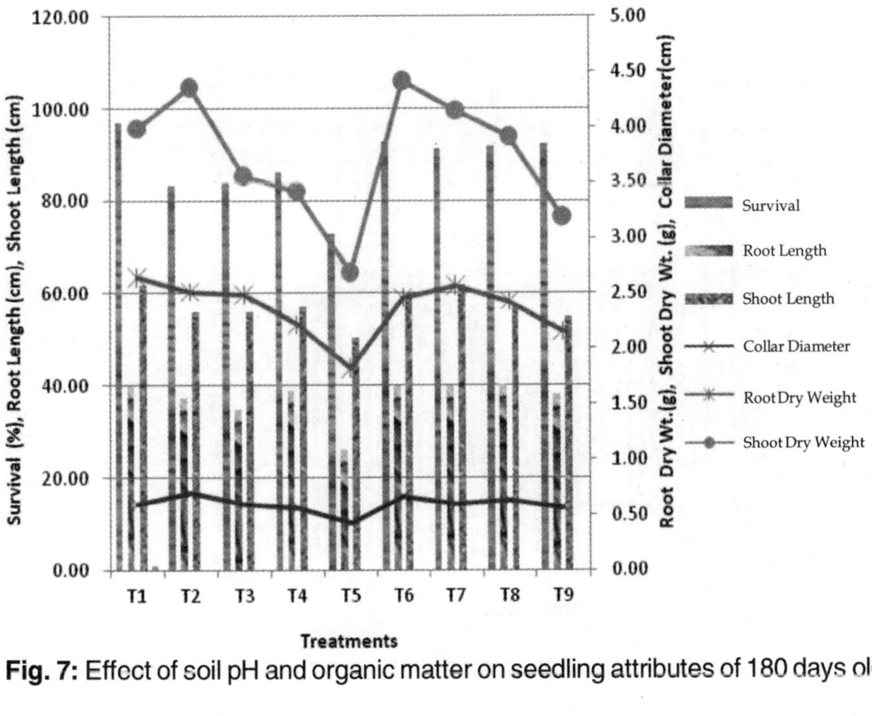

Fig. 7: Effect of soil pH and organic matter on seedling attributes of 180 days old

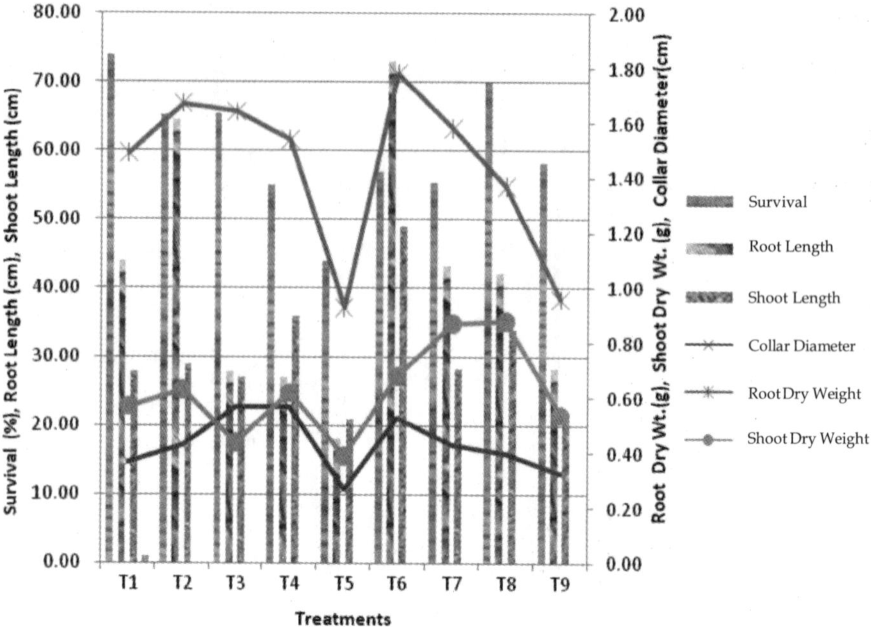

Fig. 8: Effect of soil pH and organic matter on seedling attributes of 180 days old.

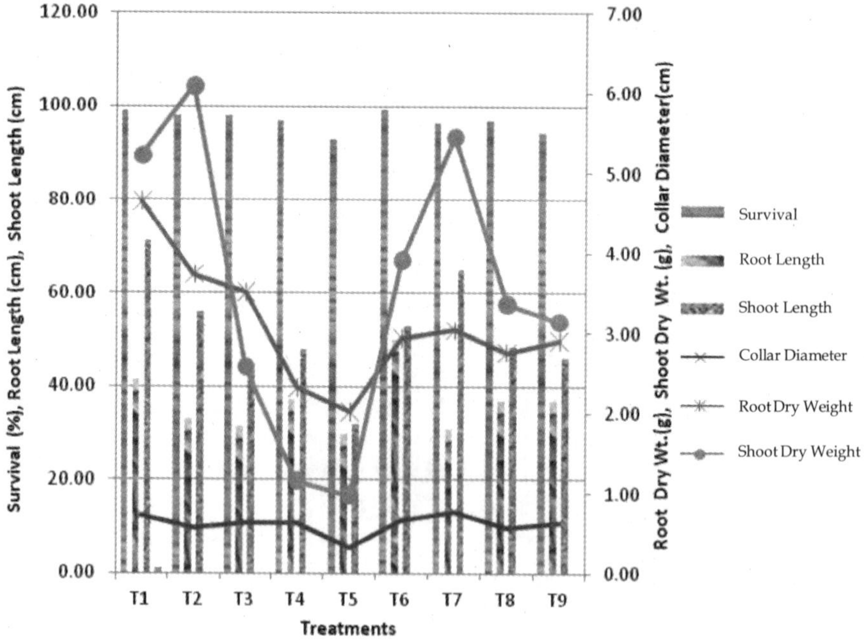

Fig. 9: Effect of soil pH and organic matter on seedling attributes of 180 days old

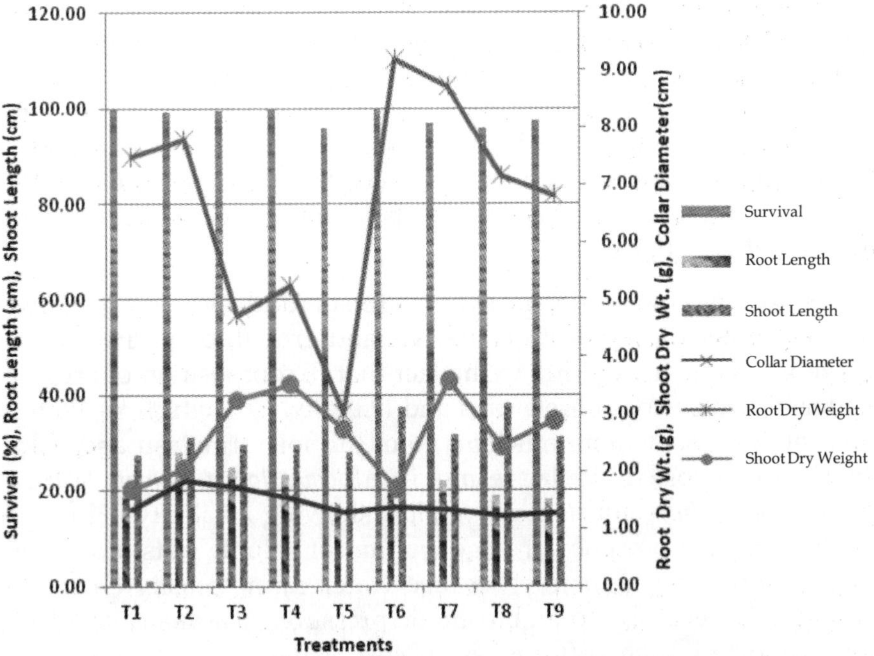

Fig.10: Effect of soil pH and organic matter on seedling attributes of 180 days old

Discussion

In this experiment, seeds were sown in soils with different pH (8.1, 9.0, 10.2 and 10.5) and in the soil with different pH mixed with sand and farm yard manure at 2:1:1 ratio were compared with control. Significant differences were observed among different treatments for all the parameters of germination and seedling growth attributes at initial stage. The results on germination and initial seedling vigour of all the species showed that the highest germination percentage in the soil with pH of 8.1 mixed with sand and farm yard manure at 2:1:1 ratio. While the lowest germination was recorded with T_5 (Soil with pH of 10.5) treatment. The lowest germination percentage noticed in the strongly alkali soil might be due to the toxic effect of sodium carbonate which would have increased the osmotic potential of the soil and decreased the water potential of soil solution. Besides the harmful effect on exchangeable sodium (49.5%), it might have the affected the radical emergence of the seedling. These results are in corroboration with the earlier findings made in Eucalyptus (Verma. 2006). Srinivasu and Toky (1996) have reported similar results in *Albizia lebbeck, Acacia nilotica, Pithacellobium dulce* and *Prosopis juliflora* in alkaline soil with pH ranged from 8.5 to 11.0. Lacey and Line (1994) observed low germination in *Eucalyptus regnans* in a soil with pH of 10.0 and the seedling survival after two weeks was zero. The presence of large amount of sodium carbonate and sodium bicarbonate in alkaline soil retards germination because of higher osmotic potential of soil solution which would have reduced the turgor pressure and in turn the entry of water into the seeds would have been prevented (Sharma *et al.*, 2007).

At 180 days old seedlings, the addition of sand and organic matter resulted in beneficial effect on survival and growth of all the species studied. This is mainly due to the fact that decomposition of organic matter dissolved the soluble salts and liberated CO_2 which facilitated the roots to absorb more nutrients from the soil. Srinivasu and Toky (1996) also reported similar results in *Acacia nilotica, Albizia lebbeck, Pithecellobium dulce* and *Prosopis juliflora* showing alkalinity pH range of 8.5-11.0. The presence of large amount of soluble salts in the soil retards germination mainly because water cannot easily enter the germinating seed due to high osmotic pressure. Sharma *et al.* (1991) have reported such differences. *Ceiba pentandra, Acacia nilotica, Pongamia pinnata, Hardwickia binata, Ailanthus exelsa, Eucalyptus tereticornis, Casuarina equsetifolia and Albizzia lebbeck* seeds and *Tectona grandis* stumps placed in higher soil pH (10.5) alone retarded the seedling growth when compared to seeds sown with a soil pH of 10.5 + sand + farm yard manure at 2:1:1 ratio (Masilamani *et al.*, 2000; Masilamani *et al.*, 2001; Masilamani *et al.*, 2002; Masilamani *et al.*, 2008; Masilamani, 2009; Masilamani *et al.*, 2009; Masilamani *et al.*, 2012: Masilamani *et al.*, 2013a; Masilamani *et al.*, 2013b). Seedling growth attributes enhanced by addition of sand and farm yard manure in varying soil pH might be due to CO_2 liberated by the decaying organic matter and by the plant roots which increases the solubility of calcium carbonate (Metha, 1986; Daji, 1992). Consequently, the concentration of Ca ions in the soil solution increases and the Ca ions on the soil solution helps to promote the reclamation of alkali soils (Metha, 1986; Daji, 1992).

The results on biochemical constituents of 180 days old seedlings of different species revealed that there was not much variation in chlorophyll content (a, b and total) between the treatments tried. The highest total N content was observed with T_1 (red earth mixed with sand and farm yard manure at 2:1:1 ratio) treatment. On the other hand, reduction of P content was observed when the soil pH was increased from 9.0 to 10.2. This might be due to reduced absorption of $H_2PO_4^-$ ion from the labile pool in the alkali soil due to fixation of soluble form of P in to poorly soluble from (Paul *et al.*, 2006). The highest K content was observed with the strong alkaline soil mixed with sand and farmyard manure at 2:1:1 ratio, whereas the normal nursery mixture recorded the lowest value for almost all the species studied. There was a significant difference in total K content among the remaining treatments. This may be due to the micro organisms which are involved in the decomposition and transformation of organic form of N in the system declined in the alkaline pH ranges (Patra *et al.*, 1996). The lowest K content was recorded in the control (red earth + sand +farm yard manure at 2:1:1 ratio) which might be due to loss of K due to higher leaching of K^+ ion in the coarse textured porous soil (Dutta and Joshi,1990).

Conclusion

In conclusion, *Acacia nilotica, Ailanthus exelsa, Albizia lebbeck, Azadirachta indica, Casuarina equsetifolia, Ceiba pentandra, Eucalyptus tereticornis, Hardwickia binata, Pongamia pinnata and Tectona grandis* had a wider range of adaptability in relation to pH. These species were able to tolerate soil pH of 10.5, when the soil is mixed with sand and farmyard manure mixture at 2:1:1 ratio, in terms of germination, seedling and chemical attributes. These species may have greater potential for wasteland afforestation and social forestry programme to meet the energy and fodder demand as well as to maintain ecological balance and environmental stability.

References

Abdul-Baki, A.A. and J.D.Anderson. 1973. Vigour determination in soybean seed by multiple criteria. *Crop Sci.*, 13: 630-633.

AOSA. 1965. Rules for testing seeds. *Proceedings of Association of Official seed Analysts*. 54:1-112.

Daji, J.A. 1992. A Text book of Soil Science. Media Promoters and Publishers Pvt. Ltd., Bombay. 380 p.

Dutta, B.K. and D. Joshi. 1990. Quantity - intensity parameters of potassium and their relationship with available forms and soil properties in dune and introduced soils. *J. Indian Soc. Soil Sci.*, 38: 404-409.

Gomez, K.A, and A.A.Gomez. 1984. Statistical Procedure for Agricultural Research. Wiley Inter - Science Publications, New York. 680 p.

Gupta, B.N., P.G. Pattanath, A. Kumar, R.C. Thapliyal and A.S. Rahuri. 1975. Rules for germination test of tree seeds for certification. *Indian For.* 101: 320-327.

International Seed Testing Association. 1993. International Rules for Seed Testing. *Seed Sci. and Technol.*, 21:177.

International Seed Testing Association. 1985. International rules for seed testing.. *Seed Sci. and Technol.*, 13 : 322-341.

Lacey, M.J. and M. A. Line. 1994. Influence of soil pH on the germination and survival of *Eucalyptus regnans* F. Muell, in Southern Tasmania. *Australian For.*, 57: 105-108.

Masilamani, P., B. Gurudev Singh, C. Chinnusamy and K. Annadurai, 1999. Influence of seed orientation and depth of sowing on germination and vigour of Anjan (*Hardwickia binata* Roxb.). *Tropical Agri. Res. Exten.* 2:76-78.

Masilamani, P., K. Annadurai, P. Saravanapandian and A. Bharathi. 2000. Effect of organic matter on germination and seedling attributes of kapok (*Ceiba pentandra* L. G aertn) in soils of varying pH. Paper presented in International workshop on agro forestry and forest products, held at Department of Forest, NEHU, Mizoram campus, Aizawl, Mizoram, India during 28-30, November 2000. P. 259-284.

Masilamani, P., K. Annadurai, P. Saravanapandian and A. Bharathi. 2001. Effect of soil reaction and organic matter on germination and seedling vigour of Babul (*Acacia nilotica* Linn.). *Madras Agricultural Journal*. 88: 287-290.

Masilamani, P, K. Annadurai and P. Saravanapandian. 2002. Effect of soil pH and organic matter on germination and growth attributes pungam (*Pongamia pinnata* Roxb). *Indian J. For.*, 25: 415-419.

Masilamani, P., K. Annadurai, P. Saravanapandian and M. Bhaskaran. 2008. Effect of soil alkalinity and organic matter on seedling emergence and growth attributes of Anjan (*Hardwickia binata* Roxb,). *Indian J. For.*, 31: 399-404.

Masilamani, P. 2009. Effect of soil pH and organic matter on germination and growth attributes of *Ailanthus excelsa* Roxb.. *Indian Journal of Agro forestry*. 2:53-56.

Masilamani, P., P. Saravanapandian and K. Annadurai. 2009. Effect of soil alkalinity and organic matter on germination and growth of *Eucalyptus tereticornis*. *Madras Agricultural Journal*. 89: 435-442.

Masilamani, P., K.Annadurai and P. Saravanapandian. 2011. Effect of soil reaction and organic matter on germination and seedling growth attributes of Neem (*Azadirachta indica A.Juss*). Paper presented in National Seminar on Abiotic Stress Management of Crop Plants held at Department of Genetics and Plant Breeding, Faculty of Agriculture, Annamalai University, Chidambaram, Tamil Nadu, during 28-29, November 2010. P. 233-242.

Masilamani, P, K. Annadurai and P. Saravanapandian. 2012. Effect of soil reaction and organic matter on germination and seedling growth attributes of Casuarina (*Casuarina equisetifolia* Forst.). Paper accepted for publication in *Seed Research.*

Masilamani, P., K. Annadurai, P.Saravanapandian and U. Arulanandu. 2013a. Study on moderating influence of FYM on pH of nursery medium for raising Teak stumps. Paper presented in National Seminar on Abiotic Stress and Crop Improvement held at Department of Genetics and Plant Breeding, Faculty of Agriculture, Annamalai University, Chidambaram, Tamil Nadu, during 22nd February 2013. P. 44-45.

Masilamani, P., P. Saravanapandian, K. Annadurai and U. Arulanandu. 2013b. Influence of soil pH and farm yard manure addition on germination and seedling growth attributes of *Albizia lebbeck* Benth.(kokko). Paper accepted for publication in *Indian Forester* June,2013 issue.

Metha, K.K. 1986. Reclamation of alkali soils of India. Oxford & IBH Publishing Co. New Delhi. 280 p.

Panse, V.G. and Sukhatme , R.V. 1985. Statistical methods for agricultural workers. Indian Council of Agricultural Research Publications, New Delhi 380 pp.

Patra, D.D., Arunprasad, Nitpendha kumar and D.V. Singh. 1996. Nitrogen Transformation in salt affected soils. J. *Indian Soc. Soil Sci.,* 44: 151-153.

Paul, R., A.K. Kapoor and S. K. Sharma. 2006. Effect of saline water, FYM and phosphorus on the performance of different varieties of Okra. *Haryana J.Agron.,* 22: 98-100.

Roy, M.M . 1986. Effect of pH on germination of *Dichrostachys cinerea* (L) Weight & Arn. *J. Tree Sci.,* 5: 62-64.

Sharma, S.D., K.G. Prasad and H.B. Singh.1991. Salinity and alkalinity tolerance by selected Eucalyptus species. *Van Vigyan.* 29:9-16.

Sharma, A., M.V. Kukkadia, D.B. Jadeja and B. G. Vashi. 1998. Effect of different salinity levels on germination and initial growth parameters of different agro forestry tree species in nursery stage. *Indian J. For.,* 21:156-159.

Sharma, Y., B. L. Verma and R. A. Singhania. 2007. Use of Agro-chemicals for minimising the alkalinity hazards and sustaining yield of mustard and cluster bean on alkaline water irrigated soils through sprinklers. *J.Water Mgt.,* 15:57-61.

Singh, P. P.S. Pathak and .M.M.Ray. 1994. Agro-forestry systems on degraded lands–An introduction. In. Agro forestry systems for sustainable land use(eds.) Panjab Singh *et al.*, Oxford and IBH Publishing Co. Pvt. Ltd. 122-135.

Srinivasu, V. and O.P. Toky. 1996. Effect of alkalinities on seed germination and seedling growth of important arid trees. *Indian J. For.*, 19: 227-283.

Verma, B.L.2006. Management of saline water in arid and semi region. Winter school on sustainable farming system for arid and semi-arid eco system held at CAZRI sub centre, Bikaner, November 04-24, 2006, pp 33-35.

❑❑❑

Chapter 13

High Temperature Stress in Crops

K. BALAKRISHNAN AND P. MASILAMANI
Dept. of Seed Science and Technology Agricultural College and Research Institute
Tamil Nadu Agricultural University Madurai, Tamil Nadu - 625 104

Heat tolerance is generally defined as the ability of the plant to grow and produce economic yield under high temperature (Wahid *el al.,* 2007). True heat tolerance at sensitive stages might be conferred by protecting structural proteins, enzymes and membranes from heat damage. The role of heat shock proteins (HSPs) and other stabilizing factors is crucial in these processes (Maestri *el al.,* 2002). Rising temperature may lead to altered geographical distribution of crops by allowing the threshold temperature for the start of the season and crop maturity to research earlier (Porter, 2005). At moderately high temperature, injuries or death may occur only after long term exposure.

Temperature is one of the most important environmental factors that govern the adaptation and speciation of plants to a wide range of environments with combinations and abiotic conditions and biotic interactions. Plants have variable complex combinations of morphological, anatomical, physiological traits evolved and biochemical traits that suit the species for its perpetuation in a particular habitat (Bowers, 1994). Plants can grow successfully in a wide range of different thermal environments, extending from temperature below freezing point in arctic region to roaring temperature above 55^{0}C of the hot desert (Table 1).

Table 1: Temperature limits of growth in different plant groups.

S.No	Group	Features	Upper Temp. limit(°C)
1	Psychrophil es z	Plants thrive at cold temperature (0 20°C), are freezing s tress tolerant. Plants of tropical and sub -tropical regions are susceptible to cold temperature.	15-20
2	Poikilotherms	Plants attain the low temperature of their environment, while still alive. Chilling sensitive plants are injured at temperature above freezing point up to 1 5°-20°C and chilling resistant plants can grow at temperature near 0°C. Temperature species are frost tolerant.	15-20
3	Mesophiles	Plants show maximum growth at moderate temperature (10°-30°C). Plants in the natural habitat as well as plant cultivated in tropical and sub - tropical regions.	35-45
4	Thermophiles	Chemotrophic bacteria	
		Plants thrive at high temperature (heat stress):	45
		Vascular plant	>100
5	Poikilohydr ous/ Resurrection plants	Plants have protoplasmic resistance to injury under extreme dehydration; assume metabolic activity on rehydration within relatively short time. Spores, seed and sporophytes of moss and grasses.	0 100

Maximum productivity of crops under optimal growth conditions may take place as a result of specific trait combinations that maximize yield. Temperature has been rated as the most important limiting factor in crop raising. Slight variation in temperature even transiently may affect the growth and yielding ability of the crop (Singh *et al.,* 1997).

Temperature Requirement

It is a measure of intensity of heat. Different crops grow at different temperature (Table 2).

Table 2: Temperature requirements of crops.

Crops	Temperature
Paddy	21°C - 35°C
Wheat	10°C - 35°C
Potato	15.5°C - 22°C
Jute	24°C - 37°C
Sugarcane	21°C - 26°C
Mustard	21°C – 25.7°C
Sesame	21°C – 27°C
Groundnut	21.°C – 26.6°C
Sunflower	28°C - 32°C
Sweet potato	21°C - 29°C
Brinjal	21°C - 29°C
Tomato	21°C - 24°C
Carrot	16°C - 18°C
Cabbage	10°C - 25°C
Cauliflower	17°C - 29°C
Lettuce	15.5°C - 20°C
Knolkhol	18°C - 25°C
Cucumber	18°C - 25°C
Watermelon	21°C - 29°C
Pumpkin	18°C - 25°C
Okra	21°C - 30°C

Air temperature has a influence on the growth of the crop to a certain extent. Generally the plant will not grow at below 4.4°C and above 48.4°C. Generally 21.2°C-26.6°C temperature is required for the proper growth and development of the plants.

Indices of Temperature Responses of Crops

Sensitive crop species are those that are inhibited by relatively low temperatures and lack the capacity to adjust their heat sensitive response to increasing ambient temperature. Low temperature sensitive crop species will be injured at tissue temperature between 30 and 40°C. Facultative heat tolerant species can adjust the cytoplasmic heat tolerance in response (above 50°C) is not commonly observed in the crop plants, except the seeds, which have evolved the unique mechanism of resumption of growth (germination) from highly dehydrated state attained during maturity. Conventionally, range of temperature variability (minimum, optimum, maximum) in the cropping season, has been defined as a the cardinal points, which are used to get an appropriate idea about the temperature requirements and the level of the temperature regimes at which it becomes stress for the growth and yield of crops (Table 3).

Table 3: Cardinal points (°C) for crops grown in warm (kharif) and cold (rabi) seasons.

Sl. No.	Response	Cool season	Warm season
1.	Minimum temperature below which growth is inhibited	0-5	15-18
2	Optimal temperature which supports maximum growth	25-30	30-37
3	Maximum temperature above which growth ceases	30-37	44-50

Heat Stress Threshold Temperature

A threshold temperature refers to a value of daily mean temperature at which a detectable reduction in growth begins. Upper and lower developmental threshold temperatures have been determined for many plant species through controlled laboratory and field experiments. A lower developmental threshold or a base temperature is one below which plant growth and development stops. Similarly, an upper developmental threshold is the temperature above which growth and development ceases (Table 4).

Table 4: Threshold temperature for some crop plants.

Crops	Threshold temperature (°C)	Growth stage
Wheat	26	Post-anthesis
Corn	38	Grain filling
Cotton	45	Reproductive
Pearl millet	35	Seedling

Tomato	30	Emergence
Brassica	29	Flowering
Cool season pulses	25	Flowering
Groundnut	34	Pollen production
Cowpea	41	Flowering
Rice	34	Grain yield

(Vijayalakshmi and Boominathan, 2011)

High-Temperature Stress on Rice

The optimum temperature for the normal development of rice ranges from 27 to 32°C. High temperature affects almost all the growth stages of rice, i.e. from emergence to ripening and harvesting (Table 5).

Table 5: Symptoms of heat stress in rice plants.

Growth stage	Threshold temperature (°C)	Symptoms
Emergence	40	Delay and decrease in emergence
Seedling	35	Poor growth of the seedling
Tillering	32	Reduced tillering and height
Booting	-	Decreased number of pollen grains
Anthesis	33.7	Poor anther dehiscence and sterility
Flowering	35	Floret sterility
Grain formation	34	Yield reduction
Grain ripening	29	Reduced grain filling

(Shah et al., 2011)

The developmental stage at which the plant is exposed to heat stress determines the severity of the possible damage to the crop (Wahid *et al.,* 2007). Flowering (anthesis and fertilization) and booting stage (microsporogenesis) are considered to be the stages of development most susceptible to temperature in rice (Farrell *et al.,* 2006). Exposure to 41°C for 4h at flowering caused irreversible damage and plants became completely sterile (IRRI, 1976). A temperature increase of 1°C shortened the number of days from sowing to heading by 4-5 days for some genotypes (Nakagawa *et al.,* 2001).

Crop Response to Temperature Stress

1. Morphological symptoms

In tropical climates, excess of radiation and high temperatures are often the most limiting factors affecting plant growth and final crop yield. High temperatures can cause considerable pre-and post-harvest damages, including scorching of leaves and twigs, sunburns on leaves, branches and stems, leaf senescence and abscission, shoot and root growth inhibition, fruit discoloration and damage, and reduced yield (Vollenweider and Gunthardt-Goerg, 2005).

High-temperature-induced modifications in plants may be direct as on existing physiological processes or indirect in altering the pattern of development. These responses may differ from one phenological stage to another. For example, long-term effects of heat stress on developing seeds may include delayed germination or loss of vigor, ultimately leading to reduced emergence and seedling establishment.

Heat stress lengthened the duration of grain filling with reduction in kernel growth leading to losses in kernel density and weight by up to 7 per cent in spring wheat (Guilioni *et al.*, 2003). and grain quality in other cereals under heat stress (Maestri *et al.*, 2002). In tomato, reproductive processes were adversely affected by high temperature, which included meiosis in both male and female organs, pollen germination and pollen tube growth, ovule viability, stigmatic and style positions, number of pollen grains retained by the stigma, fertilization and post fertilization processes, growth of the endosperm, proembryo and fertilized embryo (Foolad, 2005).

Reproductive phases most sensitive to high temperature are gametogenesis (8-9 days before anthesis) and fertilization (1-3 days after anthesis) in various plants (Foolad, 2005). Both male and female gametophytes are sensitive to high temperature and response varies with genotype; however, ovules are generally less heat sensitive than pollen. Overall, based on the available studies, it seems that plant responses to high temperature vary with plant species and phenological stages. Reproductive processes are markedly affected by high temperatures in most plants, which ultimately affect fertilization and post-fertilization processes leading to reduced crop yield.

Plant architecture can play an important role in high- temperature stress tolerance. The plant morphology is such that the panicle is surrounded by many leaves, the plant will be able to withstand high-temperature stress due to increased transpirational cooling and by preventing evaporation from the anther due to its shading by the leaves. The reduced evaporation from the anther will ensure swelling of the pollen grains, an important trait for anther dehiscence. The reduced height may also enhance resistance to lodging, thus giving more resilience to the crop against indirect stresses such as floods, which will become more frequent due to global warming.

Pollination contributing factors (pollen production, viability and reception) play a dominant role in productivity of the crop. Generally, male reproductive development in rice is known to be more sensitive to heat stress (Wassmann *et al.*, 2009). Prasad *et al.* (2006) reported that high-temperature stress during rice flowering led to decreased pollen production and pollen shed. The probable reasons were the inhibition of swelling of pollen grains, indehiscence of anthers and poor release of pollen grains (Matsui *et al.*, 2000), and thus fewer numbers of pollen grains were available to be intercepted by the stigma. Mackill *et al.* (1982) stated that the proportional fertility was positively correlated with the number of pollen grains shed on the stigma under both high and ambient temperatures. Physiologically, the decreased production of pollens at elevated temperatures may be attributable to impaired cell division of the microspore mother cells (Takeoka *et al.*, 1992). The decrease in spikelet fertility can be termed a phenotypic character of rice plant under high temperature, while the decrease in pollen germination and activity can be considered as the physiological factors responsible for this decrease (Tang *et al.*, 2008).

Temperatures higher than the optimum induced floret sterility and thus decreased rice yield (Nakagawa *et al.* 2003). A key mechanism of high-temperature- induced floret sterility in rice is the decreased ability of the pollen grains to swell, resulting in poor dehiscence (Matsui *et al.* 2000). This swelling of pollen grains in the locules is the driving force for anther dehiscence. Significant genotypic variations in high-temperature induced floret sterility exist. Matsui *et al.* (2001) found 3°C variation of temperatures causing sterility of 0-5 between the most tolerant and most susceptible cultivars among nine *japonica* cultivars. Greater increments in temperature resulted in higher proportions of sterility. Exogenous application of growth regulators has been shown recently to have some positive effects on the spikelet fertility and pollination (Mohammed & Tarpley, 2009). In fact, their exogenous application increased the level of endogenous antioxidants and thus prevented the oxidative damage to the membranes in rice (Mohammed & Tarpley, 2009).

2. Anatomical changes

At the whole plant level, there is a general tendency of reduced cell size, closure of stomata and curtailed water loss, increased stomatal and trichomatous densities, and greater xylem vessels of both root and shoot (Anon *et al.*, 2004). At the sub-cellular level, major modifications occur in chloroplasts, leading to significant changes in photosynthesis. For example, high temperatures reduced photosynthesis by changing the structural organization of thylakoids (Karim *et al.*, 1997). In general, it is evident that high temperature considerably affects anatomical structures not only at the tissue and cellular levels but also at the sub-cellular level. The cumulative effects of all these changes under high temperature stress may result in poor plant growth and productivity.

3. Phenological changes

The developmental stage at which the plant is exposed to the stress may determine the severity of possible damages experienced by the crop. It is, however, unknown whether damaging effects of heat episodes occurring at different developmental stages are cumulative (Wollenweber *et al.*, 2003). Vulnerability of species and cultivars to high temperatures may vary with the stage of plant development, but all vegetative and reproductive stages are affected by heat stress to some extent. Impairment of pollen and anther development by elevated temperatures is another important factor contributing to decreased fruit set in many crops at moderate-to-high temperatures (Sato *et al.*, 2006). The staple cereal crops can tolerate only narrow temperature ranges, which if exceeded during the flowering phase can damage fertilization and seed production, resulting in reduced yield (Porter, 2005). Under high temperature conditions, earlier heading is advantageous in the retention of more green leaves at anthesis, leading to a smaller reduction in yield (Tewolde *et al.*, 2006).

4. Physiological changes

a) Waters relations

Plant water status is the most important variable under changing ambient temperatures (Mazorra *et al.*, 2002). In sugarcane, leaf water potential and its components were changed upon exposure to heat stress even though the soil water supply and relative humidity conditions were optimal, implying an effect of heat stress on root hydraulic conductance. In general, during daytime enhanced transpiration induces water deficiency in plants, causing a decrease in water potential and leading to perturbation of many physiological processes (Tsukaguchi *et al.*, 2003). High temperatures seem to cause water loss in plants more during day time than night time.

b) Accumulation of compatible osmolytes

A key adaptive mechanism in many plants grown under abiotic stresses, including salinity, water deficit and extreme temperatures, is accumulation of certain organic compounds of low molecular mass, generally referred to as compatible osmolytes (Sakamoto and Murata, 2002).

Glycine betaine (GB), an amphoteric quaternary amine, plays an important role as a compatible solute in plants under various stresses, such as salinity or high temperature (Sakamoto and Murata, 2002). In assessing the functional significance of accumulation of compatible solutes, it is suggested that proline or GB synthesis may buffer cellular redox potential under heat and other environmental stresses (Wahid and Close, 2007).

c) Photosynthesis

The changes in various photosynthetic attributes under heat stress are good indicators of thermo tolerance of the plant as they show correlations with growth. Any constraint in photosynthesis can limit plant growth at high temperatures. Photochemical reactions in thylakoid lamellae and carbon metabolism in the stroma of chloroplast have been suggested as the primary sites of injury at high temperatures (Wise *et al.*, 2004). Chlorophyll fluorescence, the ratio of variable fluorescence to maximum fluorescence (Fv/Fm), and the base fluorescence (F0) are physiological parameters that have been shown to correlate with heat tolerance (Yamada *et al.*, 1996). Furthermore, under high temperatures, degradation of chlorophyll a and b was more pronounced in developed compared to developing leaves (Karim *et al.*, 1997, 1999). PSII is highly thermolabile, and its activity is greatly reduced or even partially stopped under high temperatures (Camejo *et al.*, 2005), which may be due to the properties of thylakoid membranes where PSII is located (Mcdonald and Paulsen, 1997).

High temperature influences the photosynthetic capacity of C_3 plants more strongly than in C_4 plants. It alters the energy distribution and changes the activities of carbon metabolism enzymes, particularly the rubisco, thereby altering the rate of RuBP regeneration by the disruption of electron transport and inactivation of the oxygen evolving enzymes of PSII (Salvucci and Crafts-Brandner, 2004b).

A well-known consequence of elevated temperature in plants is the damage caused by heat- induced imbalance in photosynthesis and respiration; in general, the rate of photosynthesis decreases while dark and photo respiration rates increase considerably under high temperatures. Also, rate of biochemical reactions decreases and enzyme inactivation and denaturation take place as the temperature increases leading to severely reduced photosynthesis (Nakamoto and Hiyama, 1999). Nonetheless, photosynthesis is considered as the physiological process most sensitive to high temperatures, and that rising atmospheric CO_2 content will drive temperature increases in many already stressful environments. This CO_2 induced increase in high-temperature tolerance may have a substantial impact on both the productivity and distribution of many crop species in future.

d) Flowering

The occurrence of flowering early in the morning was discussed many years ago as a useful phenomenon imparting heat tolerance to rice genotypes (IRRI, 1997). *A* rice floret emerges from the sheath of the flag leaf just before anthesis and opens for about one hour. in the morning (Matsui & Kagata 20036). Avoidance mechanisms also include: (i) cooling of the spikelet at flowering (ii) asynchronous tiller and panicle development (iii) asynchronous flowering time of spikelet's within each panicle and (iv) anthesis and pollination taking place within the same spikelet (self-pollination). The latter was considered as one of the major difference between heat-tolerant and susceptible varieties (Yoshida, 1981). Yield depression resulting from exposure to high temperatures during critical developmental stages is mitigated to some extent within a crop because it is composed of individuals whose tillers are not of the same age. This indicates that the panicles produced pass through their vulnerable stages at different times and some might therefore escape heat damage. It has been reported that cultivars with large anthers are tolerant of high temperature at the flowering stag (Matsui & Omasa, 2002). It can be assumed that rice cultivars with large anthers' are tolerant to temperature stress because they have a large number of pollen grains per anther, which compensates for the reduction in the number of pollen grains that germinate under high temperature (Matsui & Omasa, 2002).

The fertility of spikelet's at high temperature can be used as a screening tool for high-temperature tolerance during the reproductive phase. Selection for heat tolerance should be done for those breeding materials which can tolerate temperatures higher than 38°C (Satake & Yoshida, 1978). Exploiting the existing genotypic variation in flowering time (especially the wild-type) serves as a useful mitigation option for rising temperature as it is a relatively simple trait that needs to be focused in breeding programmes. Introgression of the early morning flowering gene from *Oryza officinalis* into *O. sativa* has recently been shown to have positive effect on reducing the spikelet fertility (Ishimaru *et al.*, 2010).

e) Assimilate partitioning

Under low to moderate heat stress, a reduction in source and sink activities may occur leading to severe reductions in growth, economic yield and harvest index. Assimilate partitioning, taking place via apoplastic and symplastic pathways under high temperatures has significant effects on transport and transfer processes in plants. It was concluded that, at least in wheat, temperature effects on translocation results indirectly from temperature effects on source and sink activities. From such results, increased mobilization efficiency of reserves from leaves, stem or other plant parts has been suggested as a potential strategy to improve grain filling and yield in wheat under heat stress. This suggestion, however, is based on present limited knowledge of physiological basis of assimilate partitioning under high temperature stress.

f) Cell membrane thermostability

The integrity and functions of biological membranes are sensitive to high temperature, as heat stress alters the tertiary and quaternary structures of membrane proteins. Such alterations enhance the permeability of membranes, as evident from increased loss of electrolytes. The increased solute leakage, as an indication of decreased cell membrane thermo stability (CMT), has long been used as an indirect measure of heat stress tolerance in diverse plant species, The relationship between CMT and crop yield under high temperatures may vary from plant to plant and invokes for study of individual crops before using it as an important physiological selection criterion.

g) Hormonal changes

Hormonal homeostasis, stability, content, biosynthesis and compartmentalization are altered under heat stress (Maestri *et al.*, 2002). Abscisic acid (ABA) and ethylene (C_2H_4), as stress hormones, are involved in the regulation of many physiological properties by acting as signal molecules. Different environmental stresses, including high temperature, result in increased levels of ABA. ABA mediates acclimation/adaptation of plants to desiccation by modulating the up or down-regulation of numerous genes (Xiong *et al.*, 2002). Under field conditions, where heat and drought stresses usually coincide, ABA induction is an important component of thermo tolerance, suggesting its involvement in biochemical pathways essential for survival under heat-induced desiccation stress (Maestri *et al.*, 2002)

A gaseous hormone, ethylene regulates almost all growth and developmental processes in plants. Heat stress changes ethylene production differently in different plant species. High temperature-induced abscission of reproductive organs relates to an increased ACC level; this is accompanied with both reduced levels and transport capacity of auxins to reproductive organs (Huberman *et al.*, 1997).

Among other hormones, salicylic acid (SA) has been suggested to be involved in heat stress responses elicited by plants. SA is an important component of signaling pathways in response to systemic acquired resistance (SAR) and the hypersensitive response (HR) (Kawano *et al.*, 1998). SA stabilizes the trimers of heat shock transcription factors and aids them bind heat shock elements to the promoter of heat shock related genes. Long term thermotolerance can be induced by SA, in which both Ca^{++} homeostasis and antioxidant systems are thought to be involved (Wang and Li, 2006).

The effects of gibberellins and cytokinins on high temperature tolerance are opposite to that of ABA. An inherently heat tolerant dwarf mutant of barley impaired in the synthesis of gibberellins was repaired by application of gibberellic acid, whereas application of triazole paclobutrazol, a gibberellin antagonist, conferred heat tolerance (Vettakkorumakankav *et al.*, 1999).

h) Secondary metabolities

High-temperature stress induces production of phenolic compounds such as flavonoids and phenylpropanoids. Phenylalanine ammonia lyase (PAL) is considered to be the principal enzyme of the phenylpropanoid pathway. Increased activity of PAL in response to thermal stress is considered as the main acclamatory response of cells to heat stress. Thermal stress induces the biosynthesis of phenolics and suppresses their oxidation, which is considered to trigger the acclimation to heat stress for example as in watermelon, *Citrulus vulgaris* (Rivero *et al.*, 2001).

Isoprenoids, another class of plant secondary products, are synthesized via mevalonate pathway. Being of low molecular weight and volatile in nature, their emission from leaves has been reported to confer heat stress tolerance to photosynthesis apparatus in different plants (Loreto *et al.*, 1998). It is proposed that endogenous production of isoprene protects the biological membranes from damaging effects by directly reacting with oxygen singlets by means of isoprene-conjugate double bond (Velikova *et al.* 2005).

5. Molecular changes

a) Oxidative stress and antioxidants

In addition to tissue dehydration, heat stress may induce oxidative stress. For example, generation and reactions of activated oxygen species (AOS) including singlet oxygen, superoxide radical hydrogen peroxide and hydroxyl radical (OH^-) are symptoms of cellular injury due to high temperature (Liu and Huang, 2000). Decrease in antioxidant activity in stressed tissues results in higher levels of AOS that may contribute to injury (Fadzilla *et al.*, 1996). Protection against oxidative stress is an important component in determining the survival of a plant under heat stress.

b) Stress proteins

Expression of stress proteins is an important adaptation to cope with environmental stresses. Most of the stress proteins are soluble in water and therefore contribute to stress tolerance presumably via hydration of cellular structures. Induction of HSPs seems to be an universal response to temperature stress, being observed in all organisms ranging from bacteria to human (Vierling, 1991).

The mechanism by which HSPs contribute to heat tolerance is still enigmatic though several roles have been ascribed to them. Many studies assert that HSPs are molecular chaperones insuring the native configuration and functionality of cell proteins under heat stress. There is considerable evidence that acquisition of thermo tolerance is directly related to the synthesis and accumulation of HSPs (Bowen *et al.*, 2002). For instance, HSPs provide for new or distorted proteins to fold into shapes essential for their normal functions. They also help shuttling proteins from one compartment to another and transporting old proteins to "garbage disposals" inside the cell.

Mechanism of Heat Tolerance

Plants manifest different mechanisms for surviving under elevated temperatures, including long-term logical and morphological adaptations and short-term avoidance or acclimation mechanisms such as changing leaf orientation, transpirational cooling, or alteration of membrane lipid compositions. In many crop plants, early maturation is correlated with smaller yield losses under high temperatures, which may be attributed to the engagement of an escape mechanism (Adams et. al., 2001). Elucidating the various mechanisms of plant response to stress and their roles in acquired stress tolerance is of great practical and basic importance. Some major tolerance mechanisms, including ion transporters, osmoprotectants, free-radical scavengers, late embryogenesis abundant proteins and factors involved in signaling cascades and transcriptional control are essentially significant to counteract the stress effects (Wang *et al.*, 2004).

The mechanisms through which the plants can cope with high temperature stress are described by Waraich *et al.*, 2012 in Fig.1. The plants can cope with the high temperature stress by physiological, morpho-anatomical and biochemical alterations. Under high temperature stress, the plants accumulate the compatible osmolytes which helps to increase the retention of water in plants for better stomatal regulation and increased photosynthetic rate. The plants also exhibit some morpho-anotomical alterations to cope with high temperature stress which includes reduction in cell size, closure of stomata, increased stomatal and trichomes densities and greater xylem vessels. The third mechanism to cope with the high temperature is the biochemical alterations. The plants increased the stress related proteins which enhance the activities of antioxidants like superoxide dismutase (SOD); catalase (CAT) and peroxidise (POD) in the plant cells. These antioxidants scavenge the ROS and reduce the photo-oxidation and maintain the integrity of chloroplast membrane and increase the photosynthetic rate.

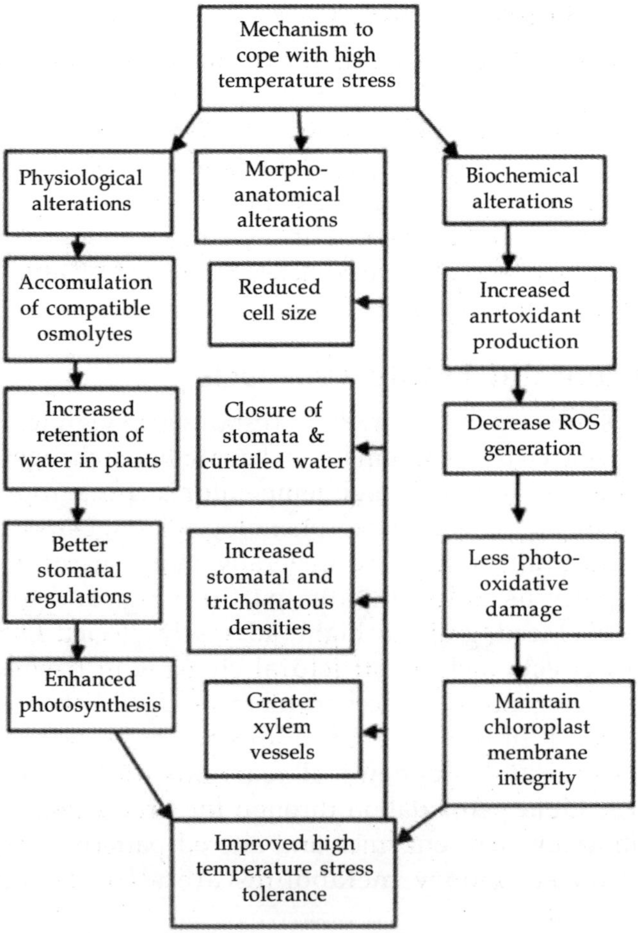

Fig. 1: Schematic representation of morphological, physiological and biochemical alteration of plants to cope with high temperature stress (Waraich et al., 2012)

The antioxidant defense mechanism is a part of heat-stress adaptation, and its strength is correlated with acquisition of thermo tolerance (Maestri *et al.* 2002). One of the most closely studied mechanisms of thermo tolerance is the induction of HSPs, which, as described in above, comprise several evolutionarily conserved protein families. However, each major HSP family has a unique mechanism of action with chaperonic activity. The protective effects of HSPs can be attributed to the network of the chaperone machinery, in which many chaperones act in concert. An increasing number of studies suggest that the HSPs/chaperones interact with other stress- response mechanisms (Wang *et al.* 2004).

Acquired Thermo Tolerance

Crop plants, like other organisms, have the ability to acquire thermo tolerance rather rapidly, may be within hours, so to survive under otherwise lethal high temperatures (Vierling, 1991). The acquisition of thermo tolerance is an autonomous cellular phenomenon and normally results from prior exposure to a conditioning pre-treatment, which can be a short but sublethal high temperature. The acquisition of high level of thermo tolerance protects cells and organisms from a subsequent lethal heat stress.

Conclusion and Future Prospects

Crop plants exhibit a variety of responses to high temperatures, which are depicted by symptomatic and quantitative changes in growth and morphology. Although high temperatures affect crop growth at all developmental stages, later phenological stages, in particular, anthesis and grain filling, are generally more susceptible. Pollen viability, patterns of assimilates partitioning, and growth and development of seed/grain are highly adversely affected. Other notable heat stress effects include structural changes in tissues and cell organelles, disorganization of cell membranes, disturbance of leaf water relations, and impedance of photosynthesis via effects on photochemical and biochemical reactions and photosynthetic membranes. Lipid peroxidation through the production of ROS and changes in antioxidant enzymes and altered pattern of synthesis of primary and secondary metabolites are also of considerable importance.

Although various signaling molecules are synthesized under heat stress, the role of Ca^{++} remains critical. A fundamental heat-stress response ubiquitous to plants is the expression of HSPs, which range from low (10 kDa) to high (100 kDa) molecular mass in different crop species. Such stress proteins are thought to function as molecular chaperones, helping in folding and unfolding of essential proteins under stress, and ensuring three dimensional structure of membrane proteins for sustained cellular functions and survival under heat stress. In addition to genetic means to developing plants with improved heat tolerance, attempts have been made to induce heat tolerance in a range of plant species using different approaches. These include preconditioning of plants to heat stress and exogenous applications of osmoprotectants or plant growth regulating compounds on seeds or whole plants.

Future research activities should be aimed at identification and breeding of heat tolerant germplasms exploiting the variation in both genotypic and morphological characters. Several approaches should be actively exploited to improve heat tolerance in current cultivars, including discovery and exploitation of new genes and alleles, improved breeding efficiency, marker-assisted selection and genetic modification. Furthermore, applications of genomics, proteomics and trascriptomics approaches to a better understanding of the molecular basis of plant response to heat stress as well as plant heat tolerance are imperative.

References

Adams, S.R., Cockshull, K.E. and Cave, C.R.J. 2001. Effect of temperature on the growth and development of tomato fruits. *Ann. Bot.* 88: 869-877.

Anon, S., Fernandez, J.A., Franco, J.A., Torrecillas, A., Alarc on, J.JS. and Anchez-Bianco, M.J. 2004. Effects of water stress and night temperature preconditioning on water relations and morphological and anatomical changes of Lotus creticus plants. *Sci. Hortic.* 101: 333- 342.

Bowen, J., Michael, L.-Y., Plummer, K.I.M. and Ferguson, I.A.N. 2002. The heat shock response is involved in thermo tolerance in suspension-cultured apple fruit cells. *J. Plant Physiol.* 159: 599-606.

Bowers, M.C. 1994. Environmental effects of cold on plants. In: R.E. Wilkinson (eds). Plant Environmental Interactions. Marcel Dekkar, New York, pp. 31-34.

Camejo, D., Rodriguez, P., Morales, M.A., DelFamico, J.M., Torrecillas, A. and Alarcon, J.J. 2005. High temperature effects on photosynthetic activity of two tomato cultivars with different heat susceptibility. *J. Plant Physiol.* 162: 281-289.

Fadzillah, N.M., Gill, V., Finch, R.P and Burdon, R.H., 1996. Chilling, oxidative stress and antioxidant responses in shoot cultures of rice. *Planta,* 199: 552-556.

Foolad, M.R. 2005. Breeding for abiotic stress tolerances in tomato. In: Ashraf, M., Harris, RJ.C. (Eds.), Abiotic Stresses: Plant Resistance through Breeding and Molecular Approaches. The Haworth Press Inc., New York, USA, pp. 613-684.

Guilioni, L., Wery, J and Lecoeur, J. 2003. High temperature and water deficit may reduce seed number in field pea purely by decreasing plant growth rate. *Funct. Plant Biol.* 30: 1151-1164.

Hall, A.E. 2001. Crop Responses to Environment. CRC Press LLC, Boca Raton, Florida.

Huberman, M., Riov, J., Aloni, B. and Goren, R. 1997. Role of ethylene biosynthesis and auxin content and transport in high temperature-induced abscission of pepper reproductive organs. *J. Plant Growth Regul.* 16: 129-135.

Ishimaru, T., Hirabayashi, H., Ida, M., Takai, T., San-Oh, Y. A., Yoshinaga, S., Ando, I,, Ogawa, T. and Ondo, M. 2010. A genetic resource for early-morning flowering trait of wild rice *Oryza officinalis* to mitigate high temperature-induced spikelet sterility at anthesis. *Annals of Bot.* 106: 515-520.

Karim, M.A., Fracheboud, Y and Stamp, P. 1997. Heat tolerance of maize with reference of some physiological characteristics. *Ann. Bangladesh Agri.* 7: 27-33.

Kawano, T., Sahashi, N., Takahashi, K, Uozumi, N and Muto, S. 1998. Salicylic acid induces extra cellular super oxide generation followed by an increase in cytosolic calcium ion in tobacco suspension culture: the earliest events in salicylic acid signal transduction. *Plant Cell Physiol.* 39: 721-730.

Liu, X. and Huang, B. 2000. Heat stress injury in relation to membrane lipid peroxidation in creeping bent grass. *Crop Sci.* 40: 503-510.

Loreto, F., Forster, A., Durr, M., Csiky, O and Seufert, G, 1998. On the monoterpene emission under heat stress and on the increased thermo tolerance of leaves of Quercus ilex L. fumigated with selected monoterpenes. *Plant Cell Environ.* 21: 101-107.

Mackill, D. J., Coffman, W. R. and Rutger, J. N. 1982. Pollen shedding and combining ability for high temperature tolerance in rice. *Crop Science.* 22: 730-733.

Maestri, E., Klueva, N., Perrota, C, Gulli, M., Nguyen, H.T. and Marmiroli, N. 2002. Molecular genetics of heat tolerance and heat shock proteins in cereals. *Plant Molecular Biol.* 48: 667-681.

Matsui, T., Omasa, K. and Horie, T. 2000. High temperature at flowering inhibits swelling of pollen grains, a driving force for thecae dehiscence in rice (*Oryza saliva* L.). *Plant Prod. Sci.* 3: 430-434.

Matsui, T., Omasa, K. and Horie, T. 2001. The difference in sterility due to high temperature during flowering period among *japonica-rice* varieties. *Plant Prod. Sci.* 4: 90-93.

Matsui, T. and Omasa, K. 2002. Rice (*Oryza sativa* L.) cultivars tolerant to high temperature at flowering: anther characteristics. *Annals Bot.* 89: 683-687.

Matsui, T. and Kagata, H. 2003. Gas exchange through the slit between the lemma and the palea in the rice (*Oryza sativa* L.) floret before anthesis. *Plant Prod. Sci.* 6: 262-264.

Mazorra, L.M., Nunez, M., Echerarria, E., Coll, F and Sanchez-Bianco, M.J. 2002. Influence of brassinosteriods and antioxidant enzymes activity in tomato under different temperatures. *Plant Biol.* 45: 593-596.

Mcdonald, G.K and Paulsen, G.M. 1997. High temperature effects on photosynthesis and water relations of grain legumes. *Plant Soil.* 196: 47-58.

Mohammed, A. R. and Tarpi. L. 2009. High night time temperatures affect rice productivity through altered pollen germination and spikelet fertility. *Agric. For. Meteorology.* 149: 999-1008.

Mohammed, A. R. and Tarpley, L. 2009*a*. Impact of high nighttime temperature on respiration, membrane stability, antioxidant capacity, and yield of rice plants. *Crop Sci.* 49: 313-322.

Nakagawa, H., Horie, T. and Matsui, T. 2003. Effects of climate change on rice production and adaptive technologies. In Rice Science: Innovations and Impact for Livelihood. Proceedings of the International Rice Research Conference. Beijing, China, 16-19, September 2002 (Eds T. W. Mew, D. S. Brar, S. Peng, D. Dawe & B. Hardy), pp. 635-658. Manila, the Philippines: IRRI.

Nakagawa, H., Takahashi, W., Hasegawa, T., Watanabe, T. and Horie , T. 2001. Development of a three-dimension simulator for rice growth and development. II. Accuracy of a rice phenology model to simulate heading stage and plat stage in leaf number. *Japanese J. Crop Sci.,* 70, 125-126.

Nakamoto, H and Hiyama, T. 1999. Heat-shock proteins and temperature stress. In: Pessarakli, M. (Ed.), Handbook of Plant and Crop Stress. Marcel Dekker, New York, pp. 399-416.

Porter, J.R. 2005. Rising temperatures are likely to reduce crop yields. *Nature.* 436:174.

Prasad, P.V.V., Boote, K.J., Allen , L.H. JR, Sheehy, J.E and Tomas, J. M. G.2006. Species ecotype and cultivation difference in spikelet fertility and harvest index of rice in response to high temperature stress. *Field Crops Sci.* 95: 398-411.

Rivero, R.M., Ruiz, J.M., Garcia, PC., Lopez-Lefebre, L.R., Sanchez, E and Romero, L. 2001. Resistance to cold and heat stress: accumulation of phenolic compounds in tomato and watermelon plants. *Plant Sci.* 160: 315-321.

Shan, F., Huang, j., Cui, K., Nie, L., Shah, T., Chen, C., and Wang, K. 2011. Impact of high-temperature on rice plant and its traits related to tolerance. *Journal Agricultural Science (Cambridge).* 149: 1-12.

Sakamoto, A and Murata, N. 2002. The role of glycine betaine in the protection of plants from stress: clues from transgenic plants. *Plant Cell Environ.* 25: 163 171.

Salvucci, M.E. and Crafts-Brandner, S.J. 2004. Inhibition of photosynthesis by heat stress: the activation state of Rubisco as a limiting factor in photosynthesis. *Physiol. Plant.* 120: 179- 186.

Satake, T. and Yoshida, S. 1978. High temperature-induced sterility in *indica* rices at flowering. *Japanese J. Crop Sci.* 47: 6-17.

Sato, S., Kamiyana, M., Iwata, T and Makitan, N. 2006. Moderate increase of mean daily temperature adversely affect fruit set of *Lycopersicon esculentum* by disrupting specific physiological processes in male reproductive development. *Ann. Bot.* 97: 731-733.

Singh, O.S., Gill, D.S. and Dhaliwal, G.S. 1997. Impact of air pollution on agro ecosystem. *Indian J. Ecol.* 24: 136-156.

Tang, R.S., Zheng, J.C., Jin Z.Q., Zheng, DD., Hung Y.H. and Chen , L.G. 2008. Possible correlation between high temperature –induced floret sterility and endogenous levels of IAA, GA's and ABA in rice (*Oryza sativa* L.). *Plant growth regul.* 54: 37-43.

Takeoka, Y., Al Mamun, A., Wade, T. and Kaufman, P.B. 1992. Primary feature of effect of environmental stress on rice spikelet morphogenesis. In reproduction adaptation to environmental stress (Eds Y. Takeoka, Al , A., Wade, T. & Kaufman, P.B). pp 113-141. Development in crop science, Vol. 22. Japan: Japan Scientific Press.

Tewolde, H., Fernandez, C.J and Erickson, C.A. 2006. Wheat cultivars adapted to post-heading high temperature stress. *J. Agron. Crop Sci.* 192: 111-120.

Tsukaguchi, T., Kawamitsu, Y., Takeda, H., Suzuki, K and Egawa, Y. 2003. Water status of flower buds and leaves as affected by high temperature in heat tolerant and heat-sensitive cultivars of snap bean (*Phaseolus vulgaris* L.). *Plant Prod. Sci.* 6: 4-27.

Velikova, V., Edreva, A and Loreto, F. 2005. Endogenous isoprene protects Phragmites australi leaves against singlet oxygen. *Plant Cell Environ.* 28: 318-327.

Vettakkorumakankav, N.N., Falk, D., Saxena, P and Fletcher, R.A. 1999. A crucial role of gibberellins in stress protection of plants. *Plant Cell Physiol.* 40: 542-548.

Vierling, E. 1991. The role of heat shock proteins in plants. *Ann. Rev. Plant Physiol. Plant Mol. Biol.* 42: 579-620.

Vollenweider, P and Gunthardt-Goerg, M.S. 2005. Diagnosis of abiotic and biotic stress factors using the visible symptoms in foliage. *Environ. Pollut.* 137: 455-465.

Waraich, E.A., Ahmad, R., Halim, A. and Aziz, T. 2012. Alleviation of temperature stress by nutrient management in crop plants: a review. *Journal Soil Science and Plant Nutrition.* 12:221-244

Wahid, A., Gelani , S., Ashraf , M. and Foolad, M.R. 2007. Heat tolerance in plants - an overview. *Environ. Experimental Bot.* 61: 199-233.

Wang, L.-J and Li, S.-L. 2006. Salicylic acid-induced heat or cold tolerance in relation to Ca2+ homeostasis and antioxidant systems in young grape plants. *Plant Sci.* 170 : 685-694.

Wang, W., Vinocur, B., Shoseyov, O and Altman, A. 2004. Role of plant heat-shock proteins and molecular chaperones in the abiotic stress response. *Trends Plant Sci.* 9: 244-252.

Wassmann, R., Jagadish, S.V.K., Heuser, S., Ismail, A., Redona, E., Serraj, R., Sing, R.K., Howell, G., Pathak and H. Sumfleth, K. 2009. Climate change affecting rice production: the physiology and agronomic basis for possible adaption strategies. *Adv. Agron.* 102: 91-133.

Wollenweber, B., Porter, J.R and Schellberg, J. 2003. Lack of interaction between extreme high temperature events at vegetative and reproductive growth stages in wheat. *J. Agron. Crop Sci.* 189: 142-150.

Yamada, M., Hidaka, T and Fukamachi, H. 1996. Heat tolerance in leaves of tropical fruit crops as measured by chlorophyll fluorescence. *Sci. Hortic.* 67: 39-48.

Yoshida, S. 1981. Fundamentals of Rice Crop Science. IRRI, Los Banos, The Philippines.

Chapter 14

Effect of Comprehensive Seed Treatments on Growth and Productivity of Cotton (*Gossypium hirsutum* L.) under Moisture Stress Condition

G. SATHIYA NARAYANAN, S. MANIMEKALAI AND M. PRAKASH
Department of Genetics and Plant Breeding, Faculty of Agriculture, Annamalai University

Abstract

Cotton (*Gossypium spp*), the 'white gold' enjoys a pre-dominant position amongst all the cash crops in India. Cotton is one of the principal commercial crops playing a key role in economic, political and social affairs of the country. Seed is a living hygroscopic material with a very complex and heterogeneous composition. Both under irrigated and rainfed conditions, suitable seed management techniques are required to encounter abiotic stresses such as excess soil moisture, salt stress, depletion of soil moisture during seed germination etc., that are anticipated during the crop growth. Enhancement in seed quality to face the challenges during seed germination and early seedling growth phases are considered as the paramount important for assured crop growth and successful production. Keeping these points in view, the present investigation was conducted to study the effect of comprehensive seed treatments on growth and productivity of cotton under moisture stress condition. The study was undertaken with the various seed treatments such as gaucho @10 gm kg^{-1}, halopolymer @ 4ml kg^{-1}, thiram @ 2 gm kg^{-1}, flyash powder pelleting @ 300 gm kg^{-1}, turmeric powder pelleting @ 200 gm kg^{-1}, arappu leaf powder pelleting

@200 gm kg^{-1}, deltamethirn @5ml kg^{-1} and vasambu rhizome powder pelleting @200 gm kg^{-1} of seeds along with control. The study revealed that the cotton seeds treated with halopolymer @ 4 ml kg^{-1} recorded higher seed quality and yield under moisture moisture stress condition, than the other treatments and control.

Keywords : Cotton, moisture stress, halopolymer, seed treatment

Introduction

Cotton is one of the principal commercial crops cultivated in the country. Cotton in a way is a gift of the Indian sub-continent to human civilization. It was in this sub -continent that the growing of cotton and its manufacture into yarn and cloth took root about five millennia ago, spreading to other countries. The area under cotton cultivation is 73.90 lakh hectares with production of 140 lakh bales during 2010-11. The productivity of lint is 322 kg/ha. This is far below the global average of 627 kg/ha. To rise up to the global standards, newer approaches on crop husbandry are warranted. Seed is one of the most vital and critical inputs for increasing agricultural production and productivity. The seeds of improved varieties have played a key role in agricultural transformation of India. The successful seed production depends upon the establishment of uniform, healthy and vigourous plants for which use of quality seed is very important. However, the availability of good quality cotton seed is often difficult because it is affected by numerous factors during and production.

Seed treatment with fungicide has a long history and is practiced for cotton seed to protect the seeds from seed mycoflora not only in storage but also to protect the germinating seedling and promote good establishment at early stages in the field as, field emergence and plant establishment are the two basic requisites in any seed or crop production programme for full exploitation of yield potential of a cultivar. Pelleting, the pre sowing treatment is the process of enclosing the seed with a small quantity of inert material just large enough to produce a globular unit of standard size to facilitate precision planting. Film coating involves the application of liquid material to the seed surface without obscuring its shape. An alternative to the chemicals, the usage of plant products having fungicidal, insecticidal and repellent properties are desirable. These botanicals are non – hazardous, non – poisonous, cheap, easily available and ecologically safe with least or no mammalian toxicity.

Various pre sowing seed treatments are being recommended by Seed Technologists, Agronomists, Microbiologists, Pathologists and Entomologists. However, presently there is no comprehensive single solution for pre sowing seed treatment available by combining various ingredients. Hence, the study was conducted to evaluating the performance of comprehensive seed treatments such as seed coating, pelleting and seed treatment for growth and productivity in cotton under moisture stress conditions.

Materials and Methods

Genetically pure linted seeds of cotton cv. LRA 5166 and Surabi were obtained from the Cotton Breeding Station, TNAU, Coimbatore. The bulk seeds were graded for uniformity using appropriate sieves. And then the seeds were given with various treatments as detailed below:

T_0 - Control
T_1 - Gaucho@10 g kg^{-1}
T_2 - Halo polymer@ 4 ml kg^{-1}
T_3 - Thiram powder 2 g @ kg^{-1}
T_4 - Fly ash powder pelleting @ 300 g kg^{-1}
T_5 - Turmeric rhizome powder pelleting @ 200 g kg^{-1}
T_6 - Arappu leaf powder pelleting @ 200 g kg^{-1}
T_7 - Deltamethrin @ 5 ml kg^{-1}
T_8 - Vasambu rhizome powder pelleting @ 200 g kg^{-1}

In pelleting treatment, the seeds were mixed with the common adhesive, maida @ 10 per cent and then sprinkled with the respective filler material (i.e.) turmaric and vasambu rhizome powder and fly ash and rolled for uniformity. Irrespective of the treatment, the dry dressing of seed was done as such without addition of water. While slurry treatment was done with addition of water @ 5 ml kg^{-1} of seed. The treated seeds were dried back to original moisture content and evaluated for the seedling quality characters *viz.*, Speed of germination, germination (%) (ISTA, 1999), root length (cm), shoot length (cm), dry matter production (mg 5 $seedlings^{-1}$) (Gupta, 1993), vigour index (Abdul-Baki and Anderson (1973)) along with untreated control.

After taking the initial evaluation, a field trial was conducted during early summer by adopting randomised block design with three replications. The crop was raised with the spacing of 60 x 30 cm and recommended package of practices were followed. The experiment was carried out under irrigated conditions and there was no rainfall during the experimental period. The moisture stress was induced by withholding irrigation at 25 DAS. The following observations viz., days to initial emergence, field emergence, days to first flowering, days to 50 per cent flowering, plant height (cm), number of monopodia per plant^{-1}, number of sympodia per plant^{-1}, number of bolls per plant, kapas yield per plant (g plant^{-1}) and seed yield per plant (g plant^{-1}) were recorded. The data were statistically analyzed as per Panse and Sukhatme (1985).

Results and Discusssion

Basu (1990) revealed that the seed treatment implies an improvement in seed performance resulting in better field performance than the corresponding untreated seed and the establishment of a seedling in the soil is an important and foremost need for the better crop production. This depends largely on the germination and vigour potential of the seeds used for sowing. Seed treatment provides an opportunity to package effective quantities of materials such that they can influence micro environment of each seed which supplies not only micro and macro nutrients, but also protects the crop from pests and diseases from the earlier stages due to the inclusion of pesticides. To increase the productivity of the cotton, various seed enhancement techniques were followed but such treatments have not come into practice among the farmers because of the non-availability of the chemicals to the marginal and small farmers and they are also expensive.

The present study reveals that the LRA 5166 cotton seeds treated with halo polymer @ 4 ml kg^{-1} recorded higher seed qualities i.e., germination percentage (95.6 per cent), root length (15.6 cm), shoot length (18.2 cm), dry matter production (0.981 g seedlings^{-5}) and vigour index (3291), when compared to control. The another variety Surabi also recorded higher germination percentage (87.6 per cent), root length (16.7 cm), shoot length (18.6 cm), dry matter production (0.955 g seedlings^{-5}) and vigour index (2273), when compared to control. These results are in conformity with the findings of Basu and Rudrapal (1979) in wheat, jute and mustard, Basu and Rudrapal (1980) in mustard and Basu and Rudrapal (1981) in mustard. The better quality of the halo polymer treated seeds of cotton might be due to the higher initial vigour of the plants induced by the halo polymer treatments and subsequent accumulation of higher nutrients in the seed.

In the present investigation, field emergence potential differed significantly among the treatments. The field emergence in LRA 5166 (89 per cent) and Surabi (87 per cent) were the highest with seeds treated with halo polymer treatment when compared to other treatments and control. The above results are in conformity with the findings of in groundnut (Anbazhagi,1997) and in sunflower (Rathinavalli,1998). The increased field emergence observed in the present study might be due to rapid conversion of the insoluble reserve substances into a soluble state and their mobilization and transport to the regions of intense growth activity by the smaller embryonic axes and larger cotyledons of the halo polymer seeds.

The seeds subjected with the halo polymer treatment also recorded higher values for the biometrical and physiological observations. The seeds of LRA 5166 recorded higher plant height (18.7 cm,70.1 cm,118.0 cm and 123 cm in 30,60,90,120 days respectively), number of monopodia per plant (1.2 plant^{-1}), number of sympodia per plant (20.3 plant^{-1}) respectively. The another variety Surabi also recorded higher plant height (18.3 cm, 69.3 cm, 115.3 cm and 121 cm in 30, 60,90,120 days respectively), number of monopodia per plant (1.2 plant^{-1}), number of sympodia per plant (18.0 plant^{-1}). The above findings are in comformity with the finding of Anbazhagi (1997) in groundnut and Rathinavalli (1998) in sunflower. This might be due to the invigorative effect of halo polymer (Chandrasenan nair, 1996) and increase in rate of photosynthesis.

In the present study, the halo polymer treatment recorded higher values for yield attributing characters. The LRA 5166 recorded more kapas yield (43.51 g plant^{-1}) and seed yield (42.7g plant^{-1}), when compared to the control. Another variety, Surabi also recorded 40.33 g plant^{-1} and 29.66 g plant^{-1} of kapas yield and seed yield, respectively when compared to control. Chandrasenan nair (1996) reported a similar increase in paddy. The increase in the yield attributing characters might be due to the increased germination and vigour of the plants induced by the halo polymer treatments.

The performance analysis of other seed treatments expressed that, among the seed coating and pelleting treatments, the order of preference would be of arappu leaf powder, gaucho, deltamethrin, vasambu rhizome and turmeric rhizome powder. The seeds treated with fly ash excelled all other treatments.

The above study revealed that among the seed treatments, seed treatment with halo polymer @ 4 ml kg^{-1} recorded higher seed yield under moisture stress condition than to other treatments and control.

Table 1: Effect of comprehensive seed treatments on initial seed quality in cotton cv. LRA 5166.

Treatments	Germination percentage	Root length (cm)	Shoot length (cm)	Dry matter production (g seedling^{-5})	Vigour index
T_0	76 .01(60.61)	10.9	11.8	0.574	1788
T_1	84.3 (63.91)	14.3	15.8	0.677	2543
T_2	95.6 (78.01)	15.6	18.2	0.981	3291
T_3	84.31 (66.70)	13.5	15.8	0.777	2597
T_4	93.62 (75.49)	14.6	17.7	0.878	3034
T_5	86.10 (68.05)	13.6	16.0	0.783	2596
T_6	90.30 (71.88)	14.3	17.3	0.845	2906
T_7	86.60 (68.62)	12.8	15.1	0.690	2523
T_8	80.61 (63.43)	13.4	15.7	0.768	2381
Grand Mean	69.52 (57.40)	13.7	15.9	0.768	2618
S Ed	0.915	0.360	0. 357	0.007	69.92
CD (p=0.05)	1.922	0.757	0.751	0.016	146.90

(Figures in parenthes is indicate arcsine values)

Table 2: Effect of comprehensive seed treatments on initial seed quality in cotton cv. Surabi.

Treatments	Germination percentage	Root length (cm)	Shoot length (cm)	Dry matter production (g seedling^{-5})	Vigour index
T_0	78.6 (62.49)	11.1	11.7	0.522	1745
T_1	84.3 (66.69)	14.1	15.4	0.645	2384
T_2	97.0 (80.40)	16.7	18.1	0.943	3326
T_3	83.6 (66.10)	14.3	16.5	0.753	2700
T_4	91.0 (76.24)	16.1	17.7	0.865	3165
T_5	83.6 (66.10)	14.6	16.3	0.764	2694
T_6	92.6 (74.29)	15.1	17.6	0.797	2875
T_7	87.0 (68.87)	13.3	15.6	0.749	2514
T_8	81.6 (64.65)	13.3	15.3	0.743	2342
Grand Me an	86.00 (65.32)	14.3	16.1	0.753	2638
S Ed	1.165	0.495	0.380	0.007	78.62
CD (p=0.05)	2.448	1.040	0.7992	0.015	165.19

(Figures in paranthes is indicate arcsine values)

Table 3: Effect of comprehensive seed treatments on plant height (cm) at different growth stages in Cotton cv. LRA 5166.

Treatments	Plant height (cm)			
	30 Days after sowing	60 Days after sowing	90 Days after sowing	120 Days after sowing
T_0	14.8	56.23	101.87	104.33
T_1	16.4	63.13	103.34	113.12
T_2	18.7	70.16	118.89	123.32
T_3	17.1	62.50	114.65	116.13
T_4	18.5	69.16	116.23	122.33
T_5	17.2	64.46	112.48	116.11
T_6	18.2	68.66	114.52	119.12
T_7	17.0	56.93	104.73	115.28
T_8	16.8	60.20	113.94	113.52
Grand Mean	17.22	63.65	111.89	116.86
S Ed	0.254	0.796	2.030	1.305
CD (p=0.05)	0.534	1.674	4.266	2.742

Table 4: Effect of comprehensive seed treatments on plant height (cm) at different growth stages in Cotton cv. surabi.

Treatments	Plant height (cm)			
	30 Days after sowing	60 Days after sowing	90 Days after sowing	120 Days after sowing
T_0	15.3	55.6	97.3	110.0
T_1	16.0	56.6	110.0	114.2
T_2	18.3	69.3	115.3	121.1
T_3	16.3	64.6	103.6	112.0
T_4	17.3	68.3	113.6	118.0
T_5	16.3	63.0	108.6	114.3
T_6	17.3	67.3	112.3	117.1
T_7	16.0	63.3	109.6	111.0
T_8	15.6	56.3	109.3	113.0
Grand Mean	16.51	62.7	108.9	114.4
S Ed	0.496	1.042	1.324	0.544
CD (p=0.05)	1.044	2.189	2.781	1.143

Table 5: Effect of comprehensive seed treatments on various growth and yield parameters in cotton cv. LRA 5166.

Treatments	Days to initial Emergence	Field Emergence (%)	Days to 1st flowering	Days to 50% flowering	Number of Monopodia plant $^{-1}$	Number of symbodia plant $^{-1}$	Number of bolls plant $^{-1}$	Kapas yield (g plant)$^{-1}$	Seed yield (g plant $^{-1}$)
T_0	4.3	65	54.0	60.0	1.0	16.8	23.0	34.7	25.3
T_1	3.8	73	51.6	57.1	1.0	18.1	23.6	36.8	26.6
T_2	3.0	87	50.0	55.3	1.2	20.3	28.0	43.5	30.1
T_3	3.3	80	52.0	57.6	1.0	18.3	24.3	39.3	26.3
T_4	3.1	86	51.0	56.2	1.2	19.4	26.6	42.7	29.8
T_5	3.5	79	52.0	57.3	1.0	18.1	23.6	38.0	27.3
T_6	3.2	85	51.0	56.0	1.1	18.9	25.3	41.5	28.4
T_7	3.6	77	53.0	58.6	1.0	18.3	24.3	38.3	26.2
T_8	3.6	76	52.3	57.0	1.0	17.8	24.0	39.2	26.8
Grand Me an	3.5	78.96	51.8	57.18	1.05	18.4	24.7	39.3	27.4
S Ed	0.099	1.507	0.222	0.272	0.031	0.260	0.471	0.459	0.332
CD (p=0.05)	0.208	3.166	0.466	0.571	0.066	0.547	0.990	0.965	0.698

Table 6: Effect of comprehensive seed treatments on various growth and yield parameters in cotton cv. Surabi.

Treatments	Days to initial emergence	Field Emergence (%)	Days to1st flowering	Days to 50% flowering	Number of Monopodia plant $^{-1}$	Number of symbodia plant^{-1}	Number of bolls plant^{-1}	Kapas yield (g plant^{-1})	Seed yield (g plant^{-1})
T_0	4.3	66	53.6	60.0	1.0	14.6	22.3	31.66	19.33
T_1	3.8	75	52.0	57.0	1.0	16.3	23.6	33.00	22.66
T_2	3.0	89	50.0	55.0	1.2	18.0	28.3	40.33	29.66
T_3	3.7	79	52.0	57.0	1.0	16.6	23.0	36.00	25.66
T_4	3.1	85	50.0	55.3	1.2	17.6	25.6	38.66	29.00
T_5	3.5	73	52.3	57.0	1.0	15.3	22.6	35.66	22.66
T_6	3.1	82	50.3	55.6	1.1	17.0	24.6	37.33	27.66
T_7	3.5	80	52.3	58.6	1.0	15.3	23.6	32.66	26.66
T_8	3.8	74	52.0	57.0	1.0	16.6	22.6	34.00	25.33
Grand Me an	3.5	52.2	78.33	57.92	1.05	16.4	24.07	35.48	25.40
S Ed	0.163	0.444	1.499	0.666	0.222	0.415	0.444	0.471	0.444
CD (p=0.05)	0.343	0.933	3.149	1.400	0.046	0.873	0.933	0.990	0.933

References

Abdul -Baki, A.A and J.D. Anderson. 1973. Vigour determination in soybean seed by multiple criteria. *Crop Sci.*, 13: 630-633.

Anbazhagi, G. 1997. Nutrient management on seed yield, quality and storability of bunch groundnut (*Arachis hypogea L.*) cv. CO 2. *M.Sc. (Ag). Thesis*, Tamil Nadu Agricultural University, Coimbatore.

Basu, R. N. 1990. Seed invigouration for extended storability. International Conference on Seed Science and Technology, Feb. 22nd, New Delhi, (India). pp. 12 - 16.

Basu, R. N. and A. B. Rudrapal. 1979. Iodine treatment of seed for the maintenance of vigour and viability. *Seed Res.*, 7: 80-82.

Basu, R. N. and A. B. Rudrapal. 1980. Iodination of mustard seed for the maintenance of vigour and viability. *Ind. J. Exp. Biol.* 18: 491- 494.

Basu, R. N. and A. B. Rudrapal. 1981. Use of chlorine and bromine in controlling mustard seed deterioration. *Seed Res.*, 9(2): 188 - 189.

Chandrasenan nair, S. 1996. Provenance on seed quality and halogenation treatment to control seed deterioration in rice. *M.Sc. (Ag.) Thesis*, Tamil Nadu Agricultural University, Coimbatore.

Gupta, P.C. 1993. Seed vigour testing. Handbook of seed testing. Quality Control and Research Dev., New Delhi. pp.243.

ISTA. 1999. International rules for seed testing. *Seed Sci. Tech., Supplement Rules*, 27: 25-30

Panse, V. G. and P .V. Sukhatme. 1985. Statistical methods for agricultural workers. ICAR Publication. New Delhi.

Rathinavalli, P. 1998. Studies on halogenation of sunflower seeds to prolong the shelf life and its resultant effect on productivity. *M.Sc. (Ag) Thesis*, Tamil Nadu Agricultural University, Coimbatore.

Chapter 15

Breeding for Salt Stress Tolerance in Crop Plants-A Review

S. LAKSHMI NARAYANAN[1], S.M.IBRAHIM[1] AND K. BALAKRISHNAN[2]
[1] Department of Plant Breeding and Genetics
[2] Department of Seed Technology
Agricultural College and Research Institute, Tamil Nadu Agricultural University, Madurai

Improving food security ought to be an issue of great importance for a country like India where one-third of the population is estimated to be poor. The World has to feed 9.2 billion people (Table 1) by 2050 (FAO, 2009). Population of India has reached 1.21 billion as per the 2011 census report which accounts for 17.5% of world's population. According to a 2005 World Bank estimate, 41.6% of the total Indian population falls below the international poverty line of US$ 1.25 a day. According to 2010 data from the United Nations Development Programme, an estimated 37.2% of Indians live below the country's national poverty line. Hence the pressure to produce and feed more is increasing day by day.

Table 1: Estimated World Population

S.No.	Year	World population (crores)
1.	1000	27
2.	1500	45
3.	1800	100
4.	1900	160
5.	2000	610
6.	2011	700
7.	2050	920

(FAO, 2009)

Abiotic Stress

Stress in plants include any factor that could lead to the death of the plant. Alternately, Stress can be defined as any environmental variable, which can induce a potentially injurious strain in plants. Exposure of plants to abiotic stresses like temperature (heat, cold, chilling/frost), water (drought, flooding), salinity, radiation (UV, ionizing radiation), chemicals (mineral deficiency/excess, pollutants, heavy metals/pesticides, gaseous toxins), mechanical (wind, soil movement, submergence) causes more than 50 percent reduction in agricultural production. Abiotic stresses adversely affect growth and productivity and trigger a series of morphological, physiological, biochemical and molecular changes in plants (Pooja Bhatnagar *et al.*, 2008). Adaptation or tolerance to these stresses is absolutely necessary to maintain growth and life cycle of plants.

Salinity

One third of earth's surface is covered by salt water and salt affects the world's land surface significantly (Flowers and Flowers, 2005). Salinity is the major factor affecting worldwide agricultural productivity, which is believed to affect nearly one-fifth of the world's irrigated land and causes 10^7 irrigated hectares to be abandoned each year (Flowers and Yeo, 1995; Nelson *et al.*, 1999 and Quesada *et al.*, 2002). Salinity refers to the quantum of soluble salts present in the soil. Salts are natural part of soils, but at elevated levels can negatively affect the plant growth and soil structure (Virtual Soil Lab Modules : http://soilweb. landfood.ubc.ca/labmodules/salinity). Salts in soil can originate from parent material; from irrigation water; or from fertilizers, manures, composts, or other amendments. The predominant salts that accumulate in soils are calcium, magnesium, sodium, potassium, sulfate, chloride, carbonate, and bicarbonate (Horneck *et al.*, 2007).

Saline soils are classified based on electrical conductivity (EC) of the soil solution which detects osmotic problems and exchangeable sodium percentage (ESP) indicative of a physical dispersion problem. Salt affected soils are classified as follows (Table 2).

Table 2: Classification of salt affected soils

Soils	ESP (%)	EC (mmhos/cm)
Non-saline/non-sodic soil	=15	= 4
Saline soil	=15	>4
Sodic soil	>15	=4
Saline-sodic soil	>15	>4

Salinity poses a great danger to the farming community limiting the agricultural production. Salinity is a silent disaster to the world of agriculture. In agriculture, the problem of salinity is becoming more severe every year as the non-saline soils and non-saline waters are increasingly exploited to accommodate the growing demand for food and plant products (Reddy *et al.,* 1997).

From the total area of 12.78 billion hectares, around 831 million hectare (6.5%) is affected by salt (Mohammad Pessarakli and Szabolcs, 2011). In India, 6.73 million ha is affected by salinity (CSSRI). The problem of salinity is getting increased because of the factors such as rise in sea levels, climate change etc.

Classification of Crop Plants Based on Salinity

Crop plants are classified in to two broad categories *viz.,* halophytes and glycophytes with reference to salt stress. A continuous spectrum of tolerance ranging from the most sensitive species which are severely affected by as little as $1/10^{th}$ of sea water (50 mol m^{-3}NaCl) to those species which are able to complete their life cycle in full strength sea water (500 mol m^{-3}NaCl.) [Garg and Gupta, 2011].

The term halophyte was described in the literature since early 1900's. Schimper (1903) defined halophytes as plants that are capable of normal growth in saline habitats and also thrive on ordinary soil. According to Stocker (1928), a halophyte is a plant that can tolerate salt concentration of more than 0.5 percent at any stage of life. Dansereau (1957) defined halophyte as the plant that can grow exclusively on salty soil. Ungar (1991) defined halophytes as those species that tolerate relatively high soil salinities and are capable of accumulating relatively high quantities of Na+ and Cl- in their tissues.

It is difficult to precisely define halophytes as opposed to glycophytes due to the variability of plant responses in dependence of a number of factors, including climatic conditions and plant phenophases. Halophytes are further categorised as Euhalophytes (Plants that grow best under certain level of salinity); Xerohalophytes (Plants that thrive under saline and arid conditions); Hydrohalophytes (Plants that thrive under saline and moist conditons (Sardo, 2005).

Glycophytes show little tolerance to salinity and do not accumulate high concentrations of ions in their growing tissues. *Salicornia europea* and *Sueda maritima* can tolerate salinities well above that of seawater (0.5 M NaCl) while glycophytes are intolerant of salinities greater than 10 mM. Obligate halophytes grow well under moderate salinity and have reduced growth under low salinity (Ungar, 1991). Facultative halophytes like *Mesembryanthemum crystallinum* gain the properties of halophytic metabolism when exposed to a pretreatement of moderate salinity. (Table 3).

Table 3: Best supported definitions of salt tolerance functional types

Functional type	Definition
Glycophytes	Plants whose growth and fitness decreases when exposed to any level of salinity greater than approximately 10mM.
Intolerant glycophytes	Plants that cannot tolerate soil salinity greater than approximately 10mM.
Halophytes	Plants that tolerate higher salinity than glycophytes (upto approximately 50mM) before a reduction in growth or fitness occurs.
Obligate halophytes	Plants that do not attain their highest fitness at low soil salinity (less than 10mM). A moderately high soil salinity (50mM) is required to attain maximum fitness.
Facultative halophytes	Plants that attain a halophytic nature only after they experience moderate salinity in the soils or as an aerosol.

(Garg and Gupta, 2011a)

Greenway and Munns (1980) adopted the ecological definition of halophytes by Jennings (1976) which states that the native flora of saline soils and have assumed the salt concentrations with osmotic potential of atleast 0.33 MPa. Plants that cannot grow in the presence of high concentration of sodium salts are called glycophytes or non-halophytes.

Only 2% of plant species are halophytes which can grow in waters of high salinity and in areas such as saline semi-deserts, mangrove swamps, marshes and sloughs, and seashores and majority of plants are glycophytes that are easily damaged by salinity. Salt effects are the combined result of the complex interaction among different morphological, physiological and biochemical processes (Reynolds *et al.*, 2001).

Salt Tolerance

Tolerance to salinity varies a great deal, almost 10 fold, amongst the crop plants and to a lesser extent amongst their genotypes. These inter and intragenic variations in salt tolerance of plants can be exploited for selecting crops or varieties that produce satisfactory growth under a given root-zone salinity. The information on crop tolerance to salinity and saline waters can be obtained from Mass (1986) and to the use of saline waters in different agro-climatic zone of India from a recent compilation by Minhas (2005) and Minhas and Gupta (1992). The ratings of some of the important agricultural crops based upon salt tolerance classification limits proposed by Maas (1986) are given in Table 4.

Table 4: Salt tolerance of some important agricultural crops.

Sensitive	Moderately sensitive	Moderately tolerant	Tolerant
EC < 1.2	1.2-3	3-6	6-10
ESP < 8	8-16	16-24	24-32
Chickpea	Broadbean	Wheat	Barley
Beans	Maize	Mustard	Cotton
Sesame	Groundnut	Cowpea	Sugarbeet
Onion	Sugarcane	Oats	Safflower
Carrot	Alfalfa	Sorghum	
Mungbean	Berseem	Soybean	
	Radish	Spinach	
	Rice	Pearlmillet	
	Cauliflower	Muskmelon	
	Chillies		

(Maas, 1986)

Salinity has three potential effects on plants viz., lowering of the water potential, direct toxicity of any Na and Cl absorbed and interference with the uptake of essential nutrients (Flowers and Flowers, 2005). Saline soils are characterised by the low nutrient ion activities and extreme ratio of Na^+/Ca^{2+}, $Na^+/K^+/Ca^{2+}/Mg^{2+}$ and Cl^-/NO^{3-} in the soil solution. Under saline conditions, nutritional disorders are developed by the halophytes.

The nutritional imbalance is due to

- Effect of salinity on availability of nutrients. Enhanced concentration of salt ions decreased micro and macro nutrients.
- Salinity affects the uptake and portioning of nutrients within the plant.
- Physiological inactivation of a given nutrient resulting in deficiency and hence requirement of the plant for that nutrient is increased.
- Salt induces growth inhibition, which then affects the respiration of roots and hence uptake of essential nutrients is decreased (Garg and Gupta, 2011).

Physiological Mechanism of Salt Tolerance

The main physiological mechanism responsible for plant growth reduction and how environment and genotype modulate plant salinity response are not yet known (Cuartero *et al.*, 2010).

Salt accumulation in the root environment lowers its water potential with an almost inevitable decrease in water potential of the plant tissues. When there is no salt uptake by the plant, the consequences will be the same as for plants grown on dry soils i.e. in environment with low matric potentials. Such plants may be of low turgor and hydration (A1) [Fig 1]. Alternatively, organic solute accumulation by the plant cells may restore turgor and hydration, a process called osmotic adjustment (A2). If this happens, considerable amount of metabolic products would be required, particularly in the case of sugars and amino acids. At external concentration of 100mmol NaCl, the cell sap would have to accumulate about 30g/l of hexose sugars or 60g/l of disaccharides. For many species, this presents about 20 and 40% sugars on a dry weight basis.

Salinity delays the onset, reduces the rate and increases the dispersion of germination events, resulting in reduced plant growth and final crop yield (Ashraf and Foolad, 2005). Seeds are particularly

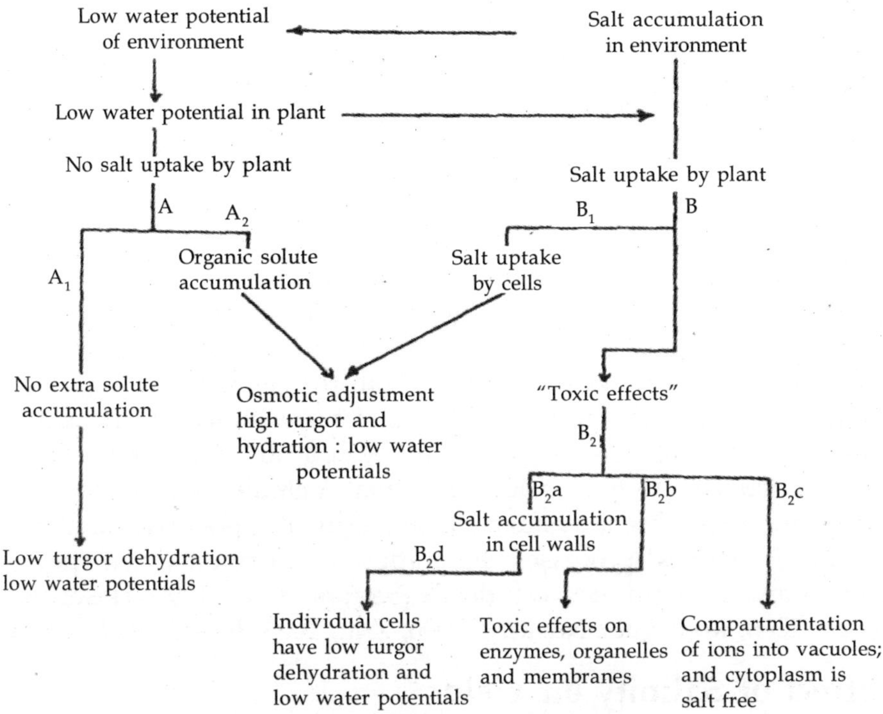

Fig. 1: Factors involved in the physiology of salt tolerance

vulnerable to stress encountered between sowing and seedling establishment while plant salt tolerance usually increases with plant ontogeny. Soil salinity may affect the germination of seeds either by creating osmotic potential external to the seeds preventing water uptake or through the toxic effects of Na^+ and Cl^- ions on germinating seed (Khajeh-Hosseini *et al.,* 2003). Salt stress affects many physiological aspects of plant growth. Shoot growth was reduced by salinity due to inhibitory effect of salt on cell division and enlargement in growing point. Salinity induces delay and decrease in sprouting percentage through lower water uptake and higher accumulation of Na^+ and Cl^- ions. Early flowering reduced dry matter, increased root: shoot ratio and leaf size caused by salinity may be considered as possible ways of decreasing yield under water stress (Sripunitha *et al.,* 2012).

Osmotic adjustment might also be induced by salt uptake (mostly Na^+ and/or Cl^-) into the cells (B1); this would avoid a considerable drain on metabolic products. However, salt uptake introduces the danger of toxicities (B2). The obvious case is the adverse effects of ions on the metabolic machinery (B2b). A more subtle alternative arises from possible high salt concentrations in the cell walls. These would adversely affect the water relation of individual cells i.e. cause their partial dehydration (B2a). However, if the salts are compartmentalised in the vacuole and the cytoplasm is relatively salt free, the toxic effects of the ions may be greatly overcome (B2c) and plant will display a higher salt tolerance as in the case of many halophytes (Garg and Gupta, 2011).

High concentrations of salts cause ion imbalance and hyperosmotic stress in plants. As a consequence of these primary effects, secondary stresses such as oxidative damage often occur under high salt stress disrupts homeostasis in water potential and ion distribution. This disruption of homeostasis occurs at both the cellular and the whole plant levels. Drastic changes in ion and water homeostasis lead to molecular damage, growth arrest and even death (Zhu, 2001). There are secondary effects of NaCl stress that include disturbance of K^+ acquisition, membrane dysfunction, impairment of photosynthesis and other biochemical processes, generation of reactive oxygen species (ROS) and programmed cell death (Serrano *et al.*, 1999; Hasegawa *et al.*, 2000; Rodriguez-Navarro, 2000; Zhu, 2003, Botella *et al.*, 2005).

Effect of Salinity on Yield

Sodium is an essential micronutrient for the normal crop growth. It is involved in the regeneration of phosphoenolpyruvate in CAM and C_4 plants. It also acts as a substitute for potassium in some circumstances. It stimulates growth, increases leaf area, stomata and improves the water balance. However, excess amount of sodium ions as accumulated in the salt affected soils are toxic to plant growth and both the quality and quantity of crop produce are getting affected (Gepstein *et al.* 2006; Blumwald and Grover, 2006 and Chandan sahi *et al.*, 2006). Salinity makes yield loss of around 20 percent in crop plants. Yield loss due to salinity ranks second among the abiotic stress factors (Table 5).

Table 5: Estimated potential yield loss due to abiotic stress

Abiotic stress factors	Estimated Potential yield loss (%)
High temperature	40
Salinity	20
Drought	17
Low temperature	15
Other factors	8

(Athar, H. R. and M. Ashraf, 2009)

Full genetic potential of crop varieties is hindered by the occurrence of abiotic stresses like salinity. Most of the irrigation source contain salt to some extent and can be measured by electrical conductivity. If the EC of the irrigation water is higher, plants find it more difficult to grow and ability of the plant to absorb water is drastically reduced. It delays and reduces the flowering of crop plants. This ultimately reflects on the crop yield. Reduction of yield and its components under salt stress condition may also be attributed to low production, expansion, senescence and physiologically less active green foliage (Schactman & Munns, 1992). The effect of both salt and water stress might lead to shrivelled seeds and consequent lower yield (Shakil Ahmed, 2009).

A typical study reported a drastic reduction in yield of barley from 29 bushels/acre to 10 bushels/acre due to increased salinity over 700 years from 1700 BC to 2400 BC (Gelburd, 1985). Yield reduction (Table 6) varies from crop to crop with changing level of electrical conductivity. Plant response to salinity depends on growth stage, soil temperature, soil moisture, and other factors (Horneck *et al.*, 2007).

Table 6: Yield reduction percentage in various crops at differing levels of EC.

Crop	Expected yield reduction			
	None	10%	25%	50%
	Electrical conductivity (EC) ds/m			
Barley	8.0	10.0	13.0	18.0
Wheat	6.0	7.4	9.5	13.0
Sugarbeet	4.0	4.1	6.8	9.6
Alfalfa	2.0	3.4	5.4	8.8
Potato	1.7	2.5	3.8	5.9
Corn(grain)	1.7	2.5	3.8	5.9
Onion	1.2	1.8	2.8	4.3
Beans	1.0	1.5	2.3	3.6
Apples, Pears	1.7	2.3	3.3	4.8
Strawberries	1.0	1.3	1.8	2.5
Sudan grass	2.8	5.1	8.6	14.0
Grapes	1.5	2.5	4.1	6.7
Broccoli	2.8	3.9	5.5	8.2
Cucumbers	2.5	3.3	4.4	6.3

(Horneck et al., 2007)

Genetics of Salt Tolerance

Different Plant species respond varyingly in their tolerance to wide range of soils with different levels of salinity/sodicity. The relative growth of plants in the presence of salinity is termed as their salt tolerance. Performance of crop plants in wide range of salt affected soils is the daunting task of plant breeders all over the world. Salt tolerance is a complex character and the mechanism of genetic control of salt tolerance is not fully understood. Salinity tolerance comes from genes that limit the rate of salt uptake from the soil and the transport of salt throughout the plant, adjust the ionic and osmotic balance of cells in roots and shoots, and regulate leaf development and the onset of senescence (Munns, 2005). Studies on inheritance of salt tolerance was initiated by Lyon (1941).

Identification of genes for salt tolerance that enhances the yield stability of crop in saline conditions has been the primary goal of many scientists (Hasegawa *et al.*, 2000; Botella *et al.*, 2005). Salt tolerance is determined by the interaction of many genes and is a complex quantitative trait (Petronia Carillo *et al.*, 2011). Salinity tolerances may increase or decrease depending on the plant species and/or environmental factors. For some species, salt sensitivity may be the greatest at germination, whereas for other species, sensitivity may increase during reproduction (Howat, 2000; Marschner, 1986, Petronia Carillo *et al.*, 2011). Several studies indicate that salt tolerance trait is governed by both additive and dominant effects (Azhar and McNeilly, 1988, Shannon, 1985).

Salinity resistance measured in terms of yield, panicle sterility, shoot dry weight, root growth and leaf necrosis showed polygenic control. In Cucumber, one dominant major gene plus several minor genes control resistance to salinity. Monogenic control of salinity resistance is also reported in soybean.

Salt tolerance determinant genes can be categorized into two functional groups. The first includes genes that encode effectors, which are responsible for processes that are necessary for stress alleviation or adaptation. The second consists of regulatory genes that control expression and activity of these effectors (Hasegawa *et al.*, 2000, Botella *et al.*, 2005). The effectors include proteins that function by protecting cells from high cytoplasmic Na^+, such as transporters, enzymes required for biosynthesis of various compatible osmolytes, LEA proteins, chaperones and detoxification enzymes. Regulatory molecules include transcription factors, and signaling intermediates like protein kinases and enzymes involved in phosphoinositide metabolism (Zhu, 2002).

Breeding for Salt Tolerance

Breeding for salt tolerance has been extensively reviewed by global researchers (Noble, 1983; Blum, 1988; Rana, 1985; Epstein and Rains, 1987; Jones, 1987). Varietal development program for salt affected soils moved very slowly because of the complex nature of the ecosystem with high temporal and spatial variability; the complex nature of the inheritance of the tolerance trait; lack of sufficient resources and research capacity and lack of precise and repeatable screening techniques. Genetic improvement of crops for salt tolerance involving traditional salt tolerant landraces were attempted but the genetic drag of local types resulted in tall, poor quality and fertilizer nonresponsive genotypes (Singh *et al.*, 2009).

Breeding procedure for the improvement of a character depends on the pattern of inheritance, number of genes with major effects and the nature of gene action. Breeder should integrate the knowledge gained on physiological traits with desirable characters to combine salt tolerance and economical yield (Ansari *et al.*, 2003).

Complete understanding of the genetic, physiological and biochemical mechanism of crop tolerance with respect to salt stress will help us to formulate strategies for identifying and developing crop varieties with higher yield potential, higher yield stability and greater production efficiency under salt affected conditions.

Genetic improvement of salt tolerance in crops can be achieved by selection, hybridisation, mutation, changing the ploidy level, use of wild source from the germplasm collections and adoption of advanced techniques like somaclonal variation, somatic hybridisation and recombinant DNA technology (Garg and Gupta, 2011).

Spikelet fertility is having highly significant negative correlation with the chlorophyll a/b ratio and shoot Na/K ratio. It indicates that the increase in chlorophyll a and Na content affects spikelet fertility, ultimately reducing the yield. Hence, selection should be done for low shoot Na and chlorophyll a content for the development of salt tolerant varieties. Chlorophyll Stability Index is also having the significant positive correlation with the productive tillers, spikelet fertility and yield. The knowledge on inter-relationship between salt tolerant traits may facilitate the breeder to decide upon the intensity and direction of selection procedure to be given on related traits for the simultaneous improvement of these traits (Shanthi *et al.*, 2012).

A lot of high yielding varieties fail miserably in salt affected areas notwithstanding the harmful effects of high salt content in the soils. Manipulation of growing environment in the salt affected areas suitable for the normal growth of the plants involve major engineering and soil amelioration process which need lot of resources that are often out of the reach of small and marginal farmers. Breeding crop varieties with in-built salt tolerance is realized as the most promising, less resource consuming /economical and socially acceptable approach. Developing crops engineered with higher salt tolerance suited to salty environments will be a boon to the farming community of high saline agriculture.

Salt tolerance of crops depend mainly on the basis of plant survival, absolute plant yield and relative yield (Garg and Gupta, 2011). Research work on unravelling the mechanisms of salt tolerance in plants was intensified from early 1980s. However, till recently salt tolerant cultivars with outstanding yield performance and commercial significance have not been released.

Attempts to improve the salt tolerance of crops through conventional breeding programmes have been met with very limited success, due to the complexity of the trait.

By genetic engineering, salt tolerant *Arabidopsis thaliana* has been developed (Inge *et al.*, 2009). To achieve this in an integrated approach involving traditional plant breeding techniques and biotechnological tools viz., molecular marker assisted selection strategy is needed for solving the problem. Entire available germplasm should be screened for salt tolerance and the existing high yielding varieties should be studied for finding out the level of salt tolerance. Novel techniques for evaluating, dissecting and mapping components of salt tolerance and subsequent transfer of this character will speed up development of salt tolerant varieties. Calculated blending of traditional breeding techniques and molecular biological tools can maximise the yield by bridging the gap in genetic yield potential of crops caused by abiotic stress. High research priority should be given without further delay to hasten the understanding of mechanism of salt tolerance at genetic, physiological and molecular level leading towards the development of cultivars specific to the salt affected areas of the farming community.

Breeding crop varieties for increased salt tolerance is now considered as a more promising, energy-efficient and economical approach than major engineering process and salt amelioration techniques which have gone beyond the limit of marginal farmers. Untiring efforts of breeder, biotechnologist, physiologist, geneticist is a prerequisite in breeding for varieties with salt tolerance. Conventional and innovative approaches of crop improvement have been looked upon to tackle the danger of salt stress. High yielding varieties are being developed rapidly with the changing pace and choice of farmers. However they underperform in stressful environment and in salt affected areas as well. The high yielding varieties developed world wide in different crops is yet to replace the local landraces for want of tolerance to the hardy environment *viz.*, salinity and sodicity.

Considerable progress has been made in developing improved varieties with tolerance of salinity, particularly for inland saline/sodic areas (Singh and Mishra 2006, Rao *et al.*, 2008 and Mackill *et al.*, 2010). Breeding methods viz., Introduction, Selection, Hybridisation, Pedigree method, Modified bulk and pedigree method and shuttle breeding method have been utilised for developing salt tolerant varieties.

Major progress has been made in breeding salt tolerant high yielding rice varieties for various inland saline, coastal saline and alkaline soils of fragile ecosystems (Mishra *et al.*, 2003). Of 32 salt tolerant rice varieties developed by Central Soil Salinity Research Institute (CSSRI), CSR 10 was the first dwarf high yielding, salt tolerant early maturing rice variety released. The recently adapted varieties in coastal saline soils are SR 26 B, CSR 1, CSR 2, CSR 3, CSR 27, CSR 13, Panvel 1, Panvel 2, Panvel 3, Pokkali 9 in Kerala only, Vytilla 1 and Vytilla 2. CSR 22, CSR 23, CSR 26, CSR 27 and CSR 30 have been identified to suit sodic soils.

Salt tolerant rice varieties like Pokkali, Vytilla 1, Vytilla 2, Vytilla 3, Vytilla 4 and Vytilla 5 have been developed by the Kerala Agricultural University to solve the coastal salinity problems of the state. SR 26 B from Orissa, Kalarata, Bhurarata, Panvel 1, Panvel 2, Panvel 3 from Maharashtra and BTS 24 from Andaman and Nicobar Islands have been developed to combat salinity problems (Deepa Sankar *et al.*, 2011). Several rice varieties have been released in Tamil Nadu for salt affected areas viz., PVR 1, IET 1444, CO 43, ASD 16, TRY 1, TRY 2, TRY 3 and AU 1 (Thenmozhi *et al.*, 2012).

In Vitro Selection for Salt Stress

Conventional approaches are not only cumbersome but also time consuming. Plant tissue culture plays an important role in the production of agricultural and ornamental plants and in the manipulation of plants for improved agronomic performance. *In vitro* culture of plant cells and tissue has attracted considerable interest over recent years because it provides the means to study plant physiological and genetic processes in addition to offering the potential to assist in the breeding of improved cultivars by increasing genetic variability (Wani *et al.*, 2010). Plant cell and tissue culture techniques offer a potent tool in developing salt tolerant lines (Stavarek and Rains, 1984). *In vitro* selection of useful variants has come to be a reliable tool in crop improvement. Morrish *et al.* (1987) opined that selection for mutants at the tissue level has the obvious advantage of saving space and time.

Plant cell and tissue culture techniques are being used for the genetic improvement and developing salt tolerant lines of rice plant throughout the world (Shanthi *et al.*, 2010, Lakshmi Naryanan *et al.*, 2012).

In recent years, traditional breeding methods have been supplemented with plant tissue culture techniques for the production of salt tolerant plants (Dix, 1993; Jaiwal *et al.*, 1997). The application of tissue culture for selecting stress tolerant mutants in rice has been discussed by Jaiwal *et al.* (1997). The approach of cell selection for salt tolerance was first applied by Zenk (1974) in *Nicotiana sylvestris.* Nabors *et al.* (1975) selected NaCl resistant cell lines of *N. tabaccum* var 'Samsum' from cell suspension culture, treated by the mutagen EMS and then grown in a medium containing 0.03 M NaCl. Cells derived from these lines even resisted concentrations as high as 0.09 M NaCl in the medium. Most of the control cultures did not grow in this saline medium. Few of them however remained alive and after several transfers started to grow fast. This growth was explained to be due to the appearance of spontaneous dominant mutations. Later, Nabors *et al.* (1980) regenerated plants from cells tolerant to 6.4 g/l of NaCl. The regenerant plants and the subsequent sexual progenies exhibited tolerance.

Croughan *et al.* (1978) isolated NaCl resistant cells of alfalfa (*Medicago sativa*) which could grow on a medium containing 0.17 M NaCl. The selected line behaved like a halophyte in some respects, including the need for salt for optional growth, maintenance of high levels of K+ in the presence of high levels of Na^+ and increased levels of No_3^- at low salt concentration, No_3^- probably replacing Cl^-. Kochba *et al.* (1980 and 1982) reported isolation of stable salt tolerant callus cell lines and embryos in *C. sinensis* and *C. auranticum*. Selection of stable, NaCl tolerant alfalfa cell lines was accomplished by McCoy (1987) by a step up selection procedure, there by cell lines originally selected for tolerance at 0.5 per cent NaCl were subsequently selected for tolerance at 1.0 per cent NaCl. In two of the seven plants regenerated from the resistant callus, whole plant salt tolerance was expressed.

Mc Hugen (1987) selected a salt tolerant flax line from cell culture, which was superior to the parent line in saline soil for germination, seedling height, flowering, seed set and seed yield, for two seed generations. The increased salt tolerance of the selected line was attributed to the increased general vigour because the salt selected line outperformed the parent line even in normal soil.

Selecting salt tolerant variants from cell cultures has been discussed by Dix *et al.* (1986) and the selection of cell lines and regenerated plants with enhanced salt tolerance has been reviewed by Meredith (1984) and Rains *et al.* (1986).

Salt tolerant somaclones has been developed in oilseed brassicas (Tyagi and Rangaswamy, 1997). In vitro selection of salt tolerant calli lines and regeneration of salt tolerant plantlets in mung bean has been achieved by Srinath Rao and Prabhavathi Patil (2012).

Breeding Through Transgenic Approach

Transgenic technology will undoubtedly aid our search for the cellular mechanisms that underlie salt tolerance, but the complexity of the trait delays engineering tolerance into sensitive species (Flowers and Flowers, 2005).

Crop improvement strategies that are based on the use of molecular marker techniques and biotechnology can be used in conjunction with traditional breeding efforts (Ribaut and Hoisington, 1998). Tremendous genetic gains in the form of varieties with resistance to pest and diseases have been achieved while the similar trend is missing in the case of abiotic stresses, thanks to the complexity and multigenic nature. Transfer of salinity tolerance trait through genetic engineering approach is being done in rice, wheat, potato, red gram, mustard, tomato and cotton.

Salt tolerance often shows the characteristics of a multigenic trait, with quantitative trait loci (QTLs) associated with tolerance identified in barley, citrus, rice, and tomato and with ion transport under saline conditions in barley, citrus and rice (Flowers, 2004). Very recently, salt tolerant genetically modified barley was developed by the research team of Australian Centre for Plant Functional Genomics, Australia.

An important cause of damage by salt stress is the generation of reactive oxygen species (ROS). Plants subjected to salt stress display complex molecular responses including the production of stress proteins and compatible osmolytes. Engineering with osmolytes such as mannitol, fructans, trehalose, proline, glycine betaine and ectoine also probably work through oxidative detoxification. These osmolytes are active in scavenging ROS. The osmolyte-producing transgenic plants are improved in tolerance not only to salts but also to various other stresses such as chilling, freezing, heat and drought, which also generate ROS (Zhu, 2001).

Breeders should use advanced tools like marker assisted selection to develop varieties with enhanced salt tolerance. A major QTL conferred salt tolerance on chromosome 1, was designated *Saltol* (Bonilla *et al.*, 2002) and has been the target of marker-assisted selection (Thomson *et al.*, 2010). On chromosome 1, QTL SKC1 was isolated by positional cloning and was determined to be a protein that functions as a Na+-selected transporter (Ren *et al.*, 2005). Besides *Saltol*, several other major QTLs/genes are being identified and targeted for marker-assisted backcrossing to combine them with *Saltol* for higher tolerance. Strategies to introgress these QTLs/genes agronomically superior backgrounds will meet the quality requirement of farmers in the salt affected areas (Singh *et al.*, 2009).

Recently, studies of salinity tolerance in crop plants ranged from genetic mapping to molecular characterization of salt induced gene products. New approaches are being developed to manipulate expression of functionally related classes of genes by characterization of signalling pathways in salt stress and characterization and cloning of transcription factors that regulate the expression of many genes that could contribute to salt tolerance. Transgenic manipulation of such transcription factors should help us understand more about multigene regulation and its relationship to tolerance. It remains a challenge to manipulate the essential protective mechanisms in plants and to utilize our biochemical knowledge for optimal molecular engineering of salt tolerance in plants (Winicov, 1998).

Conclusion

The increasing demand for food by the rapidly expanding population necessitates the extension of agriculture into salt affected soils. Soil amelioration and engineering processes consume huge resources making small and marginal farming unviable. The most cost effective and sustainable means of bringing the saline affected areas under cultivation is by developing crop varieties suitable to such areas. Contribution of breeders and physiologists will make this possible by developing varieties tolerant to salinity. To achieve higher productivity, varieties with a combination of higher yield potential and salt stress tolerance should be delivered by the breeder. The breeder has to employ advanced biotechnological tools and mix it with the traditional breeding techniques along with manipulation of physiological pathway to get the desirable level of tolerance to salinity. This will ultimately improve the agricultural production in saline problem areas and can contribute much to feed the rural and urban population.

References

Ansari, R, A. Shereen, S.M. Alam, T.J. Flowers and A.R.Yeo. 2003. Breeding for Salt Tolerance in Rice. In: *Advances in Rice Genetics. Supplement to rice Genetics IV*. (Eds). G.S. Khush, D.S. Brar and B. Hardy. Proceedings of the Fourth International Rice Genetics Symposium, 22-27 October 2000, Los Banos, Philippines. International Rice Research Institute. pp 8-10.

Ashraf, M. and M.R. Foolad. 2005. Pre-sowing seed treatment-a shotgun approach to improve germination, plant growth, and crop yield under saline and non-saline conditions. *Advances in Agronomy*, 88: 223-271.

Azhar, F.M. and T. McNeilly. 1988. The genetic basis of variation for salt tolerance in *Sorghum bicolor* (L.) Moench seedlings. *Plant Breeding*, 101: 114-121.

Blum, A. 1984. Breeding crop varieties for stress environments. *CRC Crit. Rev. Plant Sci.* 2: 119-237.

Blumwald E, Grover A. 2006. Salt tolerance. In: Halford NG. (ed) Plant biotechnology: current and future uses of genetically modified crops. John Wiley and Sons Ltd., UK, pp 206–224.

Bonilla P, J. Dvorak, D. Mackill, K. Deal and G.Gregorio. 2002. RLFP and SSLP mapping of salinity tolerance genes in chromosome 1 of rice (*Oryza sativa* L.) using recombinant inbred lines. *Philipp. Agric. Sci.*; 85:68–76.

Botella, M.A., Abel Rosado, Ray A. Bressan and Paul M. Hasegawa. 2005. Plant adaptive responses to salinity stress. In: Plant Abiotic Stress (eds.)

Mathew A. Jenks and Paul M. Hasegawa. pp 37-70, Blackwell Publishing Limited, UK.

Chandan Sahi, Amanjot Singh, Krishan Kumar, Eduardo Blumwald and Anil Grover. 2006. Salt stress response in rice: genetics, molecular biology and comparative genomics. *Funct. Integr. Genomics.* 6:263–284.

Croughan, T.P., S.J. Stavarek and D.W. Rains. 1978. Selection of a NaCl-tolerant cell line of cultured alfalfa cells. *Crop Sci.*, 18: 959-963.

Cuartero,J., Maria, C Bolarin, Vicente Moreno and Benito Pineda. 2010. Molecular Tools for Enhancing Salinity Tolerance in Plants. In: Molecular Techniques in Crop Improvement. (eds.). S.M. Jain and D.S. Brar. Springer, pp 373-406.

Dansereau, P. 1957. Biogeography, an ecological perspective. Ronald Press, New York.

Deepa Sankar P., M. A. Arabi Mohammed Saleh and C. Immanuel Selvaraj. 2011. Rice breeding for salt tolerance. *Research in Biotechnology* 2(2): 1-10.

Dix, P.J. 1993. The role of mutant cell lines in studies on environmental stress tolerance: an assessment. *Plant J.* 3: 309-313.

Dix, P.J. and K. Winter. 1986. Sodium, potassium, chloride and proline concentration of chloroplasts isolated from a halophyte *Mesembryanthemum crystallinum* L. *Planta,* 168: 421-426.

Epstein, E. and D. W. Rains. 1987. Advances in salt tolerance in plants. *Plant Soil.* 99 : 17-29.

Flowers T.J. and A.R. Yeo. 1995. Breeding for salinity resistance in crop plants: where next?. *Aust. J. Plant Physiol.* 22: 875–884.

Flowers, T.J. 2004. Improving crop salt tolerance. *Journal of Experimental Botany,* 55(396) : 307-319.

Flowers,T.J. and S.A.Flowers. 2005. Why does salinity pose such a difficult problem for plant breeders. *Agricultural Water Management.* 78: 15–24.

Garg, B.K. and I.C.Gupta. 2011. Mechanisms of Salt Injury in Plants. In: Salinity tolerance in plants: Methods, Mechanisms and Management. pp 159-186.

Garg, B.K. and I.C.Gupta. 2011a. Plant in Relations to Salinity. In: Salinity tolerance in plants: Methods, Mechanisms and Management. pp 108-158.

Gelburd, D.E. 1985. Managing Salinity, lesson from the past. *J. Soil Water Conserv.* 40: 329-331.

Gepstein, S., A. Grover and E. Blumwald. 2006. Producing biopharma ceuticals in the desert: building an abiotic stress tolerance in plants for salt, heat and drought. In: Knablein J, Muller RH. (eds). Modern biopharmaceuticals. Wiley-VCH Verlag GmbH, Weinhaum. pp 967–994.

Greenway, H. and R. Munns. 1980. Mechanisms of salt tolerance in nonhalophytes. *Ann. Rev. Plant Physiol.* 31:149–190.

H. R. Athar and M. Ashraf. 2009. Strategies for Crop Improvement Against Salinity and Drought Stress: An Overview. In: Salinity and Water Stress: Improving Crop Efficiency (eds.) M. Ashraf, M. Ozturk and H. R. Athar. pp 1-16. Springer Publications.

Hasegawa, P.M., R.A. Bressan, Zhu J-K and H.J. Bohnert. 2000. Plant cellular and molecular responses to high salinity. *Ann. Rev. Plant Physiol. Plant Mol. Biol.*, 51: 463-499.

Horneck, D.A., J.W. Ellsworth, B.G. Hopkins, D.M. Sullivan and R.G. Stevens. 2007. Managing Salt-affected Soils for Crop Production. A Pacific Northwest Extension publication, Oregon State University • University of Idaho • Washington State University.

Howat, D. 2000. Acceptable Salinity, Sodicity and pH Values for Boreal Forest Reclamation. Available at www.gov.ab.ca/env/protenf/landrec/index.html. In E. S. D. Alberta Environment, Edmonton Alberta. Report # ESD/LM/00-2. 191 pp.

Inge S. Møller, Matthew Gilliham, Deepa Jha, Gwenda M. Mayo, Stuart J. Roy, Juliet C. Coates, Jim Haseloff and Mark Tester. 2009. Shoot Na+ Exclusion and Increased Salinity Tolerance Engineered by Cell Type–Specific Alteration of Na+ Transport in Arabidopsis. *The Plant Cell*, 21 (7) : 2163-2178.

Jaiwal, P.K., R.P. Singh and A. Gulati. 1997. Strategies for improving salt tolerance in higher plants. Oxford and IBA Publishing Co. Pvt. Ltd., New Delhi. pp. 335-336.

Jennings, D.H. 1976. The effect of sodium cholride on higher plants. *Biol. Rev.*, 51: 453–86.

Jones, R. A. 1987. Genetic advances in salt tolerance. In "Tomato Biotechnology" (D. J. Evans and R. A. Jones, Eds.), pp. 125-137, A. R. Liss, New York.

Khajeh-Hosseini, M., A.A. Powal and I.J. Bingham, 2003. The interaction between salinity stress and seed vigour during germination of soybean seeds. *Seed Sci. Technol.*, 31: 715-725.

Kochba, J., G. Ben-Hayyim, P. Spiegel-Roy, S. Saad and H. Neumann, 1982. Selection for stable salt tolerant callus cell lines and embryos in *citrus sinensis* and *E. autanticum. Z. pflanzenphysiol. 106:* 111-118.

Kochba, J., P. Spiegel-Roy and S. Saad. 1980. Selection for tolerance to sodium chloride (NaCl) and 2,4-Ddichlorophenozy acetic acid (2,4-D) in ovular callus lines of *Citrus sinensis*. In plant cell cultures: Results and perspectives, (Eds.) F. Sala, B. Parisi, R. Cella and O. Ciferri, pp. 187-192. Elsevier/North-Holland Biomedical Press, Amsterdam.

Lakshmi Narayanan, S., S. M. Ibrahim, P. Jeyaprakash, M. Madhan Mohan and S. Ashok. 2012. Unravelling the potential of salt tolerance of rice (*Oryza sativa* L.) through *in vitro* screening. In: International Symposium on "100 years of Rice Science and Looking Beyond" 9-12, January 2012, Tamil Nadu Agricultural University, Coimbatore.

Levitt, J. 1980. Responses of Plants to Environmental Stresses. Vol. II Water, Radiation, Salt and other Stress. Academic Press Inc. New York, 607 p.

Lyon C. 1941. Responses of two species of tomatoes and the F1 generation to sodium sulphate in the nutrient medium. *Botanical Gazette*, 103: 107-122.

Mackill D.J., A.M. Ismail, A. Kumar and G.B. Gregorio. 2010. "The role of stress-tolerant varieties for adapting to climate change." Paper from the CURE Workshop on Climate Change, May 4, 2010, Siem Reap, Cambodia.

Marschner, H. 1986. Mineral Nutrition of Higher Plants. (Vol. 150): Academic Press Inc. London.

Mass, E.V. 1986. Salt tolerance of plants. *Applied Agricultural Research*. 1: 12-26.

McCoy, T.I. 1987. Characterization of alfalfa (*Medicago sativa* L.) plant regenerated from selected NaCl tolerant cell lines. *J. Plant Physiol.*, 123: 112-116.

McHughen, A. 1987. Salt tolerance through increased vigour in flax line (STS-II) selected for salt tolerance *in vitro*. *Theor. Appl. Genet.*, 74: 727-732.

Meredith, C.P. 1984. Selecting better crops from cultures cells. In: Gene manipulation in plant improvement, (Ed.) J.P. Gustafson, pp. 503-528. Plenum Press, New York.

Minhas, P. S. and R. K. Gupta. 1992. Quality of Irrigation Water-Assessment and Management. Indian Council of Agricultural Research, New Delhi. 123 pages.

Minhas, P. S. 2005. Crop Production in Saline Soils. In: Agricultural Salinity Management in India (eds.) N. K. Tyagi and P. S. Minhas. pp 325-350. CSSRI, Karnal, India.

Mishra, B., R. K. Singh and D. Senadhira. 2003. Advances in breeding salt tolerant rice varieties. In: *Advances in Rice Genetics. Supplement to rice Genetics IV*. (Eds). G.S. Khush, D.S. Brar and B. Hardy. Proceedings of the Fourth International Rice Genetics Symposium, 22-27 October 2000, Los Banos, Philippines. International Rice Research Institute. pp 5-7.

Mohammad Pessarakli and Szabolcs. 2011. Soil Salinity and Sodicity as Particular Plant/Crop Stress Factors. In: Hand Book of Plant and Crop Stress. (ed.). Mohammad Pessarakli. pp 3-22.

Morrish, F., V. Vasil and I.K. Vasil. 1987. Developmental morphogenesis and genetic manipulation in tissue and cell cultures of graminae. *Advance in Genetics*, 24 : 431-499.

Munns, R. 2005. Genes and salt tolerance: bringing them together. New *Phytologist.* 167(3):645-63.

Nabors, M.W., S.E. Gibbs, C.S. Bernstein and M.E. Meis. 1980. NaCl tolerant tobacco plants from cultured cells. *Z. Pflanzen-physiol.* 97: 13-17.

Nabors, M.W., A. Daniels, L. Badolny and C. Brown. 1975. Sodium chloride tolerant lines of tobacco cells. *Plant Sci. Lett.*, 4: 155-159.

Nelson, D.E., B. Shen and H. J. Bohnert. 1999. Salinity tolerance: mechanisms, models and the metabolic engineering of complex traits. In JK Setlow, (ed), Genetic Engineering, Vol 20. Plenum Press, New York, pp 153–176.

Noble, C. L.1983. The potential for breeding salt-tolerant plants. *Proc. R. Soc. Vict.* 95(3) :133-138.

Petronia Carillo, Maria Grazia Annunziata, Giovanni Pontecorvo, Amodio Fuggi and Pasqualina and Woodrow. 2011. Salinity Stress and Salt Tolerance. In: Abiotic Stress in Plants – Mechanisms and Adaptations (eds.) Arun Kumar Shanker and B. Venkateswarlu. pp 21-38 published by Intech, Croatia.

Pooja Bhatnagar, Mathur V. Vadez and Kiran K. Sharma. 2008. Transgenic approaches for abiotic stress tolerance in plants: retrospect and prospects. *Plant Cell Rep*. 27:411–424.

Quesada V, S. Garcia-Martinez, P. Piqueras, M.R. Ponce and J. L. Micol. 2002. Genetic architecture of NaCl tolerance in Arabidopsis. *Plant Physiol.*, 130: 951–963.

Rains, D.W., S.S. Croughan and T.P. Croughan. 1986. Isolation and characterisation of mutant cell lines and plants. Salt tolerance, In : Cell culture and somatic cell genetics. Vol. 3, (Ed.) I.K. Vasil. pp. 537-548. Academic Press Inc., New York.

Rana, R. S. 1985. Breeding for salt resistance: Concept and strategy. Int. *J. Trop. Agric.* III(4): 236-254.

Rao, P.S., B. Mishra, S.R. Gupta and A. Rathore. 2008. Reproductive stage tolerance to salinity and alkalinity stresses in rice genotypes. *Plant Breed.* 127:256-261.

Reddy, M. P., U. S. Rao and E. R. R. Iyengar. 1997. Carbon metabolism under salt stress. In strategies for improving salt tolerance in higher plants. Ed. Jaiwal, R. P. Singh and Anju Gulati. pp 159-190.

Ren Z.H., J.P. Gao, L.G. Li, X. L. Cai, W. Huang and D. Y. Chao. 2005. A rice quantitative trait locus for salt tolerance encodes a sodium transporter. *Nature Genet.* 37:1141–6.

Reynolds, M.P., J.I. Ortiz-Monasterio and A. Mc Nab (eds.). 2001. Application of Physiology in wheat breeding. Mexico, D.F.: CIMMYT.

Ribaut, J.M. and D. Hoisington. 1998. Marker-assisted selection: new tools and strategies. *Trends in Plant Sci.*, 3: 236-239.

Rodriguez-Navarro A, 2000. Potassium transport in fungi and plants. *Biochem. Biophys. Acta.*, 1469:1-30.

Sardo, V. 2005. Halophytes and Salt-tolerant Glycophytes a potential resource. In: Proceedings of the International Workshop, Alger, Algeria, 12-14 June 2005 (ed.) Atef Hamdy. CIHEAM. pp 87-98.

Schachtman, D.P. and R. Munns. 1992. Sodium accumulation in leaves for Triticum species that differ in salt tolerance. *Aust. J. Plant Physiol.* 19: 331-340.

Schimper, A. F. W. 1903. 'Plant Geography upon a Physiological Basis.' Clarendon Press, Oxford. 824 pp.

Serrano, R., J. M. Mulet, G. Ríos, J. A. Márquez, I. F. de Larriona, M. P. Leube, I. Mendizabal, A. Pascual-Ahuir, M. R. R. Proft and C. Montesinos. 1999. A glimpse of the mechanism of ion homeostasis during salt stress. *J. Exp. Bot.* 50:1023-1036.

Shakil Ahmed. 2009. Effect of Soil Salinity on the Yield and Yield Components of Mungbean. *Pak. J. Bot.* 41(1): 263-268.

Shannon, M.C. 1985. Principles and strategies in breeding for salt tolerance. *Plant and Soil.* 89: 227-241.

Shanthi, P., S. Jebaraj and S. Geetha. 2010. *In vitro* screening for salt tolerance in Rice (*Oryza sativa*). *Electronic Journal of Plant Breeding.* 1(4): 1208-1212.

Shanthi, P., S. Jebaraj, N. Aananthi and C. Natarajan. 2012. Relationship Between Salt Tolerant Physiological Traits and Genetic Distance in Rice. In: National Seminar on "Breeding for Abiotic Stress in Crop Plants" March 22, 2012, Agricultural College and Research Institute, Madurai. P 76.

Singh R.K. and B. Mishra. 2006. CRS 30: first basmati rice variety for sodicity stress. *Indian Farming.* 56:3-6.

Singh, R. K., B. Mishra, A. M. Ismail and G. B. Gregario. 2009. Breeding Rice for Salt Affected Areas of India. In: Progress in crop improvement research (eds.). Hossain, M., Bennett, J., Mackill, D., Hardy, B. Los Baños, Philippines, International Rice Research Institute. pp 78-90.

Srinath Rao and Prabhavathi Patil. 2012. In Vitro Selection of Salt Tolerant Calli Lines and Regeneration of Salt Tolerant Plantlets in Mung Bean (*Vigna radiata* L. Wilczek) In: Biotechnology - Molecular Studies and Novel Applications for Improved Quality of Human Life (ed.) Reda Helmy Sammour, Intech Publications.

Sripunitha, A. K. Sivasubramaniam, K. Selvarani, S. Thangeswari and J. Padma. 2012. Salt Stress and its Management. In: National Seminar on "Breeding for Abiotic Stress in Crop Plants" March 22, 2012, Agricultural College and Research Institute, Madurai. p86.

Stavarek, S.J. and D.W. Rains. 1984. Mechanisms for salinity tolerance in plants. *Iowa State J. Res.* 57: 457-476.

Stocker O. Das. Halophytenproblem. In: Frisch K.V., Goldschmidt R., Ruhland W. and Winterstein H. (eds.) Ergebnisse der Biologie. Berlin: Julius Springer; 1928.p. 266–353.

Thenmozhi,J., G.Yogalakshmi, S.M.Ibrahim and S. Lakshmi Narayanan. 2012. Importance of Rice Breeding for Saline and Sodic Soils. In: National Seminar on "Breeding for Abiotic Stress in Crop Plants" March 22, 2012, Agricultural College and Research Institute, Madurai. P 36.

Thomson M.J., A, M. Ismail, S. R. McCouch and M. J. Mackill. 2010. Marker assisted breeding. In: Abiotic stress adaptation in plants: physiological, molecular and genomic foundation. (Ed) Pareek, A., Sopory, S.K., Bohnert, H.J. and Govindjee. Springer, New York. p. 451–69.

Tyagi, R.K. and N.S. Rangaswamy.1997. In vitro induction and selection for salt tolerance in oilseed brassicas and regeneration of the salt tolerant soma clones. *Phytomorphology*. 47(2): 209-220.

Ungar, I.A. 1991. Ecophysiology of Vascular Halophytes. Boca Raton: CRC Press. 209 pp.

Wani, S.H., P. A. Sofi, S. S. Gosal, and N. B. Singh. 2010. In vitro screening of rice (*Oryza sativa* L.) callus for drought tolerance. *Communications in Biometry and Crop Science*. 5 (2): 108–115.

Winicov, I. 1998. New Molecular Approaches to Improving Salt Tolerance in Crop Plants. *Annals of Botany*. 82: 703-710.

Zenk, M.H. 1974. Haploids in physiological and biochemical research. Haploids in higher plants: Advance and potential, (Ed.) K. J. Kasha. pp. 339-354. Univ. Guelph Press, Guelph.

Zhu J.K. 2002. Salt and drought stress signal transduction in plants. *Annual Review of Plant Biology*. 53:247-273.

Zhu, J.K. 2001. Plant salt tolerance. *Trends in Plant Sci.* 6: 66-71.

❑❑❑

Chapter 16

Abiotic Stress Tolerance and Screening of Oil Palm (*Elaeis guineensis*, Jacq.) Genetic Resources

P. MURUGESAN[1], GOUTAM MANDAL[2] AND P. MASILAMANI[3]

[1] Principal Scientist, Directorate of Oil Palm Research, Research Centre, Palode, Pacha (Post), Thiruvanathapuram Dist, Kerala, gesan70@gmail.com

[2] Sr. Scientist, Directorate of Oil Palm Research, Pedavegi-534450, Andhra Pradesh

[3] Director, National Seed Research and Training Institute, Varanasi, Uttar Pradesh

Introduction

The palm oil has been important to village economies throughout its area of distrubution (Nouy *et al.*,1999). Oil Palm occupied mainly in areas parallel to the Atlantic Coast in West and central Africa extending from Cape Verde to Angola. In central Africa, the oil palm belt extends over inlands, covering Congo-K and parts of Congo-B and Angola. Oil Palm has been distributed with early trade to east Tanzania, and to the islands of Pemba, Zanzibar and Madagascar. The oil palm continues to spread eastwards in Central Africa. With the shifting of oil palm from Africa to South East Asia, it was exposed to more favourable climate. In terms of fruit production, Indonesia and Malaysia produce 81% of the world total with fruit bunch yield averaging 17-21 MT ha^{-1} (containing 3.4–4.2 ha^{-1} of mesocarp oil) (FAOSTAT, 2010) and as per recent statistics, Indonesia has attained number one posstion in palm oil production. Oil palm clonal propagation has now become a technology with commercial output of clonal plantlets constituting 5-10% of the annual oil palm planting material requirement in Malaysia (Soh, 2012).

Fig. 1: Oil Palm growth in hilly terrain, Yeroor Estate, Kollam district, Kerala state, India

Fig. 2: Oil Palm growth near river bank, Athirapalli, Kerala state, India

Fig. 3: Harvested Fresh Fruit Bunch heap in Oil Palm Mill at Jangareddygudem, Andhra Pradesh

Fig. 4: Oil Palm growth affected by water logging

Fig. 5: Screened palm grown under rain fed condition

Though oil palm is believed to be originated in wet soils on river banks, it is later considered as drought tolerant crop as they are growing under dry season of several months in traditional as well as non traditional areas. In general, oil palm grows at an altitude below 500 m, but in some instances oil palms have been found thriving on mountains in East and West Africa up to 1700 m. Oil palms are further more found on swampy river banks, in swamps on alluvial plains and low lying river islands. The recent advances in oil palm breeding programme made it possible to produce varieties with high productivity. Oil palm still responds to improved environments and there is scope for improving yields further using diverse genetic materials. Oil yield is dependent not only on genetic background but also on environmental factors such as relative humidity, water availability, soil structure, fertilizer application, agricultural management and sunshine availability (Henson and Harun, 2005). As the next step, tolerance to abiotic stresses such as water deficiency, extreme temperature, mineral deficiency, metal toxicity and ultra violet irradiation will be the target in oil palm improvement. The role of biotic factors on oil palm nutrition are expected to become more prominent as new breeds are produced with true to type inbreeds with more uniform genetic makeup (Soh, 2012). Knowledge on abiotic stress responses in oil palm is an important area which should be investigated thoroughly for application and development of drought tolerant varieties using available genetic resources.

Water Stress and Screening of Oil Palm Genotypes

Oil palm (*Elaeis guineensis*, Jacq) was once grown widely in forest and subsequently domesticated as plantation crops first in South East Asia and then in West Africa as well (Gerritsma and Wessel, 1997). With the transfer of oil palm from Africa to S. E. Asia, it was exposed to a more favourable climate. The amount of rain fall in places of origin (Nigeria) and other places and S.E.Asian countries are same but distribution is different. The climate in Benin city is comparable to Kerala state of India where December to February experience dry period. It was attributed by several authors (Soh, 2012 and Henson and Harun, 2005) that higher oil yield in South East Asian countries is primarily due to a high fresh fruit bunch yield (FFB) *i.e.* production of more bunches per palm rather than an increase in average bunch weight. Although several factors have individually contributed to the large yield improvement in the plantation industry, the great rise to a level of 5 to 6 tons of oil Ha^{-1} Y^{-1} on good estates in Malaysia is the result of optimal interaction between improved genotypes and improved cultivation methods in an overall favourable environment. Oil palm is sensitive to drought (Nouy *et al.*, 1999) and it is thus strongly affected by the increasing frequency of climate anomalies such as 'el nino' causing drought in South East Asia (Legros *et al.*, 2009 a). Legros *et al.* (2009 b) reported that main sensitive phase for drought effects was estimated at 29 months before bunch maturity, presumably associated with inflorescence sex determination. Climatic conditions prevailing in the highest yielding regions which has fairly uniform climate without any abiotic stresses are summarised by Carr (2011) which are as follows. Annual rainfall of at least 2000 mm spread evenly during the year and mean maximum and minimum air temperatures of 29-33°C and 22-24°C respectively, Relative Humidity >85 % (Equivalent to saturation deficits <0.6kPa). Bright sun shine averaging 5h d^{-1} throughout the year rising to 7h d^{-1} in some months (or solar radiation of 16-17 MJM-2 D^{-1}). However, Oil palm is cultivated successfully in West Africa, South America, India and Thailand with less favourable environments and regular dry seasons. The effect of water stress on fruit yield varies depending on its timing and severity relative to each of bunch developmental stress. It also affects distribution of yield during the year. In countries like Malaysia, yield distribution is uniform, whereas in countries with marked dry season, distinct peak lean season is observed. According to Legros *et al.* (2009 a) photoperiod sensitive phase estimated to occur nine months before bunch maturity while sensitive phase of the drought was estimated to be 29 months before bunch maturity. Large oil palm yield response in the drier environment with irrigation was reported both in Malaysia and Thailand (Palat *et al.*, 2008) and India (Suresh *et al.*, 2008, Suresh *et al.*, 2010, Mathur *et al.*, 2001, Suresh *et al.*, 2012 and Suresh and Mathur (2009). Nouy *et al.* (1999) reported that rainfall is a main limiting factor for oil palm in Benin City which has two rainy and dry seasons. Houssou *et al* (1989) reported genetic differences in survival between oil palm progenies in Benin. The results indicated that high yielding cross namely, Deli × La Me showed high death rate than Deli × Yangambi. The yield from surviving palms was not much higher from Deli × La Me than from Deli × Yangambi. Above authors could found out progenies with high yield and drought tolerance. Asemota and Conaire (2011) reported that different oil palm genotypes exhibit different response due to moisture stress in terms of membrane integrity damage and protein loss. Such differences could be exploited for selecting drought tolerant genotypes. Maillard *et al.*(1974) tried to develop methods of measuring drought resistance in Young palms. The measured root growth of seedlings under high osmotic pressure, using PEG and at tolerance of high temperature and obtained no satisfactory results. Smith (1993) found significant differences between oil palm families in stomatal openings and photosynthetic rate during the dry season in Congo and strongly suggested parameters for selecting drought tolerant genotypes. According to Nodichao *et al.* (2011), the presumed tolerant oil palm genotype has higher Total Root Length (TRL), Total Root Surface Area (TRSA) and Potential Root Water Extraction Ratio (PREP) than susceptible genotypes while the performance of the control genotype was intermediate. Earlier studies of young oil palm seedling root development in a rhizotron system showed that different progenies exhibited significant differences in root development which could be exploited in selection programme (Conaire *et al.*, 2005). These differences could also confer advantages in times of moisture deficits to progenies with profound roots. According to Corley(1996), responses to drip or micro sprinklers irrigation was less when compared to entire wetting by sprinklers or flood irrigation due to closure of stomata when atmospheric humidity is low (large negative values of vapour pressure deficit VPD). It takes two years or more to get response on effects of irrigation on inflorescence sex ratio (Corley and Tinker, 2003). Fifteen years of research on irrigation on oil palm in Thailand have shown that drip irrigation is the best systems (Tittinutchanon, *et al.*, 2009). The above authors recommended that different progenies need to be planted in areas where irrigation is not possible or irrigation responsive materials for sites where irrigation is indented to screen materials with tolerance or sensitive to drought. Muhammad *et al.* (2011) presented interesting reports that rainfall and temperature in different seasons affected yield and oil quality in oil palm of *dura* fruit form. Myristic and palmitic acid in mesocarp oil palm in the fruits harvested in dry season showed reduction in dry season compared to those from wet season. However, stearic and linoleic acid percentages increased in the samples from dry season. The study indicated that mesocarp oil yield and rainfall as well as accumulated temperature in the three months before harvesting had positive correlation while kernel oil yield showed higher concentration with accumulated temperature than amount of rainfall. Duration of bunch ripening varied with season. The duration of ripening during rainy season was 156 days whereas dry season took 165 days. Though water supply obviously plays a role on the length of ripening, other factors such as temperature and sunshine also interfere with fruit development process as per the investigations carried out in Andhra Pradesh state of India (Murugesan and Rethinam, 2000).

Screening for Cold Tolerance

It is well established fact that oil palm grows poorly at temperatures much below 20° C. Black and Sterling (1996) tested *Dura* ×*Pisifera* progenies (collected from Bamenda high lands of Cameroon) in Ethiopia at 960 m ASL and recorded precocious flowering after 12 months. Effect of low temperature and water deficiency studies showed that above factors inhibit the growth of oil palm seedlings. The relative conductivity, injury index, malondialdehyde (MDA) and proline content in the leaves increased to different degrees with the extension of low temperature and drought stress. The results obtained were utilised to initiate breeding programme used to improve drought and cold tolerance in oil palm. Similar reports have been documented by Cao *et al*. (2009) and Yang and Lin (2008). Pillai *et al* .(2000) reported that cold tolerant oil palm germplasm collected from Cameroon during 1995 and several such germplasm accessions were planted in West Bengal and Kerala states of India during 1998 and their performance in terms of Fresh Fruit Bunch yield has been reported to be satisfactory under North Eastern climatic condition (Goutam Mandal *et al*., 2012).

Water and Nutrient Stress

Cheng-Xu *et al*. (2011) stressed the need for complete knowledge on effects of water and nutrient stress on oil palm and its adoption to such condition. Legros *et al*. (2009) explained responses of oil palm to water stress. Combination of water and nutrient stress significantly increase the root/shoot ratio. Fertilization under well watered condition alleviated the nutrient stress while fertilization under water stress conditions aggravated the drought stress, water and nutrient stress had a greater nitrogen and potassium concentration. The water content of soil plays a key role in oil palm growth and development (Henson and Harun, 2005). Suriyan Cha-um *et al*. (2012) investigated physiobiochemical responses of oil palm seedlings to Mannitol and PEG induced iso-osmotic stresses. The results revealed that the water content of osmotically stressed oil palm seedlings decreased, but the proline content and the electrolyte leakage of the seedlings increased with decreasing water potential. Similar report by Nana Yamada *et al*.(2011) indicated that under drought and salt stress, proline content increases in oil palm and C DNA, genomic DNA and the promoter region of pro T were isolated from oil palm during stress. In response to water deficit, relative membrane leakage leading to the damage of the photosynthetic pigment noticed in oil palm was reported by Cha-um *et al.* (2011) and data obtained were utilized to establish multivariate criteria for water deficit tolerance screening in oil palm breeding programmes. Embryos were cultured in modified MS media for 4-8 weeks before being transferred to filter paper bridges immersed in stress inducing liquid media. Sodium chloride @ 28g/litre was added to induce water stress. Mortality of plantlets was recorded after 3 weeks. It was reported that sago palms are more tolerance to salt whereas oil palms are tolerant to water stress. In both palms, younger *in vitro* derived plantlets were more susceptible than older plantlets.

Soil Compaction, Salt Stress and Flooding

Oil palm is grown widely on waterlogged tropical peat swamp forest in Sarawak, Malaysia and concerns about green gas emissions and loss of soil carbon due to oil palm cultivation in those areas were reported by Melling *et al.* (2005). Investigation by these group in oil palm cultivated in peat soil (Typic Tropofibrist- contain 44.7% carbon and 2% nitrogen with a very high loss of ignition of about 99%) revealed that oil palm was neither a net C sink nor source but the export of Fresh Fruit Bunch without by-product utilization will cause a negative C balance of 2.01 + C $ha^{-1}y^{-1}$. Cha-um *et al,* (2010) recorded increased content of sodium ions, proline content and percentage of relative electrolyte leakage in the seedlings subjected to salt stress and advocated to do further investigations to establish criteria for salt tolerance in screening in oil palm breeding programme. Yahya *et al.* (2010) reported that soil compaction in Bernam series soil planted with oil palm does not affect FFB yield which showed increase in mean soil bulk density. Similar supportive result was reported by Haniff *et al.* (2005). The oil palm is not as tolerant of sea water as in the coconut palm and it is not a halophyte species (Munns, 1993). Yeap *et al.* (2012) reported cloning and characterisation of gene, designated as EgRBP42 from oil palm which was expressed in leaf, shoot apical meristem, root inflorescence and mesocarp with the lowest transcript level in the roots. The accumulation of EgRBP42 and its interesting transcripts were induced by abiotic stresses including salinity, drought, submergence, cold and heat stresses in leaf discs. Frequently flooded soils are poorly oxygenated, have a variable pH, low redox potential and high electrical conductivity. These conditions make the palm prone to disease (Munever *et al.,* 2001).

Summary

- Abiotic stress responses in oil palm is an important area which should be investigated thoroughly for application and development of drought tolerant varieties using available germplasm resources.
- Oil yield is dependent not only on genetic background but also on environmental factors such as relative humidity, water availability, soil structure, fertilizer application, agricultural management and sunshine availability.
- Genetic differences has been reported for drought tolerance and susceptibility in different progenies in West Africa.
- Though oil palm is believed to be originated in wet soils on river banks, it is later considered as drought tolerant crop as they are growing under dry season of several months in traditional as well as non traditional areas.
- Although several factors have individually contributed to the large yield improvement in the plantation industry, the spectacular yield rise in Malaysia is the result of optimal interaction between improved genotypes and improved cultivation methods in an overall favourable environment.
- Oil palm is cultivated successfully in West Africa, South America, India and Thailand with less favourable environments and regular dry seasons.
- Main sensitive phase for drought effects was estimated at 29 months before bunch maturity, presumably associated with inflorescence sex determination.
- Responses to drip or micro sprinklers irrigation was less when compared to entire wetting by sprinklers or flood irrigation due to closure of stomata when atmospheric humidity is low (large negative values of vapour pressure deficit VPD).
- Combination of water and nutrient stress significantly increase the root/shoot ratio. Fertilization under well watered condition

alleviated the nutrient stress while fertilization under water stress conditions aggravated the drought stress, water and nutrient stress had a greater nitrogen and potassium concentration.

- Field with compact soil planted with oil palm does not affect FFB yield which had increased mean soil bulk density.
- Frequently flooded soils are poorly oxygenated, have a variable pH, low redox potential and high electrical conductivity. These conditions make the palms prone to disease incidence.
- Significant differences were observed among oil palm families in stomata openings and photosynthetic rate during the dry season and the presumed tolerant oil palm genotype has higher Total Root length (TRL), Total Root Surface Area (TRSA) and Potential Root Water Extraction Ratio (PREP) than susceptible genotypes.
- Further intensive research is required to screen oil palm genotypes with reference to abiotic stresses.

References

Asemota, O, and B. Conaire.2011. Identification of moisture stress tolerant oil palm genotypes. *African Journal of Agricultural Research.* 5 (22): 3116-3121.

Black, G. and F. Sterling. 1996. The prospects of extending oil palm cultivation to higher elevations through using cold tolerant planting material, Planter, Kulalumpur. 72: 645-652.

Caliman, J. P. 1992. Oil palm and water deficit: production, adapted cultural techniques 1992. Oléagineux (Paris). 47 (5): pp. 205-216.

Cao, J.H., X. P. Li, W.F. Lin, G. S. Xie and J. M. Chen. 2009. Survey on resistance of 12 species oil palm cultivated in the field. *Chin. J. of Trop Agric.* 29: 1-6.

Carr, M. K. V. 2011. The water relations and irrigation requirements of oil palm (*Elaeis guineensis*): a review, *Exp. Agric.* 47(4):629-652.

Cha-um, S., N. T.Yamada, T.Takabe and K.Chalermpol. 2011. Mannitol induced water deficit stress in oil palm (*Elaeis guinensis*, Jacq.) seedlings. *J. of Oil Palm Research*, 23:1193-1201.

Cha-um, S., T. Takabe and C. Kirdmanee. 2010. Ion contents, relative electrolyte leakage, proline accumulation, photosynthetic abilities and growth characters of oil palm seedlings in response to salt stress. *Pak. J. of Botany*, 42(3): 2191-2020.

Cheng-Xu, S., H. X. Cao, H.B. Shao, X.T. Lei and Y. Xiao. 2011. Growth and physiological responses to water and nutrient stress in oil palm. *African J. of Biotechnology*, 10(51): 10465-10471.

Conaire, B., O. Asemoto, A.E.Isemila. 2005. Pattern of seedling root development among some oil palm genotypes. *J. Agric.Environ.Inter.Dev.*, 2(1): 55-64.

Corley, R.H.V. and Tinker, P.B. 2003. The oil palm, Blackwell Science, Oxford, 562 pp.

Corley, R.H.V.1996. Irrigation of oil palms, a review. *J of Plantation Crops*, 24 (Suppl) 45-52.

FAOSTAT (2010). http://faostat.fao.org/site/567/default.aspx#ancor.

Gerritsma, W. and M. Wessel. 1997. Oil palm: Domestication achieved? *Netherlands Journal of Agricultural Sciences*, 54: 463-475.

Goutam Mandal, R.K.Mathur, A.K.Shit, K.Sunil Kumar, R.S.N.Pillai, A.Nagaraju and P.Balakrishna. 2012. Evaluation of African germplasm. In: Global conference on Horticulture for food and livelihood option organised by ASM Foundation and OUAT, Bhubaneswar, 28-31, May 2012, pp.51.

Haniff, M.H, A.S. Rahim, O.Wahid, T.M.Tarmizi and M.M.Faiz. 2005. Effects of soil compaction on oil palm productivity. Proceedings of the PIPOC 2005, International Palm Oil Congress, pp 469-475.

Henson, I. E. and M.H. Harun. 2005. The influence of climatic conditions on gas and energy exchanges above a young oil palm stand in North Kedah, Malaysia. *Journal of oil palm Research, 17*(December), 73-91.

Houssou, M, A.Omore and J.Meunier. 1989. Breeding for drought resistance in the oil palm 1. Variability of some crosses for their productivity and their mortality. Paper presented at Conference of palms and palm products, 21-25 Nov, Nigerian Instt Oil Palm Res, Benin city.

Legros, S., I. M. Serra, J. P. Caliman, F. A. Siregar, A. Clement-Vidal and M. Dingkuhn. 2009. Phenology and growth adjustments of oil palm (*Elaeis guineensis*) to photoperiod and climate variability. *Annals of Botany*, 104: 1171-1182.

Legros, S., I. Mialet-Serra, J. P. Caliman, F. A. Siregar, A. Clement-Vidal, D. Fabre and M. Dingkuhn. 2009. Phenology, growth and physiological adjustments of oil palm (*Elaeis guineensis*, Jacq) to sink limitation induced by fruit pruning. *Ann. Bot*, 104: 1183-1194.

Maillard, G., C. Daniel and R. Oehs. 1974. Analuse des effets dela secheresse sur le palmier a' huile, *Oleagineux*, 29: 397-404.

Mathur, R.K, K. Suresh, S. Nair, K. Parimala and V.N.P Shivaramakrishna. 2001. Evaluation of exotic dura germplasm for water use efficiency in oil palm (*Elaeis guineensis* Jacq.) *Ind. Journal of Plant Genetic Resources*, 14(2): 257-259.

Melling, L., R. Hatano and K. J. Goh. 2005. Soil CO^2 flux from three ecosystems in tropical peat land of Sarawak, Malaysia. *Tellus*, 57B: 1-11.

Muhammad, S., P. Leewanich, V. Punsuvon, S. Chanprames and P. Srinives. 2011. Seasonal effects on bunch components *Elaeis guineensis* and fatty acid composition in dura oil palm. *African J. of Agricultural Research*, 6(7): 1835-1843.

Munns, R. 1993. Physiological processes limiting plant growth in saline soils some dogmas and hypotheses. *Plant cell Environ.*, 16: 15-24.

Munever, F, A. Acosta, and P. Gomez. 2001. Edaphic factors associated with bud rot in the oil palm of Colombia. *Palmas*, 22(2): 9-19.

Murugesan, P and P. Rethinam. 2000. Duration of oil palm bunch ripening under coastal climate of Andhra Pradesh India. *Int. J of Oil Palm*, 1(1&2): 65-66.

Nana Yamada, Suriyam Cha-Um, Hakutokageyama, Worrawat Proden, Yoshito Tanaka, Chalermpol Kirdmanee and Teruhiro Takabe. 2011. Isolation and charecterisation of proline/betaine transporter gene from oil palm. *Tree Physiology*, 31: 462-468.

Ndiaye, O, O. Diouf, B. Adon and S. Braconniar. 2007. Physiological criteria used in the selection of genotypes of palm oil (*Elaeis guineensis* Jacq.) for drought tolerance during tree early age. *Agromomie Africaine*, 19(1): 1-12.

Nodichao, L, J. L. Chopart, O. Roupsard, M. Vauclin, S. Ake and C. Jourdan, 2011. Genotypic variability of oil palm root system distribution in the field. Consequences for water uptake. *Plant and soil*, 341(1-2): 505-520.

Nouy, B., Baudouin, N. Diegui and A. Omore.1999. Oil palm under limiting water supply conditions. *Plantations, Recherché Development*, 6: 31-45.

Palat, T., Chayawat Nakharin, J. H. Clendon and R. H. V. Corley. 2008. A review of 15 years of oil palm irrigation research in Southern Thailand. *The Planter*, 84 (989): 537-546.

Pillai, R.S.N., G. Black and H. Paul Closen, 2000. Collection of Oil Palm (*Elaeis guineenisis*, Jacq.) germplasm from Africa. *Int. J of Oil Palm*, 1(1&2): 23-37.

Smith.B.G. 1993. Correlation between vegetative and yield charecteristics and photosynthetic rate and stomatal conductance in the oil palm (*Elaeis guineensis*, Jacq). *Elaeis*, 5: 12-26.

Soh, A.C. 2012. The future of oil palm clones in Malaysia. *Journal of Oil Palm & the Environment*, 3:93-97.

Suresh, K. and R. K. Mathur. 2009. Drought tolerance in Oil Palm (*Elaeis guineensis,* Jacq.). *Int. J of Oil Palm,* 6(2): 1-6.

Suresh, K., G. Nagamani, D. L. Kantha and M. K. Kumar. 2012. Changes in photosynthetic activity in five common hybrids of oil palm (*Elaeis guineenisis* Jacq.) seedlings under water deficit. *Photosynthetica,* 50(4): 549-556.

Suresh, K., G. Nagamani, K. Ramachandrudu and R. K. Mathur. 2010. Gas-exchange characteristics, leaf water potential and chlorophyll a fluorescence in oil palm (*Elaeis guineensis* Jacq.) seedlings under water stress and recovery. *Photosynthetica,* 48(3): 430-436.

Suresh, K, R. K. Mathur and M. K. Babu. 2008. Screening of oil palm duras for drought tolerance-stomatal responses, gas exchange and water relations. *Journal of Plantation Crops* 36 (3): 270-275.

Suriyan Cha-um, T Takabe and C Kirdmanee. 2012. Physio biochemical responses of oil palm (*Elaeis guineensis* Jacq.) seedlings to mannitol- and polyethylene glycol-induced iso-osmotic stresses. *Plant Production Science.* 15 (2): 65-72.

Tittinutchanon, P., C. Nakharin, J. H. Clendon and R. H.V.Corley. 2009. A review of 15 years of oil palm irrigation research in Southern Thailand. In: Proceedings of National Conference on oil palm (Oil palm for farmers prosperity and edible oil security). pp146-154.

Yahya, Z., A. Husin, J. Talip, J. Othman, O. H. Ahmed and M. B. Jalloh. 2010. Soil compaction and oil palm (*Elaeis guineensis*) yield in a clay textured soil. *American J. of Agricultural and Biological Sciences,* 5(1): 15-19.

Yang, H.G. and W. F. Lin. 2008. Effect of low temperature stress on some physiological characteristics of oil palm (*Elaeis guineensis,* Jacq.) seedlings. *Chinese Agricultural Science Bulletin,* 3: 2009-24.

Yeap, W.C, T. E. K. Ooi, P. Namasivayam, H. Kulaveerasingam and C. L. Ho. 2012. EgRBP42 encoding an hnRNP-like RNA binding protein from *Elaeis guineenisis* Jacq. is responsive to abiotic stresses. *Plant cell Rep,* 31: 1829-1843.

❑❑❑

Chapter 17

Water Stress Induced Anatomical Changes in Plants

K. BALAKRISHNAN AND P. MASILAMANI
Dept. of Seed Science and Technology
Agricultural College and Research Institute, Tamil Nadu Agricultural University,
Madurai, Tamil Nadu -625 104

Crops are continuously exposud to abiotic stresses, for example, drought, salinity, extreme temperatures, chemical toxicity, oxidative stress, etc. Factors controlling stress conditions alter the normal equilibrium, and lead to a series of morphological, physiological, biochemical and molecular changes in plants, which adversely affect their growth and productivity (Fig.1). The average yields from the major crop plants may reduce by more than 50% owing to stresses. However, plants also have developed innate adaptations to stress conditions with an array of biochemical, anatomical and physiological interventions that involves the function of many stress-associated genes.

Crop plants experience water stress either when the water supply to their roots becomes limiting or when the transpiration rate becomes intense. Water stress is primarily caused by the water deficit, *i.e.* drought or high soil salinity. In case of high soil salinity and also in other conditions like flooding and low soil temperature, water exists in soil solution but plants cannot uptake it - a situation commonly known as 'physiological drought'. Drought occurs in many parts of the world every year, frequently experienced in the field grown plants under arid and semi-arid climates. Regions with adequate but non-uniform precipitation also experience water limiting environments. The general effects of drought on plant growth are fairly well known. However, the primary effects of water deficit at the anatomical levels are not considerably understood yet and such understanding is crucial.

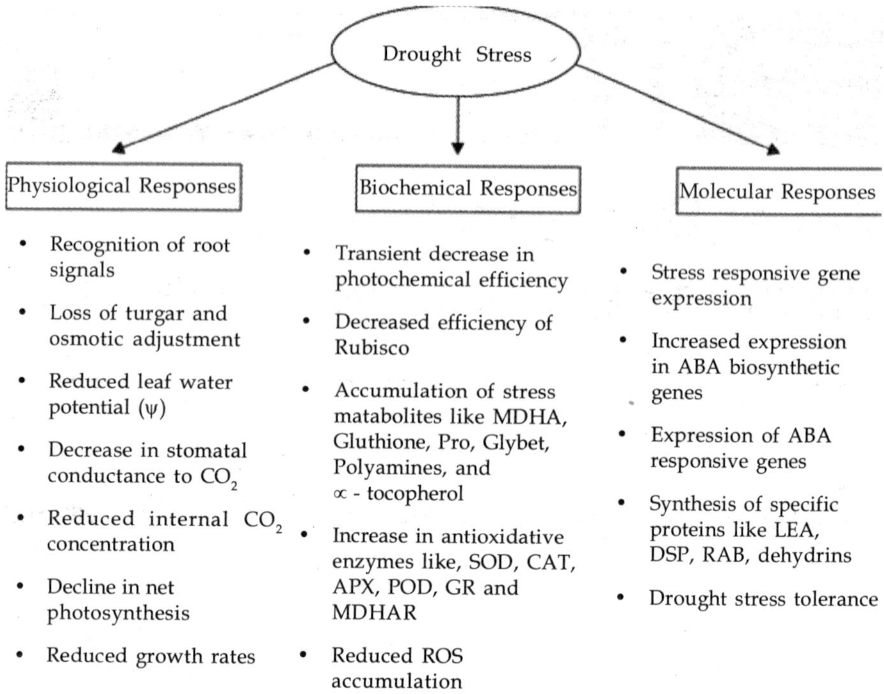

Fig. 1: Physiological and molecular basis of drought stress tolerance (Reddy et al., 2004).

Stress with a constraint or with highly unpredictable fluctuations imposed on regular metabolic patterns causes injury, disease, or aberrant physiology. Stress is the altered physiological condition caused by factors that tend to alter equilibrium. Environmental stresses trigger a wide variety of plant responses, ranging from altered gene expression and cellular metabolism to changes in growth rate and plant productivity. Plant reactions exist to circumvent the potentially harmful effects caused by a wide range of both abiotic and biotic stresses, including light, drought, salinity, and high temperatures. Among the environmental stresses, drought stress is one of the most adverse factors to plant growth and productivity (Fig. 2).

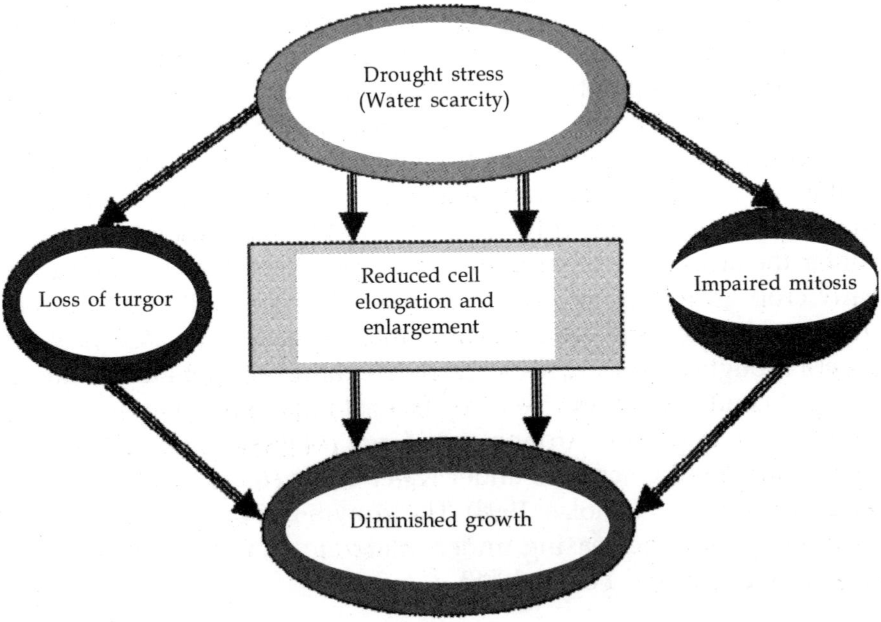

Fig. 2 : Some important causes of growth reduction in plants under drought stress.

Effect of water deficit on morphological parameters

Growth is one of the most drought-sensitive physiological processes due to the reduction of turgor pressure. Cell expansion can only occur when turgor pressure is greater than the cell wall yield threshold. Water stress greatly suppresses cell expansion and cell growth due to the low turgor pressure (Karthikeyan *et al.*, 2007). Water is being used economically for growth processes. Water Use Efficiency (WUE) maintains the better metabolic status of the plant, thus better growth potential in multipurpose agro forestry tree species is achieved. Various internal and external factors influence growth besides its genetical make-up. Growth is an important tool for assessing crop productivity in various crops. Osmotic regulation can enable the maintenance of cell turgor for survival or assist plant growth under severe drought conditions in pearl millet.

Root growth

Crops are exposed to a variety of environmental stresses, viz., drought, salinity or low temperature which constitute some of the most serious limitations to crop growth in *Helianthus annuus* (Mahajan and Tuteja, 2005). Root characteristics, especially root length, root length density, and the number of thick roots, are important for a plant to have comparatively well-established above ground parts by exploiting the available water, as in rice (Manivannan *et al.,* 2007). Drought avoidance due to a profound root system that enhances the ability of a plant to capture water is a fundamental adaptative mechanism to drought (Liu *et al.,* 2005). A prolific root system can confer the advantage to support accelerated plant growth during the early crop growth stages and extract water from shallow soil layers that is otherwise easily lost by evaporation (Martinez *et al.,* 1986). More severe drought stress suggests that the dynamics of root growth under drought conditions might be a key factor to the understanding of the contribution of roots to drought avoidance (Mathews *et al.,* 2003). The root-to-shoot ratio increases under water-stress conditions to facilitate water absorption (Nicholas, 1998). The growth rate of wheat and maize roots was found decreasing under moderate and high water-deficit stresses (Noctor and Foyer, 1998).

The root-to-shoot ratio increases under water-stress conditions to facilitate water absorption and to maintain osmotic pressure, although the root dry weight and length decrease as reported in some plants like sugar beet and populus. Higher root-to-shoot ratio under the drought conditions has been linked to the ABA content of roots and shoots. Water stress is linked to decrease in stem length in plants such as Albizzia, Erythrina, Eucalyptus and Populus, with up to 25% decrease in plant height in citrus seedling. Decreased leaf growth, total leaf area and leaf-area plasticity were observed under the drought conditions in many plant species, such as peanut and rice.

Cytological changes

Every plant organ is ideally designed to fulfil metabolic and physiological processes in specific environmental conditions. In environments characterised by arid conditions, plant survival depends upon the ability to harmonise structure and function to withstand desiccation without permanent damage. These plants show structural alterations that are mainly related to water saving (e.g. water storage and reduction of water losses) and mechanical reinforcement of tissues (e.g. thickening and straightening of cell walls) (De Micco and Aronne 2007). Mechanical reinforcement may prevent phenomena such as collapse and damage of cells, associated with irreversible wilting, and hence the loss of functionality during water deficit.

In the majority of the plant species, water stress is linked to changes in leaf anatomy and ultrastructure. Shrinkage in the size of leaves, decrease in the number of stomata, thickening of leaf cell walls, cutinization of leaf surface, and underdevelopment of the conductive system, increase in the number of large vessels, submersion of stomata in succulent plants and in xerophytes, formation of tube leaves in cereals and induction of early senescence are the other reported morphological changes.

Water stress greatly suppresses cell expansion and cell growth due to the low turgor pressure. Osmotic regulation can enable the maintenance of cell turgor for survival or to assist plant growth under severe drought conditions in pearl millet (Shao *et al.,* 2008). The leaf growth was more sensitive to water stress in wheat than in maize; *Vigna unguiculata* and sunflower (Manivannan *et al.,* 2008). Production of ramified root system under drought is important to above ground dry mass and the plant species or varieties of a species show great differences in the production of roots. The importance of root systems in acquiring water has long been recognized. A prolific root system can confer the advantage to support accelerated plant growth during the early crop growth stage and extract water from shallow soil layers that is otherwise easily lost by evaporation in legumes (Johansen *et al.,* 1992). The development of root system increases the water uptake and maintains requisite osmotic pressure through higher proline levels in *Phoenix dactylifera* (Djibril *et al.,* 2005). The root dry weight was decreased under mild and severe water stress in *Populus* species (Wullschleger *et al.,* 2005). An increase in root to shoot ratio under drought conditions was related to ABA content of roots and shoots (Manivannan *et al.,* 2008).

Anatomical Changes in Leaf

Water absorbed by plant roots is lost in the atmosphere mostly through leaf transpiration. Transpiration rates vary widely among plant groups: from 10 to 0.1 g of water $dm^{-2}h^{-1}$ in hygrophytes and xerophytes respectively. Plants in arid and semiarid environments show leaves with xeromorphic traits designed to reduce transpiration to a minimum under drought conditions. The reduction of transpiration can be reached through various means including leaf shedding as well as decrease of leaf number, leaf size and branching. In extreme conditions, shedding of twigs, branches or larger portions of plants also occurs. In plants that appear leafless under drought conditions, photosynthesis is achieved thanks to green branches, stems or retained petioles. Sclerophylly of plants is also considered an adaption to drought: hard leaves do not suffer from permanent damage due to wilting and can completely recover when favourable conditions are restored. Sclerophylls are widely distributed in arid and semiarid environments together with seasonally dimorphic species. Sclerophyllous leaves are characterised by reinforcing tissues (e.g. thick-walled epidermal cells, sclereids, etc.) which prevent the collapse of the whole structure when water availability is low, thus reducing the risk for mechanical damage. Under drought conditions, a sclerophyllous leaf slightly reduces its volume thanks to thick cuticle and thick-walled epidermal cells, but thin-walled mesophyll cells severely shrink resulting in an increase of intercellular spaces. This allows photosynthesis to remain active also in conditions of severe water stress when other leaf types wilt (Shields 1950). Seasonally dimorphic species are characterised by a seasonal reduction in their transpiring surface: larger winter mesomorphic leaves growing on dolichoblasts (long twigs) are shed at the beginning of the arid season and are replaced by smaller summer xeromorphic leaves on new brachyblasts (short twigs) (Aronne and De Micco, 2001).

Adaptation to drought conditions is achieved also through the low ratio of the external leaf surface to its volume. This is generally also accompanied by the occurrence of a compact structure made of small mesophyll cells with thick cell walls, reduced intercellular spaces and a more compact network of veins. The increased leaf stiffness, although limiting gas exchange, might reduce water losses enhancing high leaf tissue density and dry mass per area (Niinemets, 2001). More recently, the decrease of total diffusion conductance throughout the intercellular spaces from substomatal cavities to chloroplasts has been linked not only to reduced aquaporin conductance, but also to precise anatomical traits such as the reduction of chloroplast surface area exposed to intercellular space per unit leaf area (Tosens *et al.*, 2012). The presence of additional layers of palisade parenchyma at the expense of spongy tissue is also considered a way to increase the path of water through intercellular spaces to reach stomata; this would be a strategy to increase water use efficiency (ratio of carbon dioxide fixed to water lost). In xeric leaves, additional layers of palisade parenchyma can also develop adjacent to the lower epidermis leading to isobilateral anatomical structure which, together with steep leaf inclination, allows the optimisation of light interception in the early morning hours and in the late afternoon. The increase of mesophyll thickness enhances the photosynthetic capacity if it is accompanied by an increase in the number of chloroplasts exposed near the surface area facing the intercellular spaces (Oguchi *et al.*, 2005).

Xeromorphic leaves are also characterised by the presence of a thick cuticle. The hydraulic permeability of the cuticle depends on its thickness, chemical composition and crystal forms of cuticular waxes embedded in the cuticle or deposited over it. Water permeability of cuticle is lower in xeromorphic than in mesomorphic leaves and it has been demonstrated that the properties of this barrier at leaf interface with atmosphere are genetically controlled. Cuticular water permeability also depends on relative humidity, decreasing as environment dries out thanks to chemical properties of cutin and wax domains (Bargel *et al.* 2004). Moreover, different cuticle composition can determine changes in reflectance thus indirectly affecting lamina heating (Monneveux and Belhassen, 1996). Indeed, the occurrence of cuticular waxes, by increasing leaf reflectance at the visible and infrared light wavelengths, can reduce photo inhibition of photosynthesis and transpiration rates, having a positive effect on water use efficiency. Besides, UV radiation is also attenuated by the presence of flavonoids in the cuticle matrix or at the surface of epicuticular waxes (Bargel *et al.*, 2004). Phenolic compounds can also filter excess radiation either in form of vacuolar accumulations in glandular leaf hairs, in epidermal and parenchyma cells, or linked to membranes throughout mesophyll cells as also shown in cortical cells of green photosynthesising twigs. Such phenolics also have an indirect role in adaptation to drought since, being feeding deterrents, they protect plants from permanent damage due to grazing and pathogen attacks. This is critical since development of spare organs would require an extra-additional energetic cost which is often unaffordable under drought conditions.

In arid and semiarid environments, vessel size may also be limited by the need to reduce intervessel pitting and embolism by air seeding which is a phenomenon described as gas being drawn through pit membrane pores (Sperry *et al.*, 2006). The porosity of pit membranes between adjacent vessels might be designed to solve the conflict between the functional requirements to minimise vascular resistance, which favours thin, porous membranes, and to limit embolism spreading, which requires robust membranes and smaller pores (Fig.3). Indeed, a decrease in total pit area per vessel causes a decrease in the average size of membrane pore, and hence an increase in safety from cavitation. In the light of these considerations, traits such as pit membrane permeability, pit membrane area between vessels and pore size can be considered important adaptive traits potentially driving ecological differences between species (Jansen *et al.*, 2009).

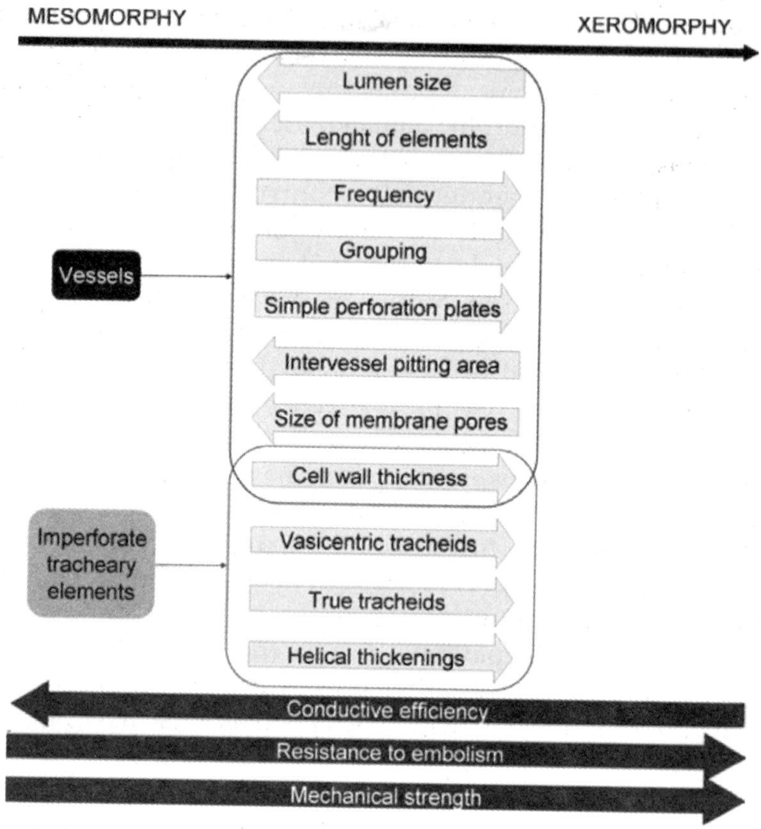

Fig. 3 : Schematic view of the direct and inverse proportionality between the main attributes of xylem elements and functional traits of efficiency/safety against embolism and impolism.

Different combinations of these traits might play a role in species geographical distribution in mesic and xeric areas.

Thickness and density of vessel sculpturing in Acer species has been shown to be correlated with resistance to cavitation (*Lens et al.*, 2011). Apart from the increase of mechanical strength, helical sculpturing has a role in preventing the incidence and spreading of cavitation, because it increases wall surface and thereby water bonding. In woods of xeric environments, helical thickenings increase mechanical strength which can be also guaranteed by the presence of vessels and imperforate tracheary elements with very thick walls and narrow lumen (Sperry, 2003). Besides, other features, such as length of conduit elements, play an important role in defining efficiency of water transport and resistance to negative pressures, but, being correlated to vessel lumen diameter, are less studied. In general, adaptation to drought can be achieved by means of various combinations of wood anatomical traits. The lack or the low expression of a specific trait can be compensated by the occurrence of other features addressed to the same goal; for example, greater wall thickness in any species may compensate for a lesser amount of imperforate tracheary elements when vessel frequency is very high as shown in *Cistus monspeliensis* L. (De Misince *et al.* 2008).

The presence of small diameter roots under reduced water availability is considered as a strategy aimed to maximise absorptive surfaces, thus increasing rates of water and nutrient uptake. The control of water loss is also exerted by the presence of specialised tissues such as a rhizodermis with thickened outer cell walls, a well-developed suberised exodermis, often accompanied by many layers of thin- or thick-walled suberised cells. The presence of suberized layers of cells, at the periphery of the root, represents an important mechanism not only in the selection of nutrient uptake but since it also regulates the inverse flux of water that, in extreme drought conditions, could pass from the root to the soil (Hose et al. 2001). After crossing epidermis and exodermis, water has to pass through the cortical parenchyma. The reduced number of cortical layers is considered of adaptive advantage under drought conditions.

The development of endodermis with thicker cell walls and the formation of additional layers of cells with suberised walls around the stele are considered means to prevent the desiccation of meristematic tissues such as the pericycle and other tissues inside the stele (North and Nobel 1992). The relevance of the apoplastic barriers in controlling water flow depends on root developmental stage: in young unstressed roots, most of the radial hydraulic resistance is more evenly distributed across cortical layers, while in water-stressed old plants, it is mainly due to exodermis and endodermis (Steudle, 2000).

Once the water has entered the stele, it needs to be transported efficiently throughout the plant to maintain a continuous water supply to the leaves. In xeric conditions, plants have evolved morpho-functional traits that change root hydraulic conductance as a mechanism for regulating transpiration complementary to those played by aerial organs (Trubat *et al.* 2006). It is recognised that the efficiency of water transport and the hydraulic safety of vascular system are fundamental for the survival of plants in arid environments. Indeed, in the context of climate changes, considering an increase in the frequency of drought and of overall higher temperatures, vegetation of arid and semiarid environments have to cope with increased xylem vessel cavitation.

Chartzoulakis *et al.*(2002) conducted an experiment on two cultivars of avocado viz., Fuerte' and 'Hass' under two irrigation regimes for 6 months in order to evaluate the growth response and leaf physiological and anatomical changes induced by moderate water stress. Irrigation was applied when soil water potential reached at −0.03 and −0.5 MPa for the wet and dry treatments, respectively. They found that cross-sections of leaves showed that the palisade parenchyma is composed of two successive and distinctive layers, the first one being in contact with upper epidermis (palisade parenchyma I, PPI) and the other with the spongy parenchyma (palisade parenchyma II, PPII) .There were significant changes in leaf anatomical characteristics induced by water stress. Water stress resulted in a significant decrease of the thickness of almost all histological components of the mesophyll, as well as of the entire lamina thickness. In stressed plants of 'Fuerte', the chlorenchyma cells are denser than those in well irrigated ones. As a consequence, the amount of intercellular spaces of water- stressed leaves was lower than in control leaves (Table 1), while the PPII cells have remarkably increased in volume and number. The PPII cells of stressed 'Hass' are peg-like and not so densely arranged as in the relevant stressed 'Fuerte'.

Table 1 : Anatomical changes in cross-sections of fully expanded leaves of two avocado cultivars exposed to intermittent water stress.

LLeaf histological component	Fuerte		Hass	
	Watered	Stressed	Watered	Stressed
Thickness (mm)				
Upper epidermis	17.4 ± 1.5[a]	15.7 ± 1.4***	14.9 ± 0.9	15.9 ± 1.9
Palisade I	64.8 ± 8.7	51.7 ± 4.0***	72.8 ± 7.1	51.6 ± 6.4***
Palisade II	31.0 ± 4.3	29.3 ± 4.3	34.0 ± 2.9	26.5 ± 4.3***
Spongy	60.3 ± 6.9	53.8 ± 6.8*	65.3 ± 8.7	55.8 ± 7.3**
Lower epidermis	14.6 ± 2.5	13.6 ± 1.9	14.0 ± 2.3	14.6 ± 2.5
Total thickness	193.4 ± 16.6	167.6 ± 7.9*	203.1 ± 8.0	168.6 ± 10.5***
% of total leaf area intercellular spaces	22.12	14.21**	24.07	13.18***

Conclusion

Drought is a worldwide problem, constraining global crop production and quality seriously, and recent global climate change has made this situation more serious. Drought is also a complex physico-chemical process, in which many biological macromolecules and small molecules are involved, such as nucleic acids, proteins, carbohydrates, lipids, hormones, ions, free radicals and mineral elements. In addition, drought is related to salt stress, cold stress, high-temperature stress, acid stress, alkaline stress, pathological reactions, senescence, growth, development, cell circle, UV-B damage, wounding, embryogenesis, flowering, signal transduction, and so on. Therefore, drought is connected with almost all aspects of plant biology. Anatomical changes induced by water deficits in higher plants were better-observed indicators, which can be directly applied to agriculture and handled. To aim at exploring efficient anatomical indices, much information has been documented, but more attention should be paid to link them with physiological and molecular one.

References

Aronne G. and De Micco V. 2001. Seasonal dimorphism in the Mediterranean *Cistus incanus* L. subsp. incanus. *Ann Bot.* London. 87:789–794.

Bargel H, Barthlott W, Koch K, Schreiber L. and Neinhuis, C. 2004. Plant cuticles: multifunctional interfaces between plant and environment. In: Hemsley A, Poole I (eds.) The evolution of plant physiology. Elsevier Scientific Publishing Company, Amsterdam, pp 273–295.

Chartzoulakis K, A. patakas. G. Kpfidis, A. Bosabalidis and A. Nastou. 2002. Water stress leaf anatomy, gas exchange, water relations and growth of two avocado cultivars. *Scientia Horticulture.* 95 : 39-50.

De Micco V. and Aronne G. 2007. Anatomical features, monomer lignin composition and accumulation of phenolics in one-year-old branches of the Mediterranean *Cistus ladanifer* L. *Bot. J. Linn. Soc.* 155:361–371.

De Micco V, Aronne G. and Baas P. 2008. Wood anatomy and hydraulic architecture of stems and twigs of some Mediterranean trees and shrubs along a mesic – xeric gradient. *Trees- Struct Funct.* 22: 643-655.

Djibril, S., O.K. Mohamed, D. Diaga, D. Diégane, B.F. Abaye, S. Maurice and B. Alain. 2005. Growth and development of date palm (*Phoenix dactylifera* L.) seedlings under drought and salinity stresses. *African J. Biotechnol.*, 4: 968–972.

Fahn, A. 1964. Some anatomical adaptations in desert plants. Phytomorphology. 14:93–102.

Field, TS, Zwieniecki, MA, Donoghue, MJ and Holbrook NM. 1998. Stomatal plugs of *Drimys winteri* (Winteraceae) protect leaves from mist but not drought. *Proc. Natl. Acad. Sci. USA.* 95:14256–14259.

Hose, E., Clarkson, DT., Steudle, E. and Hartung, W. 2001. The exodermis: a variable apoplastic barrier. *J. Exp. Bot.* 52:2254–2264.

Jansen, S., Choat, B. and Pletsers, A. 2009. Morphological variations of intervessel pit membranes and implications to xylem function in angiosperms. *Am. J. Bot.* 96:409–419.

Johansen, C., B. Baldev, J.B. Brouwer, W. Erskine, W.A. Jermyn, L. Li-Juan, B.A. Malik, A.Ahad Miah and S.N. Silim, 1992. Biotic and Johansen, C., B. Baldev, J.B. Brouwer, W. Erskine, W.A. Jermyn, L. Li-Juan, B.A. Malik, A. Ahad Miah and S.N. Silim. 1992. Biotic and abiotic stresses constraining productivity of cool season food legumes in Asia, Africa and Oceania. *In*: Muehlbauer, F.J., W.J. Kaiser (eds.), Expanding the Production and Use of Cool Season Food Legumes, pp: 75–194. Kluwer Academic Publishers.

Karthikeyan, B, C.A. Jaleel, R. Gopi, and M. Deiveekasundaram. 2007. Alterations in seedling vigour and antioxidant enzyme activities in *Catharanthus roseus* under seed priming with native diazotrophs. *J. Zhejiang Univ. Sci.* B8, 453–457.

Lens, F, Sperry JS, Christman MA, Choat B, Rabaey D. and Jansen S. 2011. Testing hypotheses that link wood anatomy to cavitation resistance and hydraulic conductivity in the genus *Acer*. *New Phytol*. 190:709–723.

Li C, and K. Wang. 2003. Differences in drought responses of three contrasting *Eucalyptus microtheca* F. Muell populations. *For. Ecol. Manage*. 179: 137–385.

Liu F, R. Christian, J.A. Shahanzari, M.N. Andersen, and E.E. Jacobsen. 2005. ABA regulated stomata control and photosynthetic water use efficiency of potato (*Solanum tuberosum* L.) during progressive soil drying, *Plant Sci*. 168: 831-836.

Manivannan, P., C.A. Jaleel, R. Somasundaram and R. Panneerselvam, 2008. Osmoregulation and antioxidant metabolism in drought stressed *Helianthus annuus* under triadimefon drenching. *Comp. Rend. Biol.*, 331: 418–425.

Mahajan S, and N. Tuteja. 2005. Cold, salinity and drought stresses - an overview. *Arch. Biochem. Biophys*. 444: 139–158.

Martinez J.P, J.F. Ledent, M. Bajji, J.M. Kinet. And S. Lutts. 1986. Effect of water stress on growth, Na+ and K+ accumulation and water use efficiency in relation to osmotic adjustment in two populations of *Atriplex halimus* L. *Plant Growth Regul*. 41: 63–73.

Mathews M.A, E. Nanvolkenburg. and J.S. Boyer. 2003.Acclimatization of leaf growth to low water potentials in sunflower. *Plant Cell Environ*. 7: 199–206.

Monneveux P. and Belhassen E. 1996. The diversity of drought adaptation in the wide. *Plant Growth Regul*. 20:85–92.

Niinemets, U. 2001. Global-scale climatic controls of leaf dry mass per area, density, and thickness in trees and shrubs. *Ecology*. 82:453–469.

Nicholas, S. 1998. Plant resistance to environmental stress. *Curr. Opin. Biotechnol*. 9: 214–219.

North, GB. and Nobel, PS. 1992. Drought-induced changes in hydraulic conductivity and structure in roots of Ferocactus acanthodes and Opuntia ficus-indica. *New Phytol*. 120:9–19.

Noctor, G. and C.H. Foyer. 1998. Ascorbate and glutathione keeping active oxygen under control. *Ann. Rev. Plant Physiol. Plant Mol. Biol*. 49: 249–270.

Oguchi, R., Hikosaka, K. and Hirose, T. 2005. Leaf anatomy as a constraint for photosynthetic acclimation: differential responses in leaf anatomy to increasing growth irradiance among three deciduous trees. *Plant Cell Environ*. 28:916–927.

Reddy, A.R., K.V. Chiatanya and M. Vivekanandan. 2004. Drought induced responses of photosynthesis and antioxidant metabolism in higher plants, *J. Plant Physiol.* 161: 1189–1202.

Shao HB, Chu LY, Jaleel CA. and Zhao CX. 2008. Water-deficit stress-induced anatomical changes in higher plants. *C R Biologies,* 33:215–225.

Shields, LM. 1950. Leaf xeromorphy as related to physiological and structural influences. *Bot Rev.* 16:399–447.

Sperry, JS. 2003. Evolution of water transport and xylem structure. *Int. J. Plant Sci.* 164: S115–S127.

Sperry JS, Hacke UG and Pittermann, J. 2006. Size and function in conifer tracheids and angiosperm vessels. *Am. J. Bot.* 93:1490–1500.

Steudle, E. 2000. Water uptake by roots: effects of water deficit. *J. Exp. Bot.* 51:1531–1542.

Tosens T, Niinemets U, Vislap V, Eichelmann H. and Castro Díez P. 2012. Developmental changes in mesophyll diffusion conductance and photosynthetic capacity under different light and water availabilities in *Populul tremula*: how structure constrains function. *Plant Cell Environ.* 35:839–856.

Trubat R, Cortina J. and Vilagrosa A. 2006. Plant morphology and root hydraulics are altered by nutrient deficiency in *Pistacia lentiscus* (L.) *Trees-Struct Funct.* 20:334–339.

Wullschleger, S.D., T.M. Yin, S.P. DiFazio, T.J. Tschaplinski, L.E. Gunter,M.F. Davis and G.A.Tuskan. 2005. Phenotypic variation in growth and biomass distribution for two advanced-generation pedigrees of hybrid poplar. *Canadian J. For. Res.,* 35: 1779–1789.